To Amos;
Who by example,
taught me more than
I'm capable of learning.

Joe Hanson

35 –
Fishing

SHRIMP AND PRAWN FARMING IN THE WESTERN HEMISPHERE

SHRIMP AND PRAWN FARMING IN THE WESTERN HEMISPHERE

State-of-the-Art Reviews
and Status Assessments

Edited by

Joe A. Hanson

and

Harold L. Goodwin

Dowden, Hutchinson & Ross, Inc.
Stroudsburg Pennsylvania

Library of Congress Catalog Card Number: 77-22280
ISBN: 0-87933-307-3

79 78 77 1 2 3 4 5
Manufactured in the United States of America.

Library of Congress Cataloging in Publication Data

Main entry under title:
Shrimp and prawn farming in the Western Hemisphere.
Bibliography: p.
1. Shrimp culture—Congresses. 2. Penaeidae—Congresses. 3. Macrobrachium—Congresses. I. Hanson, Joe A., 1928- II. Goodwin, Harold Leland, 1914-
SH380.6.S47 595'.3843 77-22280
ISBN 0-87933-307-3

Exclusive distributor: **Halsted Press,**
A Division of John Wiley & Sons, Inc.
ISBN: 0-470-99288-3

PREFACE TO THE VOLUME

As we thumb through the pages of the final manuscript, we feel that things are better than we'd hoped. You see, the two workshops from which the two books of this volume emanated were quite different in format and the decision to combine the two books in a single volume was a late one. Thus, we had feared that differences between books would be more apparent than they now strike us as being.

Nevertheless, we do feel some compulsion to explain the inconsistency of format which is apparent. The marine shrimp workshop was the first of its kind. As such, it followed a panel discussion format with many speakers in each session. While there were semiformal presentations, these were not papers in the strict sense of the term. Consequently, the path we pursued in producing the proceedings was to summarize the recordings of the sessions, flesh these out from the literature, and then send drafts off to the participants for review, correction, and expansion. Thanks to a heart-warming level of cooperation from a large number of the participants, we think that this approach in the end resulted in a rather complete and comprehensive review.

On the other hand, the prawn workshop which resulted in the second book of this volume was the second of its kind in as many years. Thus it was appropriate to have that workshop be a series of invited papers followed in most cases by questions and comments from the floor. In this way, we prepared an authoritative compendium with somewhat less effort.

So, the reader may notice some differences between the texts of the two books of this volume; differences which reflect the difference in approaches to the two workshops. Also, the sharp-eyed reader will notice a difference in the approaches to citations: the shrimp text contains numbers and the prawn text employs the more classical approach. This, too, reflects the difference in the two workshops and the ensuing compilation of the proceedings. We apologize to those who may be put off by this and hope that it will nevertheless cause no inconvenience to anyone.

Be that as it may, we find that we are struck too with the clear value of assembling these two books into a single volume. There are many similarities between shrimp and freshwater prawn farming; so many, in fact, that an interest in one predisposes one to an interest in the other. And, of course, the very fact of the high degree of similarity serves to accentuate those differences which do appear.

For example, in the freshwater prawn arena, *Macrobrachium rosenbergii* is almost alone as the species of choice for commercial culture while nearly a dozen penaeid species are enjoying active culturing efforts. Too, cultured marine shrimp can be sold in a well-established market while there is as yet no such market in the United States for freshwater prawns. Yet another difference is that marine shrimp farmers tend to talk in terms of harvesting 15-25 count animals (whole) while the prawn people tend to think in terms of 7 to 10 count "miniature lobsters." In the reproductive realm, there are few problems remaining at the moment in spawning *M. rosenbergii* in captivity while the shrimp people are just now beginning to get a handle on the life cycles of most penaeids. Going hand in hand with this last situation is the fact that genetic research has already begun on prawns but is still a fond future hope in the marine shrimp arena. Temperature tolerances are at some variance, too; however, it is important to remember that the single prawn species is a tropical one and the several penaeid species under culture originate naturally in a broader latitude band. Too, it seems likely that there are some differences in product processing problems; shrimp processing equipment and techniques are well established *and* few think that shrimp technology will handle the larger and harder-shelled prawns. Finally, since shrimp require coastal land and the prawns can settle for adequate quality fresh water anywhere within their climatic belt, it may turn out that there are some significant differences in initial land costs between the two forms too.

On the side of similarity, for one thing, our degree of ignorance concerning nutrition and feeding seems to be just about comparable in the two forms; fortunately, the two books take somewhat different tacks in looking at this same problem. For another thing, the problems and perceived opportunities for intensive versus extensive culture are very similar in the two forms and, not surprisingly, the weight of effort being devoted toward these two ends of the culture density spectrum with respect to both forms is just about the same. Also, as do all aquaculturists, shrimp and prawn farmers all face the same government and legal problems; however--another bonus of putting these two books together--the subject is treated differently in each book, with the penaeid book taking a broad view and the prawn book focusing down on FDA regulations. Thus, these two chapters are quite complementary as are many other chapter siblings in the two books.

With all this in mind, we find that we have no hesitation in recommending both books to people interested in either form of crustacean. There is amazingly little redundancy and a pleasant level of complementarity. Yes, putting the two together was a good idea.

Even though acknowledgments are given in the prefaces to each of the two books, we would like here to offer an overall credit. While we breathe a sigh of relief as this goes off to the publisher, we do at the same time recognize that we were afforded an opportunity to play the role of facilitators for the fine communities of enthusiastic professionals who are in fact the authors of this volume. We can single out John and Suzanne Huguenin for their support to the shrimp book and Paul Sandifer and William Trimble for their contributions to the prawn book. And this is a good place to recognize Dorothea Rosinsky who typed and retyped the manuscript as well as finalizing many of the drawings and photographs. But, in truth, these two books are of and by the shrimp and prawn people of the Western Hemisphere. We hope they are pleased with them.

Harold L. Goodwin
Joe A. Hanson
The Oceanic Institute
March 1977

CONTENTS

SHRIMP AND PRAWN FARMING IN THE WESTERN HEMISPHERE

BOOK I

MARINE SHRIMP FARMING IN THE WESTERN HEMISPHERE

A State-of-the-Art Review and
Status Assessment

Compiled and Edited by

Joe A. Hanson
John E. and Suzanne S. Huguenin
Harold L. Goodwin

From the contributions and with the cooperation of the participants in the workshop on the culture of penaeid shrimp held in Galveston, Texas, October 8-11, 1975

The Oceanic Institute
Waimanalo, Hawaii

January 1977

The work was supported in part by the National Sea Grant Program, NOAA, Department of Commerce, under grant #04-5-158-13 and by The Oceanic Institute, Waimanalo, Hawaii.

PREFACE

Early October is a pleasant time in southern Texas. The days may still be a little warmer than just right, but the nights are pleasantly cool, the beer is cold, and succulent shrimp can be ordered in dozens of tantalizing dishes in Galveston's very good restaurants. The hundred or so participants in the first Western Hemisphere workshop devoted solely to the culture of marine shrimp were even treated to the frozen and breaded 25-count (tails) variety by the Galveston Holiday Inn where the workshop took place.

This workshop on the culture of penaeid shrimp took place over October 8, 9, 10, and 11, 1975. It was part of a broader contract between The Oceanic Institute of Hawaii and the Office of Sea Grant of the National Oceanic and Atmospheric Administration. In this contract, The Oceanic Institute provides planning and advisory services to Sea Grant within the realms of aquaculture, education, and marine science and technology in general. Harold L. Goodwin convened and chaired the workshop and Joe A. Hanson acted as rapporteur. The participants came by invitation; they were selected to achieve the broadest and deepest possible representation of geography, skills, experience, and viewpoints within the bounds of a manageable workshop.

Although there were Japanese representatives present, essentially all work discussed in the workshop is taking place now in North and Central America, save one. Some of the most advanced penaeid research and development currently underway anywhere is being done by the Centre Oceanologique's AQUACOP group in Tahiti and we were fortunate in having representatives of that organization present. So, with thanks to the French, we do include the Tahiti work in this report which is otherwise concerned with the Americas.

The purpose of this report is to present a comprehensive picture of marine shrimp culture in the Americas up to the present time. We hope it will be a report which will be of interest and value well beyond the edges of the "American shrimp community" as it exists today. If this report is to achieve its full potential, it must be comprehensive enough to appeal to interested persons not yet thoroughly conversant with the background and language of shrimp culture but who might wish to become so. With this in mind, we have gone beyond the strict subject matter of the workshop and into the literature and personal correspondence in order to provide a background from which to assimilate and evaluate the events of the workshop. Shigueno, in his excellent book, "Shrimp Culture in Japan" (Ref. 350), has proven an excellent basic reference for this purpose as have the proceedings of the World Mariculture Society's annual meetings. We were also fortunate in receiving an offer from John and Suzanne Huguenin to use their as yet unpublished marine shrimp culture bibliography in the present work. It represents a monumental task on their part and certainly enhances the value of this publication. In addition, they have participated enthusiastically in editing the manuscript. Lastly, although representatives of Marifarms, Inc., were unable to attend the workshop, they did provide written material and that is included with thanks.

It is, of course, not possible to include all the significant shrimp culture work in progress in the Americas in a report such as this, much less in a workshop. The work that is reviewed here was selected on the basis of its representativeness and its availability. Surely, this report is not entirely complete; yet we believe it covers the spectrum adequately. As to its accuracy, the authors must take full responsibility while gratefully acknowledging the contributions of all the workshop participants and authors from which we've drawn and especially thanking the dozens of speakers who have been kind enough to review the draft of this work, provide us with improvements and corrections, and who also supplied the illustrations contained herein. The alert reader will note inconsistencies here and there. They reflect the state of the art today or author errors; usually, we hope, the former.

Although penaeid shrimp culture is the subject of intense interest on the part of government, academia, and a surprisingly broad segment of industry, it remains true that it is not yet economically viable in the Americas. As yet, market prices in the Western Hemisphere are much lower than the premium prices obtained in Japan for the very fresh cultured product. Consequently, the distance between production costs and market prices in the west remains too short so far. There is a good deal of optimism, however; Marifarms, Inc., with its huge operation is striving courageously for economic success and a number of other commercial growers are beginning pilot operations with high hopes leavened with professional caution. With the exception of the ambitious enterprise of Marifarms, Inc., commercial pilot operations at this time are mid-intensity pond cultures with expected harvests as high as 4,000 pounds per acre of, say, 25-count (heads on) shrimp. If one assumes a $2.00 / pound market price for this product, dollar productivities per acre can be over $3,500/harvest and two harvests a year are conceivable. This, of course, explains the attraction. But as this report illustrates again and again, it is not yet a case of laughing all the way to the bank. There are difficulties with extensive[a] culture and some are convinced that the way to circumvent them is through more intensive controlled culture systems. These are still in the experimental category but they do show promise and are covered in this report.

---

[a] "Extensive" is the term usually applied to medium- to low-density culturing in ponds and embayments. "Intensive" is its antonym, usually applied to very high-density, controlled-environment systems. As antonyms, the two words, of course, represent only directions on a continuum.

We wish to thank Libby Goodwin and Zoula Zein-Eldin for their gracious and effective handling of the workshop logistics and once again to thank the participants (listed in Appendix B) for their contributions to this work. Of course, the workshop would not have occurred at all without the sponsorship of Sea Grant; the NMFS Gulf Coastal Fisheries Research Center's Aquaculture Research and Technology Division and its Chief, Dr. Richard A. Neal who hosted the event; the Texas Agricultural Extension Service, Texas A&M University; and The Oceanic Institute. Finally, the National Office of Sea Grant which sponsored not only the workshop but also the preparation of the manuscript deserves full credit for its support to marine science in general and aquaculture in particular.

J. Hanson
January 1977

I. A BRIEF HISTORICAL PERSPECTIVE

In his responses to the comment draft of this book, Capt. Paul Mulvihill rightly criticized the work for not containing a history of any kind. Recognizing that volunteers are few and far between, the senior author quickly petitioned Capt. Mulvihill to provide one. Thanks to his good memory and equally good nature, Capt. Mulvihill did so. That which appears here was written from the material he so generously sent.

Dr. Motosaku Fujinaga, Japan's acknowledged father of shrimp culture, successfully spawned and partially reared larvae of *Penaeus japonicus* more than 40 years ago in 1934. Dr. Fujinaga's subsequent success, not only in rearing the larvae and growing out the shrimp, but in bringing the art to the point that mass culture was possible, is by now well known. Whatever penaeid culture may achieve in the future, it will owe its beginning to Dr. Fujinaga's original work.

During the late 1950's, Dr. Fujinaga's work became well recognized by interested groups in the United States, and representatives of what is now the National Marine Fisheries Service in Galveston and in Miami were in more or less continuous contact with him. In 1963, Dr. Fujinaga visited the U.S. and spent some time at the Galveston laboratory. In collaboration with Dr. Fujinaga, Mr. Harry Cook and others of the NMFS Galveston laboratory experimented with the Japanese techniques and successfully spawned and reared larvae of two American species: *Penaeus setiferus* and *Penaeus aztecus*.

In about 1964, the State of Louisiana became interested in shrimp culture and funded the construction of a series of experimental ponds on Grand Terre Island. The ponds were completed during 1967-1968. At roughly this same time, the National Marine Fisheries Service Tropical Biological Laboratory in Miami became involved in rearing the pink shrimp *Penaeus durorum*.

Apparently, the early work in the U.S. provided some general encouragement: American industry began to take an interest in the late sixties. In 1968, Mr. Cook left his position with the NMFS Galveston laboratory to assist Dow Chemical of Freeport, Texas, in getting into the business of farming marine shrimp. At Dow, Mr. Cook became deeply involved in mass rearing of the larvae. During this same brief time span, Armour and Company entered into a joint shrimp-farming venture with the University of Miami. The National Sea Grant Program assisted this effort with some of the funding and the Florida Power and Light Company joined in also with the provision of land and buildings at Turkey Point, some 25 miles south of Miami.

Marifarms, Inc., was established in 1967 after a great deal of siting research conducted by Dr. Fujinaga and Dr. Mitsutake Miyamura. The site chosen was and still is near Panama City, Florida, where the company has a netted-off end of a tidal bay comprising some 2,500 acres as well as two 300-acre marine lakes. The saga of the Marifarms struggle is told in several places in the present work. It is almost certainly fair to say that the Marifarms endeavor was and still is the world's largest and most dramatic attempt to achieve commercial success with marine shrimp farming. Shortly after Marifarms was incorporated, Sea Farms, Inc., of Key West, Florida, began business as both a fishing and penaeid shrimp-farming commercial concern. Sea Farms' original facilities were at Tarpon Belly Key, Florida, and the company subsequently expanded into other locations. During this same time-frame, Marinus Laboratories, Inc., was looking into the possible development of a 20-acre island on Indian River in Florida for culturing both penaeids and pompano.

The year 1972 saw several new initiations into commercial shrimp farming. A research unit called Inmont was established by a large chemical firm at Marathon Shores in the Florida keys where postlarvae were produced and then shipped to Louisiana to be grown out in ponds especially constructed for the purpose. In that same year, American Agronomics, a citrus firm, undertook a study of the feasibility of commercial shrimp-farming on Pine Island, Florida, and Aquaculture, Inc., established a hatchery and experimental grow-out facility on Shark Key near Key West.

In 1974, Neptune Mariculture undertook to grow marine shrimp on a commercial basis in large circular tanks located on Plantation Key, Florida. By now, the National Oceanic and Atmospheric Administration's Office of Sea Grant and National Marine Fisheries Service were heavily involved in supporting and conducting research in shrimp culture. The NMFS laboratories in Galveston and in Florida lead that thrust and the Texas A&M University, the University of Miami, and several other academic institutions throughout the southeastern U.S. have applied Sea Grant funds toward the solution of the bevy of problems which face the commercial culture of penaeids. The states of the South, too, are joining in the effort with the states of Texas, Louisiana, and Florida leading the fields of interest and activity.

In the early 1970's, Ralston-Purina established the Crystal River facility described later and Groton Associates began operations in Costa Rica. In spite of all the activity described here and the other attempts that this account no doubt has missed, none of the commercial ventures thus far has turned a profit. Shrimp farming in the Americas is now just 13 years old if we date it from Dr. Fujinaga's first visit. Hopefully, this entering into the teens will be accompanied by an adolescent growth spurt.

II. HATCHERY SYSTEMS

This chapter first describes the typical larval stages of *Penaeus* very briefly. This is followed by an abbreviated synopsis of the classical Japanese hatchery practice to form a basis for interpreting more recent U.S. and French experiments. The recent experiments described in the workshop are then outlined. These descriptions are followed by some notes on larval nutrition and other notes drawn from the recent literature.

A. Larval Development of Penaeid Shrimp--An Overview

Two useful basic references for information on the larval development of *Penaeus* sp. are Shigueno (Ref. 350)[a] and Bardach, Ryther & McLarney (Ref. 349).[a] This subsection provides only a cursory overview as a setting for that which follows.

Gravid females may spawn from 30,000 to more than 150,000 eggs depending on species, age, condition, and environmental conditions. Approximately 50% of the eggs usually hatch into the first nauplius stage within 14-15 hours at 28° C. Within 30-35 hours at the same temperature, the yolk-fed larvae of *P. japonicus* pass through six molts of the nauplius stage to become nanophytoplankton-feeding zoea (protozoea in western usage). Shigueno reports essentially 100% survival through nauplius-2 for hatchery-reared *Penaeus japonicus* under "normal" conditions.

After three molts and approximately eight days, *P. japonicus* zoea, well fed in diatom-rich water at 25-28° C, enter the mysis stage. Three days later under the same conditions, mysis larvae afforded an adequate supply of diatoms

---

[a] Reference numbers indicate the position of the reference in the bibliography in Appendix D.

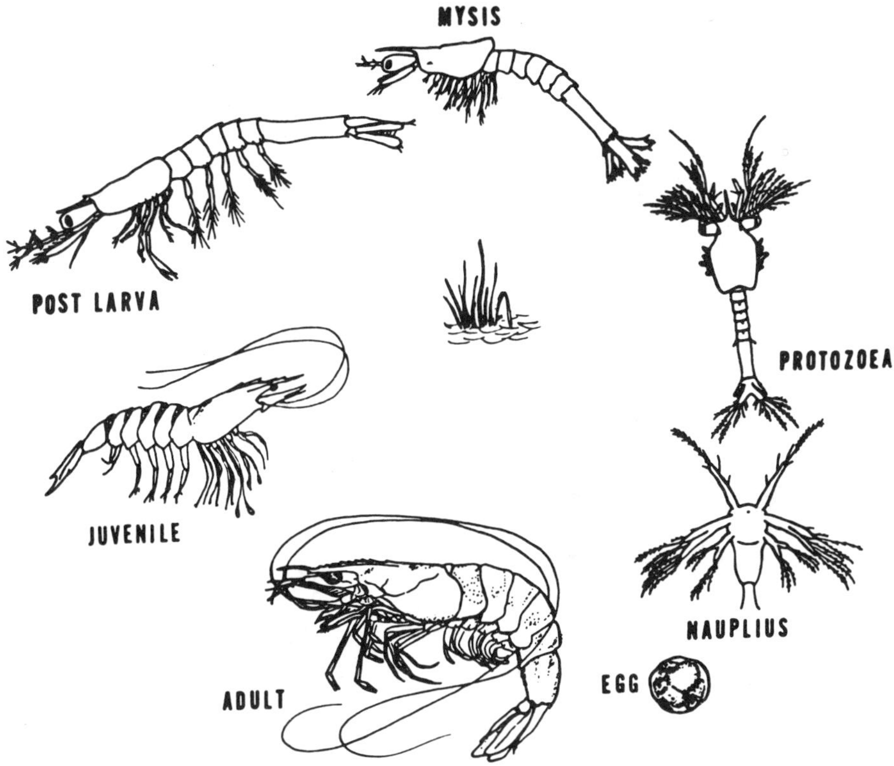

Figure II-1
Typical life stages of penaeid shrimp. Taken from Ref. 408.

and microzooplankton or a suitable surrogate pass into the postlarval (PL) stages with survival rates sometimes greater than 75%.

The times and survival rates given here are, of course, "good" average values for P. japonicus. Other species and other conditions show metamorphosis into postlarvae (PL) requiring from as few as nine to as many as 30 days and survival rates varying widely.

B. An Overview of the Japanese Hatchery Process

Once again, that given here is a cursory overview intended to provide a basis for interpretation of the workshop and recent literature information. For the complete story, the reader should consult Shigueno (Ref. 350) from which this subsection was taken.

Handling Spawners

Gravid females are selected for ripeness and good condition from freshly caught wild stocks. Thick, dark, and clearly contoured ovaries, indicating ripeness (easily visible through the translucent body) as well as lively behavior are the selection criteria. Selected females are transported either in aerated seawater at 100 shrimp to one ton of water (by boat or truck) or in chilled and dried sawdust. The latter method is less expensive and definitely inferior in terms of survival and maintenance of healthy condition. Upon arrival at the hatchery, healthy spawners are hand selected, washed, and placed in the spawning tanks.

Tanks

Hatcheries are located for availability of clear and clean seawater. The size of the unlined concrete breeding tanks ranges from 60 m^3 to more than 200 m^3 depending on the number of PL's to be produced and other hatchery management concerns. Spawner stocking density is on the order of 30 shrimp to a

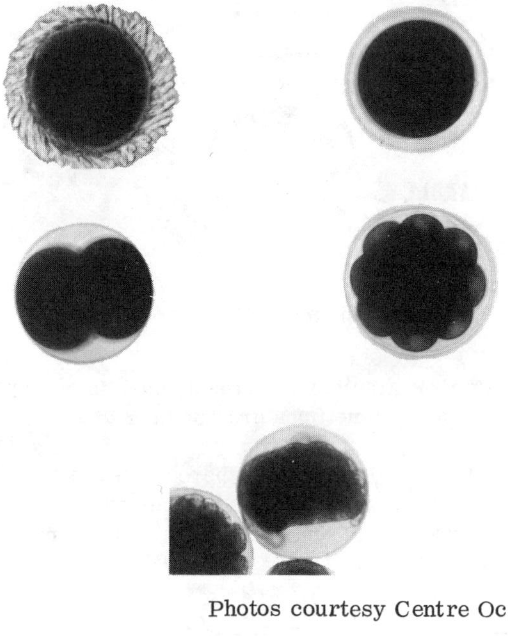

Photos courtesy Centre Oceanologique, Tahiti

Figure II-2
Shrimp eggs and their development. Top left: just-spawned egg of *P. merguiensis*. Top right, mid-left and -right: early development of *P. monodon* egg. Bottom: *P. monodon* nauplii in egg.

60 m^3 tank. The tank floors are slightly inclined to facilitate draining and have lowered catchment basins just outside the drains for collecting the PL's when the tank is drained. Aeration is provided by perforated, weighted plastic hoses scattered over the floor and fed by a roots-blower. Vinyl-coated stainless steel heating tubes fed by a small boiler are employed to maintain hatchery water temperature at about 28° C.

Spawning

Larval-rearing tanks are always drained, cleaned, and dried before receiving a new batch of gravid females. When the spawners arrive, the tanks are filled to a depth of 80 cm with seawater filtered through long mesh bags with 96 x 76 or 83 x 64 openings per 2.54 cm^2. The bags are attached to the output end of the filling pipe. Shigueno maintains that it is best to fill the tanks and bring them to the required temperature just before introducing the spawners rather than several hours before. It is necessary to block the spawning tank's outlet pipe to preclude the spawners congregating in them and suffocating.

Mature females may be expected to complete spawning spontaneously during the first two nights in the tanks; they are removed on the third day not only because spawning is essentially complete but also because the induced plankton bloom begins to cause enough turbidity to inhibit visibility and prevent sure removal of all spawners. Egg discharge is associated with nocturnal swimming activity and thus, smaller spawning tanks may be expected to inhibit spawning. (Note: This apparently has not been the case in U.S. experience with other species.) Spent spawners are removed by dip nets. The newly discharged eggs are irregular in shape and gradually become spherical. Nauplii hatch within 14-15 hours.

Nauplii and Phytoplankton

When nauplii are found swimming, usually in the afternoon of the second day, potassium nitrate (KNO_3) at 2 ppm and potassium phosphate (K_2HOP_4) at 0. 2 ppm (total volume) are applied to accelerate phytoplankton production. This application rate is maintained daily through PL_{10} (10-day-old postlarvae). With substantial sunlight, plankton growth begins to produce a brownish color in the water as early as late afternoon of the first day of application of the fertilizers. Brown turbidity, gradually increasing day by day, is an indication of normal plankton growth.

Sometimes normal phytoplankton growth is inhibited by lack of sunlight or unknown factors in the water. In the latter case, introduction of small amounts of fresh seawater may result in immediate resumption of the bloom. In the former, the clear polyethylene sheets normally employed to cover the tanks to prevent heat loss and exclude airborne detritus may be removed to admit more intense insolation. Removal of the sheets is acceptable so long as the water can be maintained at 28° C.

Nauplii grow by their own yolk and do not feed on the developing bloom. Hence, the first four days of hatchery activity in the tanks consist of two simultaneous but mutually independent processes: 1) development of the larvae to the nanoplankton-feeding zoeal (protozoeal) stage and 2) development of the phytoplankton bloom to a cell density high enough to support good growth and high survival of the zoea.

Zoea and Nutrition

The zoeal stage of *P. japonicus* goes through three molts and lasts 4-5 days. As long as the plankton bloom continues healthy, tank water is maintained as is and no additives other than the chemical fertilizers already mentioned are introduced. The zoea swim constantly forward by action of the swimming legs and drag behind them a threadlike feces.

Should the bloom fail, the Japanese introduce shrimp formula feed crumbled into 100-mesh particles. The input rate is 1.5 g/10,000 zoea/day on a quantitative basis. Qualitatively, feed can be added at a rate which results in water color (turbidity) similar to that resulting from normal plankton growth. Hirata (Ref. 350, p. 15) has achieved good results also with soybean cake which is a byproduct of the brewing of soy sauce and a good deal cheaper (in Japan) than the formula feed.

Mysis and Nutrition

The mysis stage lasts three days if conditions are normal; it entails three molts. Mysis larvae continue to feed on phytoplankton and will take microzooplankton such as rotifers and oyster eggs and larvae as well.

At the time of appearance of the mysis, new seawater is added to the tank at a rate which increases the depth by 15 cm/day. By PL_5 (postlarvae--fifth day), the depth of the water reaches 2 m where it is maintained. The planktonic bloom, if it maintains its proper composition, gradually turns the water dark coffee brown. The appearance of "green water" instead indicates a flagellate bloom; if such a bloom occurs, development is abandoned, the tank purged and sterilized, and spawning reinitiated. In tanks showing acceptable larval development, the phytoplankton forms noted in order of their relative abundance are *Melasira*, *Nitzchia*, *Skeletonema*, *Rhizoselenia*, and *Chaetoceras*. Water pH during all phases of larval and postlarval hatchery development remains between 7.5 and 8.5 without any active control being mentioned by Shigueno. Should the plankton bloom fail entirely, soybean cake or formula feed is substituted in the same fashion as that described for zoea (water color approach). Larval development rates on soybean cake may be retarded by as much as two days; survival is unaffected.

Postlarvae and Nutrition

Postlarvae of P. japonicus gradually shift from an omnivorous diet to a carnivorous one and prefer a pure animal diet by PL_{5-6} according to Shigueno. By the PL_{3-4} stage, planktonic copepods at a density of 60/l along with lower densities of veliger, trocophore, balanus larvae, noctiluca, and polychaete larvae as well as some flagellates occur. These naturally occurring forms are augmented with Artemia salina nauplii and as many as 46-84 nauplii per day may be consumed by shrimp fry in the late mysis and early PL stages. It is important that only the nauplii of Artemia be introduced (no eggs or shells) and that large nauplii not be fed to mysis and early PL shrimp.

At PL_{5-6} minced clam or formula feed replaces Artemia nauplii as the basic feed. Clam meat is expensive and of limited nutritional value after the necessary washing; Shigueno suggests that eventually formula feed may be expected to replace it entirely. During rearing from early PL to PL_{20-30}, the tank water is exchanged at the rate of 20-40% per day. There is no set formula; the rate is varied according to objective and subjective evaluations of water quality. New water is filtered through 60-mesh cloth as it is introduced.

Photo by W. D. Madden

Figure II-3
Pure algal cultures at AQUACOP, Centre Oceanologique, Tahiti.

Nutrition and Disease

Mass kills of cultured larvae in Japan are entirely too common. The disease organism is thought to be Vibrio sp. However, Shigueno feels that it is likely that such epizootics occur primarily as a result of poor nutrition.

At present, the Japanese do not practice pure phytoplankton culture and so species compositions in the tank-reared phytoplankton community vary widely and indigestible forms such as Navicula and Coscinodiscus may become dominant; this results in poor larval nutrition. Setaguchi et al (in Shigueno, Ref. 350, p. 39) have undertaken formula feed experiments to counteract the problem with considerable success.

Also, the Japanese are now reevaluating pure culture phytoplankton approaches using Chaetoceros sp. and chemically sterilized seawater. Apparently the original cause for the Japanese swing away from pure culture approaches was caused by their inability to achieve stable Skeletonema costatum cultures at temperatures above 25° C as well as the impracticality of thermally sterilizing the large volumes of water involved in their methods. Chaetoceros and chemical sterilization promise to overcome these drawbacks. Setaguchi (ibid.) reports survival rates between 64.8% and 83.4% with combined feeding of pure Chaetoceros sp. and formula feed H-105 and rates between 54.2% and 82.9% on the formula feed only. Zoea (protozoea) and mysis stages of P. japonicus were prolonged two or more days on the formula feed alone, but the results are considered encouraging nonetheless.

Harvesting and Shipping of Postlarvae

At about PL_{20-30}, the P. japonicus PL's begin to settle and congregate on the sides and bottom of the tanks. This, of course, corresponds to the wild transition from free-swimming to sand-burrowing. The PL's should be harvested at this stage since the concrete tanks tend to injure the telson and walking legs and thus invite disease.

In harvesting, the tank is first slowly reduced to a water depth of 0.5 m by means of a siphon and strainer. Then a course mesh net bag is tied to the outlet pipe. The outlet pipe stopper is then removed and the PL-laden water escapes through the mesh bag in which the young animals are entrapped. At intervals, the flow is stopped and the PL's transferred to water-filled buckets.

Shipping for short distances is done by truck in canvas bags with aeration at a temperature of 27° C and a density of about one-half million PL's to 1 m^3 of water. This approach is acceptable for up to three hours of driving. For longer distances, polyethylene bags containing 8 liters of water, 4 l of pure oxygen, and about 6,000 fry at 18° C are employed. Cooling from 28° C to 18° C is gradual.

This method is acceptable for up to 12 hours of transport by truck or by air. Shipping in live fish holds of boats has been tried with limited success. It has been found that when temperatures are high (28-29° C), densities also are high, and times are long (16+ hours), several molts and a high incidence of cannibalism may reduce survival by as much as 40%.

C. Hatchery Techniques Described in the Workshop

Mr. Cornelius R. Mock first discussed progress with the closed-cycle hatchery systems being developed by the NMFS Gulf Coastal Fisheries Research Center's Aquaculture Research and Technology Division at Galveston, Texas. Mr. Don Sweat of Sea Farms, Inc., then described progress with the system his organization is employing in Honduras, and Dr. Alain Michel commented upon recent work of AQUACOP, the aquaculture team of the Centre Oceanologique du Pacifique in Vairao, Tahiti. This is a laboratory created in 1972 by the Centre National pour l'Exploitation des Oceans (France). Additionally, Mr. Bruce Hysmith of the Texas Parks and Wildlife Department sent data on the hatchery experience of the Department's Marine Fishery Research Station and we have added exerpts from Marifarms' literature.

Mr. Cornelius R. Mock of the NMFS Gulf Coastal Fisheries Research Center

Since 1972, the Aquaculture Research and Technology Division of the NMFS Galveston Center has been working with the University of Arizona's Environmental Research Laboratory (ERL) on the rearing of *P. aztecus* and *P. setiferus* in closed-raceway inflated "greenhouses" of the type initially pioneered by ERL for growing produce on desert seacoasts (Ref. 442). The greenhouse grow-out facilities are described later under Grow-out Systems.

Initially, Gulf seawater was employed directly in the hatchery which supports this work and now supplies PL's to other workers. It was found that the quality of this water was sufficiently poor that the production of diatom blooms of consistent composition and intensity was essentially impossible. Larval rearing now is accomplished in autoclaved tapwater with Instant Ocean.

Mr. Mock pointed out that the amount of feed demanded by penaeid larval cultures, be it cultured diatoms or artificial feed, is a function of water volume rather than the absolute number of larvae; so the optimization problem is one of maximizing the number of larvae per unit of water volume while holding other parameters such as metabolite and oxygen concentrations within acceptable limits. So far, *P. setiferus* and *P. aztecus* densities of approximately 250 postlarvae per liter of seawater have been achieved with fair consistency in the Galveston work.

Pure *Skeletonema* cultures are maintained separately and fed to the larvae at rates that maintain appropriate food cell densities. At P_2 (2-day-old protozoea), *Tetraselmis*, also maintained separately in pure cultures, is introduced and at P_3, *Artemia salina* nauplii are fed. When this method works normally, larval growth is exceptional with *P. setiferus* attaining the PL_1 stage in nine days. However, if an excess of *A. salina* nauplii are introduced, they compete with the shrimp larvae for *Tetraselmis* and increase the metabolite load in the closed system.

Scientists at the Galveston laboratory have observed blooms of a dinoflagellate, *Oxyrrhis*, when feeding *Tetraselmis*. Blooms of *Oxyrrhis* may reduce growth and survival in the shrimp larvae by consuming *Tetraselmis* cells. However, it has been found that, if *Skeletonema* is reintroduced, the dinoflagellates die out and the protozoeal shrimp will return temporarily to the *Skeletonema* for food. Once the dinoflagellate bloom has subsided, *Tetraselmis* can be reintroduced and larval development proceeds as if the bloom had not occurred.

Mr. Don Sweat of Sea Farms, Inc.

Sea Farms, Inc., employs a system similar to the Japanese method using 7,000-gallon (26.5 m^3) tanks with a water depth of 1.5 m. When this system was initiated in Key West, seawater derived from wells was employed and no trouble was experienced in maintaining stable *Skeletonema* populations in the larval-rearing tanks. However, when a hatchery system was established in Honduras, water quality was found to be less stable and it was necessary to resort to growing the diatoms separately under more controlled conditions. The diatom-rich water subsequently was introduced into the larval-rearing tanks following the Galveston method.

During the late protozoeal to early PL stages, the shrimp are fed *Artemia* nauplii presently; Sea Farms, Inc., is experimenting with a variety of pelletized feeds as well. The latter are reduced to a flour consistency in a blender prior to feeding to the young shrimp. The animals are fed once a day through mysis and then in the morning and evening thereafter. Minced clams are fed sparingly as a supplement and seem to give animals so fed an edge over non-clam-fed young. But clams are expensive and dangerous; if not completely washed, water fouling is likely. The shrimp species being cultured are *P. occidentalus* and *P. stylirostris* and larval development times range from 14-18 days in water temperatures of 29-30° C and salinities of about 30 ppt. The developing shrimp are taken off *Artemia* nauplii as quickly as possible since the nauplii are expensive and difficult to obtain.

As yet, Sea Farms, Inc., has experienced no dinoflagellate blooms. Tank water is not changed except for the addition of the diatom water. Feed density generally is monitored by sampling.

Mr. Alain Michel of AQUACOP, Centre Oceanologique, Tahiti

In Tahiti the first problem is that the available ocean surface water is very low in nutrients and therefore in desirable phytoplankton species. Most of the time, artificially nutrified water stays clear for many days and whenever good blooms do begin, they are prone to swing to dinoflagellate dominance. In 1976 Centre Oceanologique hopes to circumvent this problem by using the artificial upwelling approach Dr. Oswald Roels used at St. Croix. This involves dropping a pipe down the island slope into Intermediate Water (600 m) and raising this nutrient-rich and hopefully dinoflagellate-free water to the surface.

The Centre Oceanologique employs the Galveston approach to larval rearing of penaeids using 0. 5 and 2. 0 cubic meter tanks and producing unispecific cultures of *Phaeodactylum*, *Cylindrotheca*, and *Tetraselmis*. Densities of 100-150 PL/l have been achieved but not yet consistently. Recently two serious problems have occurred: a fungus and a bacterial attack. The first was due to *Sirolpidium* and was combated with Treflan, a herbicide added to the larval-rearing tanks at a concentration of 10 μg/l (= 0. 01 ppm), three times a day as a prophylactic measure and six (or more) times per day as a therapeutic. The second problem was a bacterial necrosis of the appendages beginning during the protozoeal stages. It was able to destroy a complete batch of larvae in 48 hours. Water treatments and antibiotics have been tested and AQUACOP is now able to stop this disease. Five species have been reared thus far. They are *P. merguiensis*, *P. aztecus*, *P. japonicus*, *P. monodon*, and *M. ensis*.

Photo by W. D. Madden

Figure II-4
A view of AQUACOP's hatchery.

Artemia are difficult and expensive to obtain from Tahiti and this poses a larval nutrition problem. Other organisms are being tried and it has been found that rotifers can be used during the mysis and early postlarvae stages.

Mr. Bruce Hysmith of the Texas Parks and Wildlife Department

Mr. Hysmith writes (personal communication) that his experience with *P. aztecus* is that the nauplii pass through five molts and require 40 to 48 hours to reach the protozoeal stage. The protozoea of *P. aztecus*, however, require only four days for the three molts which bring them to the mysis stage. Mr. Hysmith reports that *P. aztecus* mysis stages are fed *Artemia* nauplii, rotifers, and *Brachionus* sp. in his laboratory.

Dr. Paul F. Bente, Jr., formerly of Marifarms, Inc., Florida

The following was exerpted from Dr. Bente's keynote address to the Symposium on Marine Chemistry in the Coastal Environment of the American Chemical Society's April 1975 meeting:

> "Three concrete, 28,000 gallon tanks, approximately 30 ft. by 30 ft. by 6 ft. deep, the largest size that had been used successfully in Japan, were installed for the hatching tests. After a few abortive starts in the summer of 1968, the hatchery turned out 7.5 million baby shrimp of the pink variety. In 1969, about 14 million white, 5 million pink, and 70 thousand brown shrimp were hatched. Fortunately, the hatching of each variety sufficiently resembled the process for the kuruma shrimp that the technicians were able to adapt procedures to achieve good yields.
>
> "On the strength of this progress, 32 more of these tanks were constructed for commercial work. During the 1970-74 seasons it turned out a total of about 1.5 billion baby shrimp. Each year 20 million were stocked in public waters, and modest quantities were provided to several research centers.
>
> "The hatching of baby shrimp is shown by Table I.
>
> "This is a fascinating, complex process involving 12 transformations after the spawned eggs have been fertilized. The literature is now replete with biological details of these changes. Excellent descriptions are given in two U.S. Patents[a] which have been dedicated to the public.
>
> "This paper is concerned only with a few of the procedures in which modifications of commercial significance have been made. In general, the hatching water must be gently aerated. It should be at 25°-28° C

[a]Patent Numbers 3,473,509 and 3,477,406.

Table I
Characteristics of Hatching Cycle

Form	Time in Form Indicated (days)	Common Food	Size at Close of Period (mm.)
Egg	1	--	0.25
Larva			
Nauplius (Stages I-V)	2	egg yolk	0.5
Zoea (Stages I-III)	4	phytoplankton (5-10 μ in size)	2.2
Mysis (Stages I-III)	3	zooplankton, artemia, yeast	5.0
Shrimp			
Baby (suspended in water)	6	artemia, pellet feed	7.5
Baby (bottom dwelling)	14	marine life incl. benthic, pellet feed	20.0[a]

[a]Weight at this size is about 0.02 gms.

and have a salinity of at least 14 parts per thousand. If salinity is low, it can be increased by addition of rock salt rather than sea salt, a simpler, cheaper procedure which the company developed as flexibility in operating procedures was explored.

"The zoea does not swim and must be kept suspended in water and fed by having its mouth collide with phytoplankton. The original process involved culturing special phytoplankton (chlorella and Skeletonema costatum) by adding nutrients, including residues obtained from the manufacture of soy sauce. One may judge what this hatching water looked like from the fact that the Japanese called the mixture "black tea." This procedure was gradually modified to culturing naturally occurring phytoplankton by adding chemical fertilizers as nutrients to enhance their multiplication. Yeast was also added as a food.

"The mysis swims after its food and lives on zooplankton and other meat. The original process added various forms of supplemental meat, such as finely ground up clams. This was modified to ground fish, and that was later displaced by additions of a live food, in the form of hatched artemia, commonly called brine shrimp. These procedures simplified operations, reduced labor, and improved yields.

"In this process the capacity of the hatching tanks is governed chiefly by the total weight of living shrimp (biomass) at the time the shrimp are finally moved to the growing areas of the farm. This varies for each variety of shrimp. White shrimp can tolerate a much higher density than brown or pink shrimp. Time for the full hatching cycle also varies and is another limitation on productivity. Details are given in Table II.

Table II
Productivity Limits of 28,000 Gallon Hatchery Tanks

Kind of Shrimp	Avg. Limiting Population[a]	Size of Shrimp at End of Cycle (mg.)	Duration of Total Cycle (days)
Brown	1,250,000	5-10	35-40
White	4,000,000	3-6	22
Pink	2,000,000	6-8	30

[a] Occasionally under ideal conditions densities 2.5 times this high have been obtained.

"The limiting or maximum number of eggs or larvae to start the process is thus determined by these facts, taking into account mortality expected as the larvae move through their many transformations.

"To increase capacity, the process has been divided into two steps, involving scale-up of an exploratory laboratory research procedure used by Cornelius Mock and coworkers of the U.S. Galveston Laboratory. The first step is conducted in a cylindrical, open top, 7,000 gallon, fiberglass tank standing on legs in one corner of the 28,000 gallon concrete tank. It is partially submerged in the water of the big tank. Water in the cylindrical tank is gently circulated--by being removed from the bottom into air lift pipes which discharge it at the top so as to cause a slow rotary stirring motion of the water column.

"This operation accomplishes several things. The circulation prevents the shrimp larvae, food particles, and general debris, such as egg shells, from settling to the bottom, a situation which generally kills off a large proportion of a hatching if it is allowed to persist. To avoid that in the 28,000 gallon tanks, technicians occasionally had to swim about with flippers to stir up the settlings accumulating on the bottom of the tank despite vigorous air agitation. The procedure also allows the first part of the process to be done in much higher density, thus enhancing the larvae's capture of the food particles and thereby reducing the total amount of extra feed required. It turns out that this also increases the survival of the larvae as they pass through the transformations.

"At the appropriate time the 7,000 gallon tank is drained into the 28,000 gallon tank which contains water prepared for the second step of the hatching cycle when the shrimp are in the post larval form. While this step is being carried out, if desired, the first step of the next hatching can commence and run concurrently. Thus, once operations are underway, the number of cycles that can be turned out is essentially doubled. Since survival is increased, the new procedure is capable of essentially doubling and at times even tripling production rate, while simultaneously reducing the amount of food used per given number of shrimp hatched. Though every tank is generally operating in a different stage of the complex hatching process, all this is done by four biologists and two helpers who are deployed to other operations when the hatchery is shut down."

A Replacement for *Artemia*?

Mr. Michel's earlier mention of *Artemia* difficulties stimulated a round-robin discussion. Although the problem is perhaps more acute today in Tahiti, it is clear to all aquaculturists that increasing demands for *Artemia* today are being accompanied by undeniably diminishing supplies and ever higher prices. And, although several competent workers are attacking the problem of *Artemia* mass culture[a], there are as yet no clear indications that the situation will reverse in the near future.

Mr. Mock reports that he has raised larvae up to PL_2 on *Skeletonema* and *Tetraselmis* alone employing the former at a density of about 150,000/ml and the latter at a concentration of about 5,000/ml.

Mr. Sweat mentioned some work with rotifers by Yang at the University of Miami that seemed encouraging.

Dr. Webber reported that, in a recent conference in Belgium, seven or eight reports on alternate larval feeds were given. These included work in progress on rotifers, protozoa, and nematodes. Of the lot, the work on nematodes looks most encouraging according to Dr. Webber.

Mr. Sweat has noticed that PL's in Honduras eat the algal mat growing on the sides of the tanks. It is not known whether or not this could form a total diet but it could be that, if lower growth and survival rates can be tolerated, this mat

---

[a] There are pilot commercial *Artemia*-rearing efforts in California (Dr. Bernard Cohlan), on Christmas Island, and at The Oceanic Institute in Hawaii (Mr. Terry Astro), to cite only three of the better-known efforts. See also References 262, 263, and 159.

could be used. In this context, Mr. Sweat points out that the objective is not high survival rates per se, but an adequate supply of PL's as inexpensively as possible.

Mr. Mock reported that the algal material which sloughs off his biodiscs produces good growth in juvenile white shrimp but he has not attempted feeding it to larvae. He also noted that, in high-density systems, it is necessary to keep all material in the tanks suspended and well oxygenated. Development of an anaerobic bottom layer will cause problems without doubt.

Alice Kenslow of Marine Nutritional Systems, Inc., mentioned that her firm is now producing algal feeds that may work with penaeid larvae and, moreover, that they are willing to work toward special penaeid larvae feeds if someone will tell them what nutritional requirements must be satisfied.

One approach may be to maintain a mature benthic and suprabenthic plankton community in separate ponds and transfer needed amounts to larval-rearing tanks. This would be an extension of the phytoplankton approaches being employed at Galveston and elsewhere.

All in all, there appears to be a rich supply of ideas for alternatives to Artemia and a relative desert of empirical work in progress. Perhaps the empiricism will begin soon if the Artemia problem worsens. Or perhaps we will take advantage of the work of other nations where "San Francisco Bay Brand" is not so easy to obtain even now.

D. Notes from Recent Literature

The World Mariculture Society's annual meetings seem to be the best recurrent forum for the most recent advances in aquaculture in general and crustacean culture in particular. Accordingly, most of the notes selected for presentation here are drawn from the published proceedings of the WMS over the past five years. For the serious student, the bibliographies cited in these papers and reproduced in Appendix D will provide a foundation for determining what is and is not known about larval penaeid shrimp culture.

Hatchery Systems

Sandifer, Zielinski, and Castro are developing "a simple airlift-operated tank for closed-system culture of decapod crustacean larvae and other small aquatic animals" (Ref. 489 and 490). The advantages of this engineering approach are: 1) water conservation and exclusion of disease and competitive organisms once the water is clean and 2) suspension and uniform distribution of feed and larvae throughout the water along with prevention of an anaerobic benthic layer. So far, these models have been used only with larvae of the freshwater prawn, Macrobrachium sp.

Mock and Murphy (Ref. 408) describe a larval-rearing method different from the Japanese approach in that 1) much smaller and separate containers are used for spawning, 2) phytoplankton cultures are reared separately under artificial light in "Instant Ocean" mixed with tapwater, and 3) larval-rearing containers are much smaller than those employed by the Japanese.

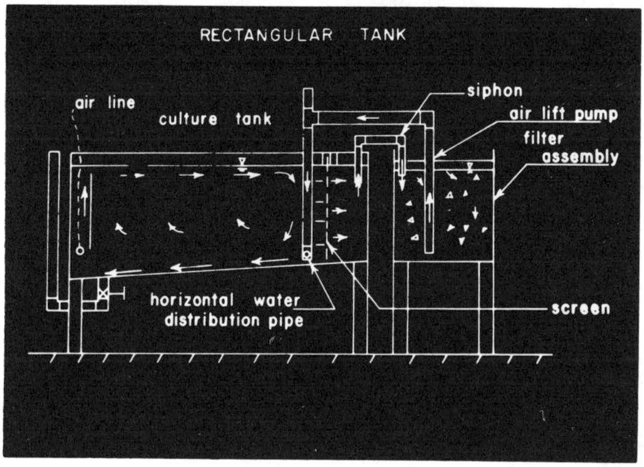

Photo courtesy of Paul Zielinski

Figure II-5
Schematic drawing of a larval culture tank designed by Zielinski and Castro, Clemson University, for the South Carolina *Macrobrachium* program.

Larval Nutrition

The central problems of larval nutrition are first, producing and maintaining adequate densities of unicellular algae for nauplius and protozoeal stages and second, finding a suitable substitute for *Artemia* nauplii as feed for mysis and postlarval stages.

Probably the solution to the phytoplankton problem lies in further developing the pure culture approach along one or some combination of three paths: 1) pure cultures maintained with the larvae, 2) pure cultures (possibly of more than one species) maintained adjacent to, but separately from, the larvae and fed at appropriate rates, and 3) pure cultures harvested elsewhere and preserved for subsequent sale to hatcheries.

Boschi and Scelzo (Ref. 444) report successful rearing of *Artemesia longinaris* larvae on pure *Skeletonema* through protozoea at temperatures of 17-20° C; this lasted 15-17 days. Mysis of the same species were fed *Artemia* nauplii during 12-14 days. Postlarvae were gradually transferred from *Artemia* nauplii to minced shrimp and special vegetable diets.

Griffith, Murphy, Kenslow, and Ross (Ref. 433) report a successful mass culture method for the green flagellate Tetraselmis that results in cell densities averaging just under 0.5 million per ml and sometimes exceeding 1 million per ml. The method has been employed using both natural and artificial seawater and both natural and artificial light. Within reasonable limits, the success of the method is independent of culture size. Approximately 60% of the culture can be harvested every four days. Mock, Neal and Salser (Ref. 210) describe an airlift pump that has been used successfully at Galveston for circulating and aerating cultures of Skeletonema, Thalassiasira, and Cyclotella as well as larval penaeids. (See Figure 4) The advantages claimed for this pump are 1) elimination of mechanical agitators, 2) total water circulation exposing all cells to the light source and preventing settling, 3) even distribution of nutrients, and 4) provision of air essential for phytoplankton growth.

There are continuous and batch methods for algal culture, and algal culture very probably can be improved technologically to enhance production rates and to manipulate quality in terms of digestibility and in terms of additives that can improve growth and survival rates of algal feeders. A prerequisite, however, is much better knowledge of the nutritional demands of algal feeders.

Thus far the scientific literature is notably lacking in reports of successful work leading to replacements for Artemia nauplii. Paulsen, at the Oceanic Institute in Hawaii[a], is beginning work with copepods and rotifers in tropical waters with particular interest in finfish larval nutrition and there is the work mentioned by Dr. Webber earlier in this chapter and that reported by Shigueno, Mr. Sweat, and Marifarms relating to the use of soybean cake and formula feeds in powdered form. So far, that seems to be the sum of it.

[a]No published results at the time of this writing.

III. OBTAINING SPAWNERS

Recent results with induced ovarian maturation in penaeid shrimp, though far too preliminary to be used as a foundation for authoritative prediction now, do lead to some hope that shrimp breeders will succeed in closing the life cycle of several species reliably within the next few years. Until then, the gravid females of all species are in the oceans and shrimp hatchery operators will be obliged to go there to get them and bring them back to the hatchery for spawning. This session of the workshop was led by Mr. Don Sweat. Formal discussants were Dr. Ausbon Brown of the NMFS Galveston Center, Dr. Hoyt Holcomb of Texas A&M University, and Mr. Jerry Broom of Groton Associates in Costa Rica. Also speaking from the floor were Mr. B. R. Salser of ERL, Dr. Spencer Malecha of the University of Hawaii, Dr. Alain Michel of AQUACOP, Mr. Wallis Clark of the Galveston Center, and Mr. Mock, also of the NMFS Galveston Center.

The Japanese approach as reported by Shigueno has already been discussed in Chapter II. Throughout the world where penaeid shrimp are pond cultured in so-called LDC's (lesser developed countries), a method is employed whereby tidal flows infuse young PL's into estuarine ponds for rearing. This method, described well by Bardach et al (Ref. 349), results in low-density growout and labor-intensive harvesting. Almost certainly, it is practical only where labor is quite inexpensive. At any rate, the method obviates the need for high-density larval rearing--which is the subject underlying the demand for spawners.

One final word before beginning: the literature is notably innocent of descriptions of techniques for the capture and transport of penaeid spawners. For this reason, this chapter, after an introductory overview furnished by Marifarms, Inc., contains only that information offered in the workshop itself. Note also that some of the experiences related herein have not yet been verified by adequate replication and so are too preliminary to warrant publication. Their presentation here must not, therefore, be taken as an assertion of their scientific verification; nor may it reflect in any way on the person who shares it informally with his colleagues.

A. Mother Shrimping by Marifarms, Inc.

This is another exerpt from Dr. Paul F. Bente's keynote address to the Symposium on Marine Chemistry in the Coastal Environment of American Chemical Society's April 1975 meeting. Dr. Bente relates Marifarms' experience with the collection of spawners as follows:

> "After hatching was demonstrated, the company made many exploratory trips, chartering commercial trawlers, to learn where in the Gulf and when during the calendar year each variety of shrimp in the gravid condition could be caught in sufficient numbers to operate commercially. Gravid shrimp are females carrying roe sufficiently developed so that spawning can be induced in a hatchery. Fishing is good when 5 percent or more of a normal catch are gravid.
>
> "Though many sites were found where mother shrimping can be done successfully, those most suited to the company's location are the waters off Mobile during February through April for brown shrimp, Appalachicola during May and June for white shrimp, and both locations during April through September for pink shrimp. During the seasons of 1970 through 1974, the company procured a total of 154,000 mother shrimp, equivalent to about 13,000 pounds, heads on basis.
>
> "Mother shrimping in itself is a highly technical operation, for which the crew must be specially trained and the trawler specially equipped to hold the shrimp alive in aerated, chilled water, and then to package and ship them successfully to the hatchery by chartered plane or, if the run is short, by company truck.
>
> "Handling white mother shrimp poses a special problem. The sack of sperm acquired by the mother shrimp during copulation is carried outside the shell. (Pink and brown mother shrimp carry it inside the shell.) If the gravid white shrimp becomes agitated enough to snap its tail, as it does in trying to escape, it generally loses the spermatophore making fertilization of the subsequently spawned eggs impossible. Despite such odds, by careful handling, the hatching of white shrimp was quite successful. These odds were finally reduced by placing each mother shrimp, as soon as it was caught, into a perforated cylinder, capped on each end, and of a diameter that prevented each shrimp from flipping its tail. After transportation the shrimp was released safely in the hatching tank.
>
> "When a gravid shrimp is placed in the hatching tank, spawning almost invariably takes place during the next night. Since this happens in the dark, no one sees what goes on. By special techniques, the Japanese government has captured the spawning process on movie film. This shows that as the mother shrimp swims around it releases a stream of roe that is dispersed in the water by the gentle paddling action of its 10

swimmerets. The spawning also ruptures the spermatophore so that the released eggs are fertilized in the sea water."

B. Mr. Don Sweat of Sea Farms, Inc., Honduras

Mr. Sweat felt strongly that anyone who wishes to collect spawners on commercial scales is more than well advised to obtain his own boat. The gatherer of gravid penaeids already must contend with such incompatible and uncontrollable variables as weather, lunar cycles, and the natural unpredictability of the shrimp themselves. Mr. Sweat emphasized that there certainly is no point in compounding these problems with a captain who may or may not feel like working when one wants to collect and a boat owner who seems to think that a struggling shrimp hatchery should pay off his mortgage. In Honduras, Sea Farms operates its own boat with a crew of four. Total experience thus far is just less than one full year so the preliminary findings related here are yet to be verified.

The species available to Sea Farms' Honduras operation are *P. occidentalis*, *P. stylirostris*, and *P. vannamei* with the latter appearing to make up about 10% or less of the shrimp population in the areas fished. The drags are of short duration (about 30 min.) and done typically in 30-40 feet of water. The catch is dumped on deck and all four crew hurriedly go through it, picking out any animal which appears attractive and putting it into an ambient water container (∝ 30° C). This serves to wash off bottom slime and detritus and acts as a primary sorting operation as well.

The next phase involves a re-examination of the animals remaining after the first sort and picking out all females which are ready to spawn and in good shape. The criteria used by Mr. Sweat are olive green ovarian color, (for white shrimp) presence of a spermatophore or remnant thereof between the 3rd and 4th walking legs, and healthy condition.

All the gravid shrimp thus selected are transported in separate tubes sized to immobilize the animals without over-constraining them. Tube transport along with reduced temperature is credited with further reducing the likelihood of shrimp spawning in transit and/or injuring themselves or, in the case of whites, throwing off the spermatophore. Tubes may be made of PVC or clear lucite or Dupont's "Vexar" screening rolled into a tube. The latter provides all necessary immobilization without restricting water flow in 2 of the 3 available dimensions as the solid plastic tubes do. Mr. Sweat offered a note of caution at this point: if all or most animals placed in a particular tube die, it very likely is toxic and should be discarded. Like aquarium glass, plastic tubes sometimes exhibit unexplained toxicity.

For transport, *P. occidentalis* in tubes are placed directly in seawater chilled to 19° C. The shrimp are not acclimated gradually and the survival rates are around 80%. _P. stylirostris_, on the other hand, do not tolerate rapid cooling from the ambient 30° C to 19° C; 21°-22° C being the transport temperature finally selected for this species. On the boats, the tubed shrimp are held in cooled and aerated tanks. At the dock, the tubed spawners are transferred to oxygenated plastic bags for transport to the hatchery which is about two hours away by truck.

As for seasonality of catches in Honduras, Mr. Sweat reports that gravid _P. occidentalis_ seem to be available year round although the best months appear to be January and April so far. _P. stylirostris_ has been slow all year and just began to pick up in August. _P. vannamei_ is such a small percentage of the penaeid community in the Sea Farms collecting area that the trawler crews see very few gravid females. Best catches of gravid _P. stylirostris_ in this region have occurred during the day on new moons with peaks around noon and termination around 6 p.m. For _P. occidentalis_, on the other hand, best catches have occurred during the hours of darkness under a full moon. It has been Sea Farms' experience that spawners caught at times other than the favored lunar phases may appear to be fine to the eye but perform poorly or not at all once in the spawning tanks.

C. Dr. Ausbon Brown, Jr., of NMFS, Galveston

Obtaining spawners for the aquaculture operation at the NMFS Laboratory, Galveston, has for the most part been successful. The Laboratory has been fortunate in securing at least two commercial shrimp boats for this purpose. The greatest success has been in obtaining gravid brown shrimp (_Penaeus aztecus_) as opposed to limited success in obtaining gravid white shrimp (_Penaeus setiferus_) and gravid pink shrimp (_Penaeus duorarum_). It is especially difficult to determine the white shrimp's readiness to spawn because the developing stage of the ovary in white shrimp superficially resembles the mature stage (Ref. 519).

In the past, the determination that white shrimp were ready to spawn was based entirely upon ovarian coloration. However, another criterion has been added now; this is the presence or absence of a sperm packet (Harvey Persyn, personal communication to Dr. Brown). During the years that Dr. Brown has collected gravid white shrimp, they were fished during the daylight hours because of their nonburrowing habit during these hours. White shrimp collected during daylight hours off the coast of Freeport, Texas, have been spawned and raised to postlarvae. Recently, however, gravid white shrimp have been caught during the nighttime hours from 7:00 p.m. to 11:00 p.m. along the Galveston coast. The nighttime collections were based primarily on the presence of the sperm packet, with a consideration of the ovarian color and, invariably, the animals have spawned.

The equipment used in transporting spawners and for collecting them on board ship has not changed significantly. Ice chests are used primarily for transportation purposes, limiting the number of spawners to 10 or less in the smaller ice chests, and 20 or less in the larger ice chests. Aeration is supplied by portable battery-powered aerators. Depending upon the vessel, there may be a box with constant water flow, into which the suspected spawners are placed until they can be culled. The animals thought to have the least chance of spawning are placed in ice chests for the trip back to shore, and the animals with the best chance of spawning remain in the constant-flow boxes.

On vessels without the constant-flow boxes, the suspected spawners are placed directly into the ice chests and culled. Other precautions are taken, especially in the collection of white shrimp; these include bags of ice to reduce the temperature and hopefully retard the spawning of the shrimp during the trip back to shore. If there is no constant-flow box on board (and, in the collection of white shrimp, this has been the rule), the water in the ice chests is changed several times before reaching shore.

D. Dr. Hoyt Holcomb of Texas A&M University

During June 1975, an intensive program for acquiring gravid female *P. setiferis* was conducted. All in all, seven boats varying in size from 40-foot commercial trawlers to small outboards participated for about 15 days. The waters fished were those from very near the beach out to the 10-fathom line. During June, no more than two gravid females were taken in any one night.

At the beginning of July 1975, fishing was shifted down the coast. On the night of July 3, four spawners were taken. On the next night in the same area, 76 spawners were taken from 36-40 feet of water between 3:45 and 9:15 p.m. On July 9 in the same area, 62 spawners were taken between 4:00 and 9:30 p.m. The July 3 catch hatched about 720,000 eggs leading in two weeks to approximately 480,000 PL's. The July 9 catch suffered mistakenly high temperatures during transport back to Galveston and several spawned on the way. Consequently, only about 200,000 nauplii were hatched.

For live female transport, Dr. Holcomb presently employs a Ralston/Purina-type styrofoam box about 18" x 24". In these are placed plastic bags containing seawater chilled to 18° C, 5-6 newly caught gravid females, and about 30% pure oxygen. The shrimp are cooled to 18° C gradually over a period of 45-60 minutes. Although this is the method used so far, Dr. Holcomb suggests that one female per bag would be best because it is possible for the creatures to spawn even at 18° C and if one of the animals in a bagged group does, it very likely will kill the rest.

E. Dr. Alain Michel of AQUACOP - Centre Oceanologique, Tahiti

Dr. Michel explained that obtaining spawners in Japan and the United States is quite different from the Tahitian problem, since AQUACOP's gravid shrimp must be flown in to Papeete from New Caledonia or the Fiji Islands and then trucked several hours to the Centre Oceanologique.

The first attempt at air transport employed oxygenated plastic bags containing one shrimp each at 22° C. All the animals arrived dead due to repeated jumping in the relatively large space of the inflated bags. The next attempt employed the same bags, but the shrimp were enclosed in a roll of soft plastic screen to restrict their movements; all arrived in Tahiti in good condition. The species with which AQUACOP is working presently include: *P. merguiensis*, *P. aztecus*, *P. japonicus*, *P. monodon*, and *M. ensis*.

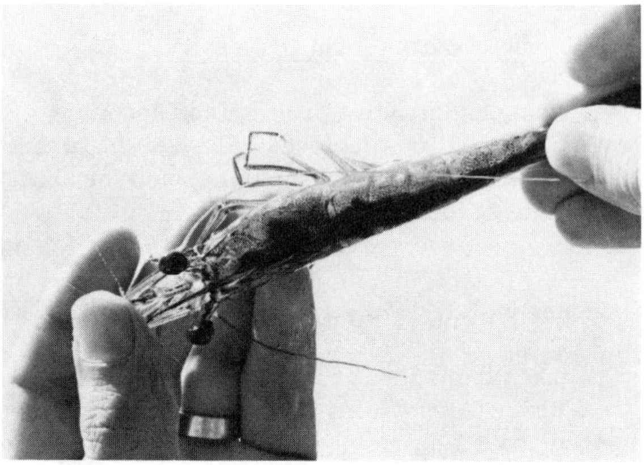

Photo courtesy Centre Oceanologique, Tahiti

Figure III-1
P. merguiensis ready to spawn.

F. Mr. Jerry Broom of Groton Associates, Costa Rica

The area in which spawners are collected by Groton Associates in Costa Rica is relatively small and narrow since there is a steep dropoff in that location. At this time, Groton is leasing boats and having its share of the sorrows mentioned earlier by Mr. Sweat. Usually, Groton fishes in 10 fathoms or less for gravid females.

Mr. Broom suggested that the presence of a spermatophore or remnant thereof alone is a sufficient criteria for selecting white spawners from a haul. The olive green ovaries may well not be a necessary criterion since Mr. Broom has found that, with indications of the sperm plug present, ovaries mature very rapidly after capture. No full sperm plugs have been observed by Groton's operators on white shrimp.

This past year has been an all-around disastrous fishing year for white shrimp in Costa Rica and so very little collecting data is available. But P. californiensis has been abundant. After capture, the spawners are transported to shore in clear plastic tubes about 6" long immersed in boxes containing aerated water at 20–21° C. It has been noticed that spawns occurring after the full moon are poor compared to those that occur before and during a full moon.

G. Mr. Billy Ray Salser of the Environmental Research Laboratory (ERL) of the University of Arizona, Puerto Peñasco, Mexico

Mr. Salser reports that the species available to ERL in the upper Gulf of California are P. srylirostris, P. californiensis, and some P. vannamei; but not enough of the latter species to be significant. Most of ERL's work is with P. californiensis.

Mr. Salser noted that P. stylirostris catches by ERL were best between 4:00 a.m. and 2:00 p.m. (as contrasted with 9:00 a.m. - 6:00 p.m. in Honduras) and that all catches have been in May with none later. Water depths in which spawners were collected were 3-4 feet.

Since the shrimping grounds are only 20–30 minutes by boat plus another 5-10 minutes by car from the ERL Puerto Peñasco laboratory, ERL has not troubled itself to develop anything like an interesting spawner transport technology. Females are hauled to the spawning tanks in such containers as ice chests and plastic trash cans--cooled and aerated.

In reviewing the reports contained in this chapter, it becomes clear that the technology or art of successful spawner collection involves: 1) finding adequate locations, 2) determining the best times of year, month, and day to collect, 3) identifying animals ready to spawn, and 4) transporting them unspawned and in good condition back to the hatchery. This chapter shows that methods for successful spawner collection can differ markedly, depending on the particular circumstances.

In closing this topic, it seems useful to refer to the opening. If expectations for closing the life cycle are well grounded, then the technology or art of catching and transporting spawners as it has developed so far may, in a few years, represent a rustically nostalgic period. But, at this moment, successful collection of adequate numbers of gravid females is an absolute prerequisite to mass hatchery culture of penaeid shrimp of all species. The next chapter has more to say concerning the control of penaeid life cycles.

IV. PENAEID LIFE CYCLE CONTROL

In general, the shrimp culture community is optimistic about its ability to control the life cycles of commercially valuable penaeids within the not too distant future. Ovarian maturation, fertilization, spawning, and subsequent hatching of eggs and development of larvae using only females reared in captivity is occurring already. Some species mature and spawn spontaneously under some culture conditions. But for others, unilateral eyestalk ablation seems to be the surest route to success. But it is by no means yet foolproof. As yet reliability is far too low. However, success of a sort is being had with:

1. *P. stylirostris* in Crystal River, Florida

2. *P. stylirostris* and *P. setiferus* in Corpus Christi, Texas

3. *P. aztecus* and *P. setiferus* in Galveston, Texas

4. *P. merguiensis*, *P. aztecus*, *P. japonicus*, *P. monodon*, and *M. ensis* in Tahiti

5. *P. monodon* in the Philippines

6. Verbal reports of a "red prawn" in India

Thus, seven captive species are known to be maturing and spawning to some extent in at least five research locations. Background references on the search for control of penaeid life cycles will be found in the bibliography under "Reproduction and Maturation" and points brought out in the workshop are summarized in this section.

A. Dr. Wallis Clark, Jr., of the NMFS, Galveston Laboratory

Dr. Clark led this session. In his introductory remarks, he observed that there appear to be three approaches to maturation of penaeid shrimp in captivity. They are:

- Maturation in pond cultures.
- Maturation under controlled rearing conditions after eyestalk ablation.
- Laboratory studies which manipulate individual females under careful control to stimulate ovarian growth, vitellogenesis (yolk formation), and, finally, nuclear maturation followed by stripping of eggs and fertilization.

Dr. Clark stated that scientists of the NMFS Galveston Laboratory have been taking an essentially cellular approach to the biology of ovarian maturation in penaeid shrimp for the past one to two years. At present the research is focused upon isolating and defining the hormonal mechanisms responsible for maturation and spawning as well as looking into the links between nutrition and vitellogenesis.

Dr. Clark defined nuclear maturation as the change from 4n chromosomes to 1n as determined by DNA concentration. So far, no true nuclear maturation has been demonstrated in females in which ovarian maturation was induced, except in some unexplained instances. The best guess of the moment is that the process is controlled by a small polypeptide of nervous system origin. According to Dr. Clark, some evidence suggests that lipids are particularly important in stimulating yolk laydown in ovarian oocytes.

As of now, Dr. Clark reports that the Galveston laboratory can stimulate female brown shrimp (*P. aztecus*) to mature and spawn at will, although he did not state exactly how this is done. So far, success with the white shrimp (*P. setiferus*) has been more limited but fertilized eggs are being obtained.

B. Mr. Alain Michel of AQUACOP, Tahiti

Mr. Michel then reported on recent life cycle work by AQUACOP in Tahiti. Since there are no species of penaeid shrimp appropriate for commercial culture indigenous to the waters of Tahiti, it has been particularly necessary for the Centre Oceanologique to achieve maturation and spawning in captivity. As of now, spawning experiments have been conducted with members of five different species, all of which came to the laboratory in Tahiti as postlarvae or juveniles. All species are reared either in "small" 12 m^2 tanks or in larger ones ranging from 400-1,200 m^2. The water used is clear and has tropical oceanic characteristics (salinity = 34 $^o/oo$, 26-30° C, and pH = 8.3). The animals are fed a dry preparation produced in the laboratory.

Under these conditions, Mr. Michel reports, P. merguiensis matures spontaneously throughout the year and spawning has been observed to occur at a rather small size (6 g) in this species. P. japonicus and M. ensis also have been maturing spontaneously in Tahiti while P. aztecus and P. monodon have not yet undergone ovarian maturation without unilateral eyestalk ablation (Ref. 460). This is done by pinching off one of the eyestalks. Survival is as high as 95%. Since the water is quite clear in AQUACOP's facilities, maturing females may be observed and hand netted easily in the 12 m^2 tanks. In the larger tanks, the same operation is accomplished by diving. After the mature females are retrieved, spawning takes place in special one-cubic-meter tanks where an airlift device achieves recovery of the eggs in four baskets fitted with 207-micron plankton mesh. With this approach, AQUACOP, as of August 1976, has achieved the sixth generation of P. merguiensis, the third generation of P. aztecus, the second generation of P. japonicus and P. monodon, and the first generation of M. ensis. According to Mr. Michel, production of commercial quantities of fertile eggs from captive-reared females appears to be a strong likelihood in the near future.

In answer to questions from the floor, Mr. Michel mentioned that the number of eggs produced by captive-reared females in Tahiti (40,000 eggs per 22 g female P. aztecus and 50,000 eggs per 50 g female P. japonicus) was lower than that said by the literature to be spawned from wild-caught females. This, however, is not the case with P. monodon and P. merguiensis; the egg rates being 150,000 per 80 g female and 30,000 per 25 g female, respectively. Hatching

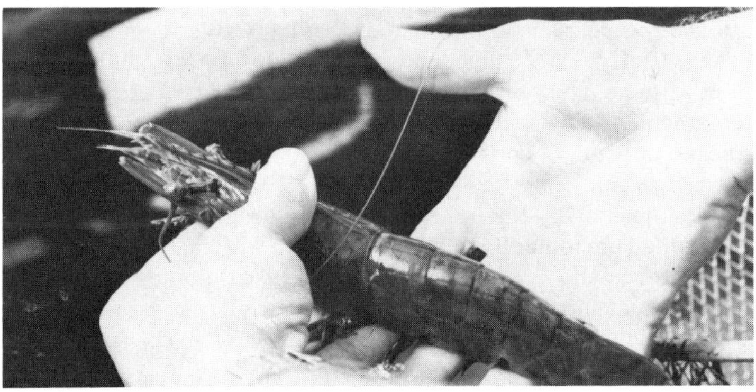

Photo courtesy of Centre Oceanologique, Tahiti

Figure IV-1
Female P. monodon with one eyestalk ablated.

rates average about 80% which compares favorably with results obtained from wild stock. There have been no attempts by AQUACOP to induce polyploidy and as yet, no attempts at hybridization. Mr. Michel also mentioned that unilateral eyestalk ablation does not always work; the ablation must be done on healthy animals and, to obtain a prompt response, probably at the right intermolt stage. Some females mature 3 or 4 days after the ablation; others molt first and mature after. With P. monodon, all the ablated females have matured and, in 3 months, 18 spawnings have been obtained from 6 females giving a total number of 3,3 x 10^6 eggs. This is the proof that females can mature more than one time in a short period of time.

Unilateral eyestalk ablation in Tahiti was begun when it was first noticed that all maturing P. aztecus in the stock had only one eye. Later work in the Philippines by SEAFDEC with P. monodon (Ref. 478) shows heavy (usually total) mortality when bilateral ablation is practiced (ed).

C. Dr. Fred Conte of Texas A&M University, Corpus Christi

Dr. Conte then reported on a joint maturation and reproduction study being conducted by Texas A&M University in Corpus Christi, Texas, and the National Marine Fisheries Service in Galveston, Texas. In May, 1974, postlarval P. stylirostris were stocked at 112,500/ha in an 0.2 ha pond near Corpus Christi, Texas. At the end of 139 days, shrimp averaging 150 mm and 30 grams were harvested. Ovarian development was not observed in the females, but the males exhibited small developing spermatophores. Five hundred of these shrimp were transported to Galveston, Texas, and overwintered in a heated raceway (Ref. 210).

In April 1975, 100 shrimp were returned to Corpus Christi, and along with 17 adult P. setiferus captured offshore, were stocked in a 0.1 ha clay-bottom pond. The pond received heated water from the Barney M. Davis Generating Station's cooling lake and ambient temperature water from the generating station's intake canal (Ref. 515). The pond was kept in a static condition with no flow-through of water. The shrimp were fed Ralston-Purina Marine Ration 20. Plankton samples taken from the pond were monitored throughout the study for the presence of penaeid eggs and larvae.

In late June 1975, the female shrimp averaged 52 grams and 190 mm. The pond was seined periodically with a 50 m net and from July through September, Penaeus setiferus and P. stylirostris exhibited all stages of gonadal development. From mid-August through September, six P. stylirostris with attached spermatophores were captured, placed in tanks and spawned. The fertilized eggs underwent cleavage to the morula stage.

From the number of females exhibiting gonadal development, and the sexual activity observed, it is probable that numerous spawnings occurred in the pond.

Plankton samples taken from the pond revealed a massive population of copepods, chaetognaths, and other zooplankton, but no evidence of penaeid eggs or larvae was found. Salinity ranged from 41.0 to 50.0 ppt and this, along with the large grazing population of zooplankton, would be detrimental to the survival of penaeid eggs or larvae.

Periodically, mature male and female P. stylirostris were confined in net cages and placed in the pond. The female shrimp used were of three categories: those exhibiting gonadal development; those lacking gonadal development; and those lacking gonadal development but with one eyestalk ablated.

The general condition of the shrimp maintained in net cages held in the pond deteriorated with time. The animals became soft, remained in this condition, and chitinoclastic infection developed on the abdomen, pleopods, uropods, and telson. Female shrimp exhibiting gonadal development when placed in the cages soon reabsorbed their ovaries. Female shrimp lacking gonadal development when placed in the cages never exhibited ovarian development. Female shrimp that lacked gonadal development, but had one eyestalk ablated, exhibited gonadal development within 4 to 6 days. No copulation or evidence of spermatophore transfer was observed in the cages.

This data shows that mature, sexually active shrimp can be obtained from pond systems in temperate climates. It is obvious that before industry can employ these methods, the techniques and systems will have to be refined and operated to produce spawners in sufficient numbers at the beginning of the growing season.

D. Dr. William MacGrath of Ralston-Purina Reported from the Floor

The Ralston-Purina Crystal River facility is employing unilateral eyestalk ablation techniques and finds that they do work in Crystal River's closed systems. Dr. MacGrath also noted that, while no one is sure what the effective ratio of males to females for a high probability of spermatophore transfer may be, it certainly is higher than 1:1 and more likely is on the order of 4:1 or possibly even 8:1. He did not elaborate.

E. Dr. Clark Concluded

He noted that, so far, all the attention is being given to the females. Accordingly, we have little knowledge of the male maturation process as it may affect fertilization and hatching and even (possibly) the final stage of female ovarian maturation. Dr. Clark emphasized that it is important that we gain knowledge concerning spermatogenesis and the mature sperm. Such information will be crucial to studies concerning cryo-preservation of sperm, in vitro fertilization, and hybridization. Dr. Clark reviewed the theory that active eyestalks

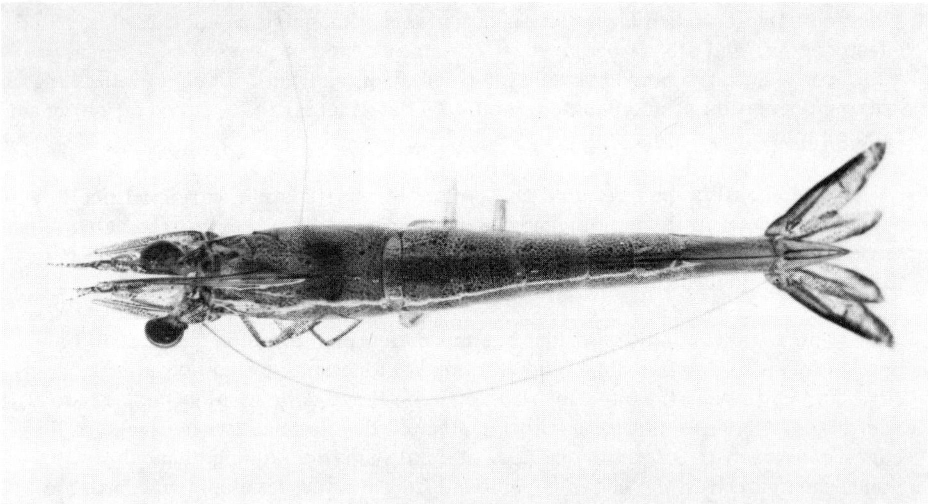

Photo courtesy of Centre Oceanologique, Tahiti

Figure IV-2
A maturing female member of AQUACOP's sixth captive generation of *P. merguiensis*.

produce or stimulate a polypeptide of neural origin and that this hormone normally inhibits release of the steroid hormone ecdyson which, when it reaches adequate levels in the blood, stimulates or permits ecdysis and possibly ovarian development; at least, increased levels of ecdyson parallel the onset of ovarian maturation in many instances. Thus, it appears that penaeids have an eyestalk-controlled polypeptide which suppresses release or production of a steroid which plays a role in ovarian development; this condition continues until either eyestalk ablation or some combination of external (perhaps optical) stimulii reduces the polypeptide levels.

Since different species inhabiting the same region appear to breed at different times of the day, month, and year (see Chapter III), and since there are reports of spontaneous maturation in turbid ponds, deeper holding facilities, and in ERL's polyethylene-covered raceways, investigators may wish to turn some research attention to light levels, photoperiodicities, and wavelengths as a manipulatory mechanism potentially more effective than pinching off eyestalks. At any rate, the present eyestalk ablation approach seems to be working to an almost acceptable level now and commercial-scale life cycle control may be within reach rather soon.

V. GROW-OUT[a] SYSTEMS FOR PENAEID SHRIMP

The workshop session concerned with the subject of grow-out systems included four presentations, two concerned with intensive culture approaches and two with extensive pond grow-out experiments. It was not practical to examine all of the pond activities underway in the Americas at this time. Therefore, two which were felt to be broadly representative of the state of today's art were selected. Conversely, the two intensive grow-out experiments reviewed here are the only ones underway in the Americas at this time which are concerned directly with penaeids. Also, although Marifarms, Inc., was not represented in the workshop, we have taken the liberty of including a brief description of that enterprise since it is unique in the Americas and by far the largest single shrimp-farming effort in the Western Hemisphere.

The workshop participants, it must be recalled, are members of a rather close community. Before gathering in Galveston, most already knew each other personally and each was familiar with the work of the others. It therefore usually was not appropriate for panelists to spend very much time on background information and so the contents of the presentations tended very much to consist of updates on recent progress. This was both economical and consistent with the objectives of the workshop. It even would make acceptable text for a newsletter. It is not, however, appropriate to a full report on the state of the art which is intended to serve persons outside as well as inside today's penaeid culture community. Accordingly, this chapter begins with broad examinations of where the extensive and intensive penaeid culture art appears to stand today and follows this with summaries of the four presentations and the brief Marifarms description. These are then supplemented by additional information and observations which surfaced during the workshop discussion period.

[a]Hyphenation of this term as opposed to use as a single word or as two words follows the preferred usage as given in Webster's 3rd International Dictionary--Unabridged; page 31a, paragraphs 3.7 and 3.7.1.

A. Pond Grow-out Systems

Fourteen interrelated aspects of pond grow-out systems are discussed in this subsection. With some exceptions, this background is general and is intended to provide a basis for using the workshop information contained later in this chapter only. For the details of specific approaches being followed by others in given situations, readers should refer to the bibliography.

Pond Size and Depth

For the most part, ponds in use in the Americas to date range from 1/4 acre up to 20 acres. Depths range from one to several feet. These last two statements exclude the much larger Marifarms, Inc., operation which is discussed later. As a rule, smaller ponds are easier to handle and, should an epizootic strike, less stock is lost. It is a sort of how-many-eggs-in-a-basket tradeoff: many small ponds usually will cost more to build and maintain than a few large ones and yet less of the total stock is lost if catastrophy strikes only one of a multitude of small ponds. On the other hand, smaller ponds usually will be subject to greater fluctuations in temperature, salinity, and pH and so may be more prone to stressing shrimp physiologically and thereby inviting disease. Benthic preparation (tilling, grading, sterilizing, and liming predators) may be somewhat easier with smaller and shallower ponds; so, generally, is predator control, harvesting, feeding, chemical disease control, aeration, fertilization (if used), monitoring of the biomass and the survival rate, and monitoring of growth rates. In general, monitoring the stock is more accurate in smaller ponds because a convenient-sized sample simply represents a larger percentage of the population in the smaller pond than in the larger. Also there is a tendency for population distributions to be more variable in larger ponds. There are few data now which suggest differences in growth rates, survival rates, or feed conversion rates as functions of pond size alone. It would seem that, today,

Figure V-1
An aerial view of the Texas Parks and Wildlife Department's Marine Fisheries Research Station on the north shore of Matagorda Bay. The station has 40 acres of land and 23 acres of water. The 21 ponds range in size from 1/4 to 4 surface acres.

decisions on pond size cannot yet be made on the basis of hard and fast rules amenable to mathematical analysis. Rather, the decision is a subjective one which, in the final analysis, usually is a tradeoff between the lower construction costs of larger ponds and the expected easier management of smaller ponds.

Mr. Hysmith writes (personal communication) that his group recommends ponds of 5 surface acres, 210 feet wide by 1,050 feet long with bottoms sloping from 2 feet at the shallow end to 5 feet at the deep end, and with an average depth of 3 feet. An 18-inch-diameter pipe with a gate valve for draining and harvesting should be installed in the deep end. The long narrow design is advantageous for distributing fertilizer and feed. Flow-through and pond level control are achieved in the ponds recommended by Mr. Hysmith with an 8-inch vertical pipe in the deep end which is connected to a horizontal section by a swivel joint. Inflow occurs at the shallow end. Mr. Hysmith further recommends that the long axis of these ponds should be about 30° off from the prevailing winds. This allows wind-induced agitation and aeration to occur without excessive erosion of the leeward levee. Mr. Hysmith also mentioned that research at the Texas Marine Fisheries Research Station indicates that shrimp stocked at equal densities grow faster and to a larger size in 1-acre ponds than they do in 1/4-acre ponds. He did not cite the species, the stocking densities, or the grow-out period. Lastly, Mr. Hysmith mentioned that supplemental aeration in the ponds he described did not appear to increase growth or yield significantly (Ref. 523).

Water Exchange Mechanisms

Penaeid shrimp live in saline water ranging from 15-50 $^{o}/oo$. Water can be pumped from wells, from estuaries, from the open ocean, and in the case of the AQUACOP, Tahiti plans, from the deep nutrient waters of the tropical seas. Or water at elevated temperatures can be had from seaside electrical power stations. Or, where the topography is appropriate and the tidal differences are great enough, water can be exchanged as tides rise and fall.

With few exceptions, pumping one's own water results in a high level of control over the water supply and is to be preferred if it can be afforded. Pumped saline well-water is attractive since it tends to be of comparatively consistent salinity, temperature, and pH; and concentrations of pathogens and predators usually are quite low. But saltwater wells are relatively expensive, too, since they cost tens of dollars a foot to dig in most locations. However, pumping water directly from oceanic or estuarine Surface Water usually involves a sizable capital outlay also and it brings in predators and competitors that must be killed or filtered out at considerable expense. Near-shore seawater also is more prone to quality fluctuations than is most well-water.

Obtaining heated outlet water from electrical generation plants sometimes has the attraction of demanding little or no pumping nor any outlay of capital for wells or intake pipes. It also has the satisfaction of utilizing heat energy wasted by the power plant to stimulate protein production and it thereby can eliminate needs for supplementary heating facilities and thus may lower capital and operating costs. However, such water is available only while the power plant is up and running and not while it is shut down for maintenance, repair, or in response to decreased power demands. In addition, power plants often employ poisons to prevent fouling and the shrimp farmer must somehow assure himself that concentrations of such chemicals in his ponds will not reach harmful levels. The obvious approach to protection, of course, is to cause the power plant effluent to bypass the shrimp pond when it may be contaminated; this requires that the farmer know in advance when the water will be contaminated and know it long enough in advance to take the necessary actions. Penaeids seem to prefer their water between 26° and 30° C. If the power plant/shrimp farm complex is in a region with wide seasonal temperature variations, there may be some chance of the culture medium getting too warm during the warm months if the only source of water is the heated power plant effluent. If this condition exists, an auxiliary source of cooler water may be required and this indicates an auxiliary seawater supply system. Thus, the capital cost savings associated with the use of heated power plant effluent may, in some cases, be more apparent than real.

Obtaining water exchange from tidal fluctuations, where it can be done, certainly should be an inexpensive approach. However, it also tends to admit more predators and competitors and probably disease organisms. Moreover, it is appropriate only to areas where tidal changes are adequate year round. Locating ponds so as to permit tidal exchange of culture water invites disaster by inundation in regions subject to severe storms.

Adequate water exchange in ponds obviates the need for aeration and, if input water quality is good, it also eliminates or vastly reduces requirements for active water quality control. It does, however, essentially prohibit fertilizing ponds and may result, depending upon the exchange rate and the biomass as well as the feeding rate, in either beneficial or detrimental removal of uneaten feed. At the same time, a flow-through approach can reduce hazards from lowered salinity or salinity layering caused by heavy rains or terrestrial runoff of fresh water. Note that, where static saline ponds are employed, make-up water to replace that lost to evaporation should be fresh so as to avoid ever-increasing salinity.

Climate

Unless a source of heated water is available or the design of the operation is such that heating water is economically feasible, ponds must be located in climatic regions which allow water temperatures to remain between, say, 25° and 30° C most of the year. Freezing temperatures for extended periods in winter may kill penaeids or require their relocation to warmer waters for overwintering.

There are observations in the literature that indicate that low-density penaeid shrimp given appropriate burrowing conditions can survive freezing winters (Ref. 438). Probably this does not extend to pond culture, however. Temperatures between 10° or 12° C and 25° C result in slowed growth and yet survival rates may be higher between 20° and 25° C and lower at 10°-12° C than they are within the preferred 25°-30° C range, the range where optimal growth rates are obtained.

The other most significant climatic variable is precipitation. Heavy rains can lower salinities detrimentally and regions subject to severe storms such as hurricanes are invitations to pond inundation, berm erosion, berm breakdown, and ground conditions which prohibit effective pond repair and maintenance. At present, the economics of penaeid culture do not usually admit the construction of an infrastructure of asphalt roads and concrete berms that would allow rolling stock to operate in extreme climatic conditions in and around a shrimp farm. In addition, other less common natural and man-made climatic conditions such as deposition of dust, silt, toxins, and so forth should be avoided.

Pond Material and Lining

Generally, ponds will be built of whatever earth is ambient in the area. So the site must be selected with this in mind. If the soil is not appropriate for pond and berm construction, extra expense will be incurred in pond linings and berm structural materials. Soil that is too soft usually is also too porous; water will be lost through seepage and shrimp will burrow and be difficult to harvest. Moreover, there will be a tendency for a sub-benthic anaerobic layer to form with concomitant buildup of hydrogen sulfide and other toxic substances. Even if this does not occur, some evidence suggests that shrimp allowed to burrow do not grow as rapidly as those not so allowed. Alternately, although burrowing is natural, shrimp reared in plastic-lined raceways do not appear to suffer the loss. Probably, burrowing is predator-avoidance behavior and not only unnecessary where predation is absent but also a form of behavior which inhibits maximum feeding and thereby reduces growth rates.

Berms built of crumbly soil or soil vulnerable to erosion will be costly to maintain and will gradually fill in ponds with loose material which aggravates the conditions discussed in the previous paragraph. Generally, compact soils with a fairly high clay content are most appropriate. Too clayish soils are difficult to work with bulldozers and draglines, and too porous and loose soils exhibit the difficulties just described.

Bottom Preparation and Maintenance

Pond bottoms should be sloped toward the drains. The Japanese also place several trenches in the bottom which lead toward the drains. This approach facilitates removal of sulfurous reducing compounds when ponds are drained and facilitates harvesting if it is done by the pond-draining method.

Penaeids are intolerant of even low levels of sulphur compounds such as H_2S. In any pond, buildup of a sulphurous reducing benthic layer is a constant threat that is aggravated if the bottom is soft and porous. Soils with high iron oxide contents can combat this buildup as can broadcasts of iron oxide at a rate of one kilogram per square meter (Ref. 350).

In many pond cultures, total draining for harvesting followed by drying and tilling of the bottom soil to aerate it is common practice. This practice, sometimes combined with applications of lime, can assist in reducing predation and competition, too.

Fertilization

Fertilizing penaeid grow-out ponds with standard commercial fertilizers or, as in Tahiti, with nutrient-rich oceanic Intermediate Water, can stimulate the growth of plankton and of attached algae mats and their accompanying populations of copepods, amphipods, isopods, and other nutritious animalcules. The evidence suggests that the benthic layer (called "lab-lab" in Asia) can be a whole diet for shrimp and that it definitely increases growth rates of shrimp being grown out on formula feeds. Moreover, zooplankton is highly important in young shrimp diets.

Like most blessings, however, fertilization is mixed. Since it causes phytoplankton blooms, too, it results in turbid water; and turbidity inhibits monitoring of sexual maturation, biomass, and survival rates. Since the algae produce oxygen during daylight and consume it during darkness and since the zooplankters consume oxygen constantly, fertilization rates must be balanced so that shrimp do not suffer from supersaturation[a] of oxygen during daylight and oxygen depletion at night. Moreover, it is not a simple calculation to determine rates of fertilization; as the animals grow rapidly on prepared diets augmented by lab-lab, their rate of metabolite production increases exponentially and this, too, is a fertilizer which alone or in combination with commercial fertilizers can stimulate plankton growth <u>and</u> produce toxic effects in the shrimp.

For example, Mr. Hysmith writes (personal communication) that he and his colleagues at the Texas Marine Fisheries Research Station have found cottonseed meal to be a superior fertilizer for stimulating zooplankton growth in saltwater. Prior to filling, the pond bottom is broadcast with 150 pounds per acre of cottonseed meal. After filling, a golden-brown bloom is evident within three

---

[a] ERL has found this to be a potential problem in larval and young juvenile shrimp

days. To maintain the bloom, 50 pounds of the meal per acre are added weekly. However, when the water temperature reaches 28° C, some care must be taken in adding the meal because dissolved oxygen becomes a potential problem. At this time it becomes important to monitor temperature and D.O. closely and to adjust fertilization rates accordingly. The cottonseed meal is used for young shrimp and its application is continued until commercial feed replaces it (Ref. 469).

Feed, Feeding, and Feed Conversion Rates for Grow Out

As long as it is not taken overly seriously, it may be useful to categorize shrimp grow-out feeds into three sets: 1) natural feeds (lab-lab and other benthic biota and detritus) which may occur in extensive grow-out systems, 2) prepared natural feeds (trash fish, clams, mussels, and seafood processing offal), and 3) formula feeds.

It is usually held that early larval penaeids are herbivorous, becoming omnivorous as they move through the PL and juvenile stages and then gradually becoming mostly or wholly carnivorous as they mature. Whether or not they represent an analog of wild diets, most formula feeds contain a high percentage of vegetable matter. Crude protein content may vary between 20% and 50%. The protein may be of animal or vegetable origin or both, but a flavoring and scent of seafood (essence of squid works very well) appears necessary for gustatory attractiveness. A binder which allows the pellets to remain whole in water for 24 hours or so is necessary as, so it appears, is a fair percentage of starch. A later section is devoted to nutrition, so it is appropriate to say little more about formula feeds here other than that so far, they remain fairly expensive and that they frequently produce lower growth and feed conversion rates than do the other two categories of feed--when conditions are appropriate to the other two categories.

Lab-lab is largely the sole feed enjoyed by shrimp cultured in the ancient Asian tradition, although it is occasionally augmented by whatever animal protein trash might be convenient. So long as population densities remain rather low, the Asians seem to achieve good growth and survival rates with this approach and production reports range between a few and 1,000 kg/ha/yr. It is interesting to note that the ERL Puerto Peñasco controlled environment, intensive culture experiments have indicated that it is valuable to leave lab-lab growth on the sides of the plastic-lined raceways. When this is done, penaeids maintained on formula feeds forage repeatedly in the lab-lab and grow faster than those denied this supplement.

Processed natural feeds tried on shrimp include trash fish and seafood offal, mussels, clams, crabs, barley, wheat, and a variety of other substances. All the animal meats just mentioned produce good growth when fed at rates high enough to maintain good growth but low enough to prevent leavings and waste from contaminating the water and/or the benthic realm. They are, however, usually malodorous and messy to handle and may contain skeletal material and bivalve shells that may accumulate on pond bottoms and become a source of trouble--and there are few cheap and easy methods of separating the meat from the hard parts without losing a large part of the meat's nutrient value. Moreover, processed natural feeds pose refrigeration problems, are not always available, and their prices fluctuate.

Herbaceous grow-out feeds alone seem not to result in acceptable growth and survival rates except where they are broken down bacterially in the ponds into fertilizers that result in adequate growths of plankton and lab-lab. If they find acceptable application at all, it is likely to be with rather extensive grow-out approaches.

At the moment, it appears that pond shrimp-farming developments in the Americas are likely to gravitate toward formula feeds augmented by natural pond productivity. Note that, when this approach is used (as it is in most pond cultures whether it is intentional or not), feed conversion rates calculated on the basis of weight of shrimp produced per weight of formula feed introduced, are quite truthfully false; how much of the nutrition is coming from the natural productivity of any given pond cannot be known accurately at present. Thus we find feed conversion rates reported in the literature ranging from 1.1:1 (Ref. 416) to 10:1 and more.

Feeds and feeding rates must, of course, be approached with the realization that uneaten feed of most sorts can contaminate the water, invite disease, and break down into plant nutrients which stimulate plankton and lab-lab growth. Ponds with high enough water turnover rates can avoid the turbidity and produce good lab-lab mats at the same time while avoiding oxygen balance and disease problems in the bargain. Also, it should be mentioned that growth rate and suvvival rate do not necessarily go hand in hand. Ponds with high individual growth rates very often show lower survival rates than ponds with more moderate growth rates.

Control of Predators and Competitors

If water is drawn from oceanic surface waters, be they pumped or tidally exchanged, there is no good method for prohibiting the entry of predators and competitors. Screen and mesh barriers can keep the adults, juveniles, and most of the larvae of predators and competitors out but some microscopic eggs and larvae are bound to get in sooner or later. Moreover, mechanical barriers must be cleaned frequently; usually by hand. Most screening materials deteriorate more or less rapidly and therefore require continual vigilance. Saline well-water usually is free of predator/competitor larvae.

When ponds are drained, drying, tilling, and liming of the bottom soil helps to eliminate spore that may remain there. If adult finfish predators/competitors can be kept out of ponds, applications of rotenone or saponin in proper proportions will eliminate the juveniles without harming the shrimp. Malachite green is also a possibility but it can kill diatoms and inhibit lab-lab growth. Some species of shrimp have been found to be several times more resistant to the above poisons than are finfish; if competitors are other decapods, this approach may be somewhat less appropriate.

Predatory birds are a constant bother in exposed ponds and some reptiles and mammals (including _Homo sapiens poacherii_) may be expected to prey upon unprotected ponds via the terrestrial entry. Steeply sloped banks and adequate water depths will frustrate wading birds. And clean banks will discourage snakes while making the larger mammalian predators at least more visible. Yet, all in all, some loss from predation should be factored into production estimates unless the farmer is so affluent that he can afford protection from invasion by land, by sea, and through the air.

Disease Prevention

There are no chemical or biological disease prevention measures being employed in pond cultures today aside from the mechanical and water quality factors mentioned briefly in this chapter. Stressed animals of any species are usually less disease resistant than animals maintained under nonstressful conditions. Abrasion of exoskeletons invites infection and ponds usually are kept free of abrasive substances. Jumping resulting from sudden disturbances or poor water quality can damage exoskeletons, too. Oxygen stress from dissolved oxygen levels being either too high (see ERL footnote under "Fertilization" above) or too low can be fatal in itself and probably serves at nonlethal levels to lower disease resistance. High-temperature effects on shrimp have received little scientific attention since they are seldom a danger but it is known that survival rates tend to decline while growth rates increase at temperatures in the low 30° C range (Ref. 81). Penaeids seem highly tolerant of pH variations within the neutral to alkaline realm but probably are less tolerant of pH's much lower than 7.0. Thus, acidic water may be an invitation to disease. More discussion of diseases and treatments is to be found in Chapter VI.

Aeration

It was mentioned earlier that ponds with higher water exchange rates may not require aeration. But, if water exchange rates are very low, aeration should be considered. The advantages and limitations of fertilizing to achieve biological oxygenation along with food production have been discussed already. Mechanical aeration can be achieved with surface agitators and various bubble-producing devices. They are all somewhat expensive in terms of capital and operating costs. Shallower ponds with higher surface-to-water and surface-to-biomass ratios can reduce and perhaps eliminate the need for active aeration, particularly in breezy locations. In ponds without active aeration mechanisms, temperature or salinity stratification of contained water may result in lower oxygen tensions in water near the bottom where the shrimp spend most of their time. Moreover, animals suffering oxygen depletion will not necessarily move to the higher water. Evidence suggests that they do this only during darkness and then they may behave in a counterproductive way. In this realm, too, smaller and shallower ponds are easier to monitor and manage.

But even the question of aeration is not straightforward. Mr. Hysmith, for example, has mentioned that, in his experience, aeration has not resulted in significant increases in growth and survival and often is accompanied by the growth of blue-green algae mats.

Natural Disasters

It so happens that many of the considerations important in pond culture of penaeids combine with land and labor costs to make locations in lower latitudes particularly attractive as sites for shrimp farming. And many such areas have histories of intense tropical storms with torrential rains. It goes almost without saying that the climatic history as well as past incidents of such natural spectaculars as earthquakes, volcanism, and tsunami (tidal wave) inundation should be considered in the selection of sites for shrimp farming. The assessed probability of the occurrence of such phenomena should then be balanced against such factors as land and labor costs, ambient temperatures, availability of good water, and so forth.

Nursery Ponds

Smaller ponds which are stocked at high densities with early PL's to be grown out to about 1- to 3-gram juveniles frequently are called nursery ponds. PL's can be stocked in nursery ponds at very high densities since they will be removed while still small. Such juveniles, when subsequently stocked into grow-out ponds, seem to have higher survival rates than do early PL's stocked directly into grow-out ponds. Ralph Latapie writes (personal communication) that one reason for the apparently higher survival rates of juveniles is that their counts can be more accurate than counts of PL's. He cites an instance in which a careful count of PL's showed a bag of 4,000 ± 10% to contain, in reality, only 2,700. If these had been stocked and had 100% survival, the recorded survival would have been less than 70%.

During their time in the nursery ponds, the animals are much more concentrated and so are easier and cheaper to feed and manage. In fact, since they can feed nicely on zooplankton, only fertilization may be required. Moreover, in the more extreme latitudes, it may be economically feasible to warm the nursery ponds and thereby to get a one- to two-month jump on the normal growing season and thereby increase the number of crops harvested each season. Because stocked juveniles are larger, hardier, and easier to estimate in terms of counts, this approach also results in better knowledge of correct feeding rates, biomass, and mortality. By and large, the nursery pond approach seems to give better control and better overall production and is gaining favor with commercial shrimp farmers. The actual stocking density of nursery ponds will depend upon the approach taken to feeding, the approach to water exchange and aeration, and a variety of other interrelated factors.

A final observation is that the nursery pond approach on larger scales is appropriate to culturing bait shrimp since the animals are harvested at a small size and therefore can be stocked at high densities (B. T. Hysmith; personal communication).

Stocking Density

Like stocking densities for nursery ponds, stocking densities for grow-out ponds are dependent upon a number of closely coupled factors. If only natural feeding is to be employed, the final biomass achieved before harvest must be no greater than that which can be supported by a pond's plankton and lab-lab. If supplemental feeding with prepared natural and/or formula feeds is used, stocking and ultimate biomass density may be limited by the aggression characteristics of the species in use, by considerations for oxygen tension or metabolite production rates, or a variety of these and other factors including the size at harvest. Generally, high stocking densities result in lower survival rates and lower growth rates. But calculated feed conversion rates frequently are higher and the final biomass harvested for a given cost of production may be higher too--although the individual animals may be smaller at harvest and thus bring a lower price per pound. Stocking densities of PL's mentioned in the literature (for example, see Ref. 349) range from around 100,000 to several hundred thousand per hectare and most probably the upper limits and the effects of intermediate densities remain to be explored fully. Undoubtedly there are differences in optimum densities among species, too, since behavioral observations indicate some species to be much more sociable than others.

It is obvious, but nonetheless worth mentioning, that high population densities increase water quality control and feeding rate problems and are more prone to epizootics. On the other side of the balance, optimum return on capital and labor will be achieved by employing the highest possible stocking densities consistent with good survival and growth rates.

Biomass Monitoring

The shrimp farmer interested in maximum return on investment (as if there were another kind) must monitor his standing crop in order to determine feeding rates and in order to know survival rates and the general condition of his crop. In turbid water, this can be done with samples trapped in representative locations throughout the ponds. Other workers may use hand-operated trawls to do sample pond transects (B. T. Hysmith, personal communication). These sampling methods are less than 100% satisfactory since the representativeness of the samples must be assumed and cannot be verified without draining the ponds. It is known that shrimp in large ponds seldom distribute themselves homogeneously and that their distribution at any point in time probably is influenced by several interactive variables. However, little hard data is available and it is possible to estimate probable biomass distributions in ponds only crudely at present. If clear water can be maintained in ponds, they can be surveyed visually and a higher degree of confidence achieved. But clear water means low natural productivity which, in turn, means high supplemental feeding costs. So far, the ability to do visual surveys in ponds seems to be the exception rather than the rule. High stocking densities with adequate feeding result in high rates of metabolite production. And this must result in planktonic turbidity that can only be overcome with high water exchange rates from a clear water source--which may wash any suspended feed away. Also, the propensity of some penaeid species for burying themselves in the bottom with only their beady little eyes sticking out makes even visual surveys in fairly clear water of questionable reliability if the pond has a soft bottom.

Harvesting

Harvesting shrimp from grow-out ponds is accomplished by one and sometimes a combination of four methods: 1) trapping, 2) netting, 3) trawling, and 4) draining.

In draining, the pond is so constructed that, as the water is drained, the animals follow the receding water and collect in a catchment at the outlet where they can be netted from the catchment or collected as they come through the large-diameter drainpipe. The Japanese have used small electric trawls but, although such trawls are employed for fishing in the U.S. (Ref. 374), there are no reports of their use in shrimp farming in the Americas. In electric trawls, a week current between electrodes just ahead of the trawl mouth causes the buried shrimp to jump a few centimeters out of the mud where the oncoming trawl collects them before they resettle. Nets may be static and designed to trap shrimp moving about at night; or the nets may be actively drug through the pond to entrap all shrimp not well buried in the bottom. Of the four, pond draining is the only method of closely approaching a 100% harvest with minimal damage to the animals. It also goes well with pond revitalization between crops and for now is the preferred method among shrimp farmers in the Americas.

Draining is, of course, more appropriate to smaller ponds than to larger. Harvesting smaller ponds one at a time can result in a harvest rate consistent with a processing capability that represents minimum capital and it can be adjusted to market demands. For larger ponds, trawling or netting can produce the same results and the ponds can eventually be drained at the end of the season (Ref. 430).

B. Intensive Culture Approaches

No one in the Americas is yet getting wealthy from penaeid shrimp farming in ponds, but shrimp are being produced and sold using this more extensive approach. This is not true in the case of intensive culturing approaches; these are still in the experimental stages--stages that are more heavily subsidized by government and venture capital than are most pond culture projects.

Figure V-2
Controlled-environment raceways at the shrimp research station at Puerto Penasco, Sonora, Mexico, operated jointly by the Universities of Arizona and Sonora.

The central purpose of intensive culture technology is to achieve economy of production by rearing more and better animals faster and in less space. This seems to demand tighter controls over the rearing environment which results in such systems being more or less disconnected from their ambient environments. It is true that such approaches usually demand higher capital investments than the more extensive approaches; however, so do broilerhouses. And the economic and gastronomic superiority of broilerhouses over the old-fashioned chicken yards is by now undisputed.

When and if they become successful, intensive shrimp culture approaches will widen the geographic latitude band within which the animals can be cultured commercially. They may eventually eliminate serious disease, predation, and competition problems or reduce them to a secondary concern. They will maximize feed conversion rates, growth rates, and survival rates. They will achieve 100% harvesting at rates consistent with market demands and they will produce a product which is far superior to wild or pond-reared stocks in taste, texture, and head-to-tail ratios. In short, intensive penaeid culture techniques will do for shrimp what the broilerhouse has done for chicken--if they are successful.

The next section of this chapter first discusses the two intensive culture approaches represented in the workshop and then reviews recent developments in pond grow-out systems.

C. Brief Reports from the Workshop--Intensive Grow-Out Systems

This section contains reports of work in progress at the National Marine Fisheries Service (NMFS) Gulf Coastal Fisheries Research Center in Galveston, Texas and at the University of Arizona/University of Sonora Puerto Peñasco facility.

Mr. Cornelius R. Mock, the Closed Raceway System Research at the NMFS Gulf Coastal Fisheries Research Center

By and large, Mr. Mock reports, closed raceway systems are not yet economically viable. He believes they can become economical when the cultured product is of premium quality, its production is reliable on a year-round basis, and the intensive systems can become independent of the rising costs of power purchased from the utility companies.

As of now, the raceways of the Galveston Laboratory are completely closed, hydrologically speaking. All water is recirculated and the only makeup water is tap water added to replace water lost through evaporation. The raceways are approximately 75 feet long and 10 feet wide, about 1.5 feet deep and enclosed in pairs under inflated polyethylene greenhouses of the type originally developed by the University of Arizona's Environmental Research Laboratory. Aeration is heavy and employed in such a way that it drives the water recirculation pattern. Water in the tanks flows in two opposing and parallel horizontal spiral motions down one side and back the other. This keeps feed and metabolites in suspension and prevents any tendency for an anaerobic bottom layer to build up. Filters are both mechanical and biological. Forty-inch-diameter rotating biodisks in a separate compartment at one end of each raceway are the biological filters for dissolved metabolites and inclined plate separators filter out the solids. In these tanks, PL's have been stocked at a density of $1,324/m^2$ and growth to 52 mm has been achieved in 79 days using artificial feeds developed in the Galveston laboratory.

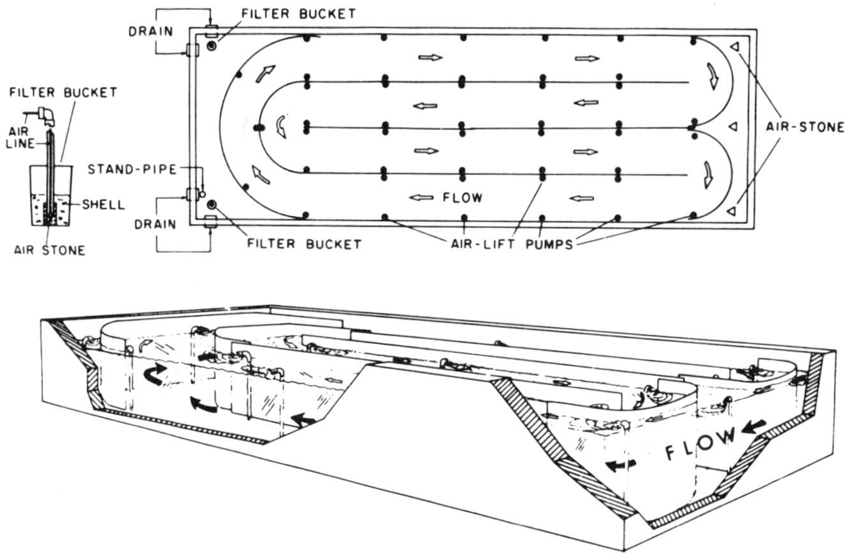

Figure V-3
Sketches showing the design approach used in the NMFS Galveston raceways. Taken from Ref. 210.

Near-future plans are to employ solar heaters for water temperature maintenance during the winter and to use the biological "wastes" produced in large quantities by the high-density biomass in the raceway as feedstock for methane generators. In this way, methane energy can be used for pumping and cloudy-day supplemental heating; solar heaters will carry the main water-heating burden and the system can be disconnected from the utility lines. If this works, and if the comparative quality and price of raceway-reared shrimp rises, a point at which closed-system grow-out will be financially attractive will be achieved.

Mr. Carl Hodges, Work in Progress by the University of Arizona's Environmental Research Laboratory (ERL) in Conjunction with the University of Sonora at Puerto Peñasco on the Gulf of California in Sonora, Mexico

The report here consists mostly of written material received from ERL after the workshop. In the workshop, Mr. Hodges spent some time discussing the economics of production in the ERL intensive system; this discussion will be found in the Production Economics chapter later in this report.

Engaged in solar energy and water purification research since the 1950's and controlled-environment vegetable production since the 1960's, ERL initiated production research in controlled-environment culture of penaeid shrimp in 1973. The objective of the program is the development of a commercially viable production system throughout the life cycle of selected species of marine shrimp. The site of the work is the coastal field station at Puerto Peñasco, Sonora, Mexico (160 air miles from ERL's main laboratories at Tucson). The facility is operated jointly with the University of Sonora.

Thirty full-time U.S. and Mexican scientists and technicians are employed on the project. Controlled-environment facilities with seawater systems include a hatchery with six 2,000-liter hatchery tanks; five air-inflated structures 30 m long, each containing two experimental raceways 3 m x 23 m (two of the ten systems can be supplied with elevated-temperature seawater); twelve 5 m^2 fiberglass tanks and one hundred seventy .25 m^2 and .5 m^2 fiberglass tanks for nutrition and pathology trials. Outdoor tankage consists of fourteen 5 m^2 tanks and three older concrete tanks of slightly larger size. Ancillary facilities include two seawater wells with a combined capacity of 1,600 gpm; an algal culture facility; a nutritional kitchen and drying rooms; pathology and general laboratories; shops, boiler-room, and storage buildings. Computerized systems for data collection and analysis are utilized.

All three species of Gulf of California penaeids are under investigation: *Penaeus californiensis* (brown), *P. stylirostris* (blue), and *P. vannamei* (white). Occasional comparison trials are made with such exotic species as *P. aztecus* from the Gulf of Mexico and *P. monodon* from the Republic of China. The hatchery operates primarily with wild gravid females captured offshore, although one of the native species (*P. californiensis*) has closed its own cycle, reproducing spontaneously in the controlled environment of the raceways.

The most significant intensive-culture grow-out to date (March 1976) has been achieved with *P. stylirostris*; animals with an average weight of 20.7 g were recorded after 24 weeks in the system, with a shrimp biomass in the raceway at harvest of 2.31 kg/m^2.

Improvements continue in raceway and tank design, water quality systems, and overall management techniques. Recent emphasis has been placed on the development of new postlarval grow-out systems, *Artemia* grow-out, and the establishment of an algal culture facility for the hatchery. An engineering prototype of a new raceway is under construction with a larger-scale systems prototype planned for 1976-1977.

In nutrition and food technology, two cardinal challenges are being faced simultaneously: the establishment of minimum and optimal nutritional requirements throughout the life cycle, and the development of formulated diets utilizing commercial feed components and marine products to achieve the requirements. Related investigations include feed stability and the use of various binding agents.

Effective feed conversion appears to be enhanced in raceways inoculated with select species of algae for the photosynthetic production of oxygen and diet supplementation.

It was anticipated initially that these experiments in intensive culture would identify new problems in the health of population-stressed animals. This expectation has been justified. To date, 13 diseases or disease-like conditions have been observed, for most of which preventive techniques and/or successful chemotherapeutic treatments have been developed by ERL's pathology section. Two of the most significant etiological agents appearing in the high-density populations have been *Fusarium solani* and a *Leucothrix* sp., the latter causing a filamentous gill disease (see Chapter VI for further details).

D. Brief Reports from the Workshop--Extensive Grow-out Systems

In the realm of extensive grow-out, Dr. William MacGrath of Ralston-Purina reported on work in progress at that company's two pilot projects and Mr. Hoyt Holcomb of Texas A&M University reported on work in progress at two of that institution's research facilities.

Dr. MacGrath, Ralston-Purina's Operations in Crystal River, Florida, and More Recent Developments in Panama

The Crystal River facility began about five years ago. Its purpose is to determine the economic viability of raising penaeids in heated power plant effluent. It is a joint venture with the Florida Power Corporation. The maximum temperature elevation of the Crystal River power plant is 14° F. By and large, the material available for forming grow-out ponds is limerock; however, the area is underlain with porous aquifer so that a number of problems have plagued the operation. Among them are unexpected recession of pond water into the aquifer, difficulty in draining ponds because of in-seepage from the aquifer, and the instability of the limerock which results in berms eroding into the ponds. Much of this has been overcome by lining the ponds with reject concrete from the power plant where new construction of a nuclear facility has been underway. Ponds are of the flow-through type and flow rates are adjusted to maintain optimum temperatures insofar as possible.

The use of power plant effluent can be used to extend the growing season of shrimp in Central Florida and can even be used to overwinter potential spawners. However, one should realize that a power plant (at least today) is in the business of generating electricity and not producing warm water. In addition, the peak use of electricity in Florida is in the summer when the warm water is not as essential as it would be in the winter when peak needs are lower.

The Ralston-Purina project at Crystal River has been successful in identifying and screening species of shrimp which have been shown to be commercially viable. Over the years, Crystal River workers have evaluated carefully 14 species of saltwater shrimp from all over the world. Production figures at Crystal River with these various species have ranged from 0 - 4,000 lbs./acre of commercial-size shrimp. Standing crops of better than 5,000 lbs./acre of 2.5-gram shrimp have been harvested.

The Ralston-Purina land in the Republic of Panama is quite barren and is in a region with tidal fluctuations as large as 23 feet. This demanded a perimeter berm to prevent tidal flooding. As the facility was constructed, ponds and their surrounding berms were formed with bulldozers backed up with draglines when needed. By and large, the bulldozers are much faster while draglines do a better job if one can afford the time. There are sixteen 1/4-acre nursery ponds in the Panamanian facility, twenty-four 2-acre grow-out ponds, three 6-acre grow-out ponds, and one 17-acre unit.

In order to maximize control, water for the ponds is pumped from the adjacent "estera" rather than reliance being placed upon the tidal flow. There are four 14-inch-diameter fiberglass intake pipes delivering between 5,000 and 12,000 gpm depending on the tide level. Intake water is screened through a graduated filter series and this turns out to be a labor-intensive approach as it demands constant manual cleaning of the screens. Shrimp in the grow-out ponds are fed formula rations with a commercially available mobile blower.

In operation, PL's are harvested from the nursery ponds through a drainage system which results in "99.9%" recovery. They are harvested in batches of 30-40 kg and weighed before being put into aerated tank trucks for transfer to the grow-out ponds. This allows the operators to monitor the productivity and survival rates of the nursery ponds, to determine early feeding rates for the grow-out ponds, and to monitor survival rates in the grow-out ponds.

Mr. Hoyt Holcomb, on the Texas A&M Agricultural Extension Sea Grant-sponsored Shrimp Research Program Conducted in Corpus Christi and Galveston, Texas

In the Galveston area facility, built in 1974, there are twenty-one 1/2-acre ponds divided by a central seawater reservoir plus two 5-acre ponds. These are all served by the reservoir which itself is supplied with seawater by a single pump drawing Surface Water from Galveston Bay. The soil is good clay pan and there have been no problems with percolation. The bottoms of all the Galveston ponds are above mean high tide. The 1/2-acre ponds have flat, sloped bottoms and can be drained completely while each of the two 5-acre ponds has a shallow internal ditch and can be drained only to the surface of the ditch. A flow-through approach is used with water in the 1/2-acre ponds turning over completely about every 45 days while the 5-acre ponds require "much longer" to achieve 100% water replacement.

The Corpus Christi facility began in 1972 with one 1/8-acre nursery pond, one 1/4-acre intermediate pond, and one 1/2-acre grow-out pond. Shrimp are moved from one pond to the next every 30–40 days. All of these three ponds, too, are above mean high tide and can be completely gravity drained.

More recently, eighteen 1/4-acre ponds have been built adjacent to the Corpus Christi Central Power and Light Company's electrical generation plant. All eighteen employ the flow-through approach and can be completely drained. Two pumping locations supply water to these ponds. One is located in the cooling water inlet canal for the power plant and the other is located in the cooling basin for the outlet water. Thus the water temperature in the shrimp ponds can be maintained considerably above ambient. In the Corpus Christi production ponds, productivity has been as great as 4,396 lbs./acre/month.

Current research being conducted at the two sites includes the following:

1. Comparison of the growth potentials of a number of native and exotic species.

2. Stocking density studies ranging between 25,000 and 150,000 per acre.

3. The maturation studies which Dr. Conte discussed in Chapter IV.

4. Development of effective systems management techniques for 1/2- and 5-acre ponds and comparisons to determine how the two differ.

5. Determination and comparison of the effects of stocking different larval and postlarval stages into grow-out ponds.

6. Analysis of the live bait shrimp market as a potential outlet for cultured penaeids.

7. Experiments with the comparative effectiveness of night harvesting methods.

E. Marifarms, Inc.

It was mentioned earlier that Marifarms, Inc., was not represented in the workshop. However, that firm's efforts most certainly are significant in the embryonic American shrimp industry. Moreover, the Marifarms operation is neither strictly a pond-oriented operation nor is it an intensive-culture operation with respect to its grow-out approach. Therefore, we have collected what information we could and summarized it under a separate heading in this chapter.

Marifarms was incorporated in 1967 and began commercial operations in 1971. In that year the firm realized a harvest of 210,000 pounds (after processing) which was sold for $215,000. The firm incurred a net loss of $2,307,580 for 1971 and an additional $551,537 for the five months ending in May 1972 according to its June 1972 prospectus. According to its 1975 annual report, the firm produced and sold $269,367 in shrimp in 1973 and $756,431 in 1974. In 1975, hatchery disease problems and Hurricane Eloise resulted in sales dropping to $228,521 according to the 1975 annual report. That report shows that the firm's annual net operating losses declined steadily from $1,894,267 in 1971 to $841,443 in 1974 and then rose to $1,409,700 in 1975 as a result of $777,050 in losses attributed to the hurricane damage.

The central grow-out "pond" employed by Marifarms is a 2,500-acre embayment with its seaward end enclosed by a long nylon net treated with an antifouling agent. There are, in addition, two 300-acre lakes fed by pumped seawater. Tidal flows and storms produce severe strains on the net which encloses the embayment, and developing a completely effective design has taken some time and incurred significant costs in capital and lost stock.

Postlarval shrimp were originally contained in pens within the larger grow-out enclosures. As the shrimp became juveniles, they were released into larger second-stage pens but not yet released into the large enclosures. When the shrimp reached adequate size, the second-stage pens were opened and the animals moved into the large embayment and lakes to mature. Timing was critical as a release too early resulted in low survival due to predation and a release too late depressed growth. The total length of the nets used in all three stages is about 12 miles. More recently, the firm has found it more effective to employ the nursery pond concepts described by Mr. Hysmith earlier in this chapter.

Water exchange for all posthatchery grow-out operations is essentially tidal in the embayment and pumped in the lakes. Although the net barrier precludes entry of most adult swimming predators, the company's literature mentions no active measures being taken to rid the 2,500-acre grow-out pond of any predators or competitors which may be already present or which may enter as larvae. In fact, the nursery pond operation appears to be partially a mechanism to reduce predation, the other reasons being: 1) that very small shrimp can escape through the larger mesh of the main barrier net in the 2,500-acre enclosure and 2) the multiple cropping mentioned later.

The large embayment and the lakes, of course, already are naturally bioproductive so that low population densities of shrimp could be grown without supplementary feeding. However, the success of Marifarms' venture depends upon increasing the natural productivity levels very markedly and this demands supplementary feeding. During 1970, the firm fed a total of 800,000 lbs. of fresh (refrigerated) trash fish which were ground and distributed by a feed boat. But supplies of trash fish were unreliable; spoilage was a constant threat; the operation was messy and malodorous; and laboratory tests indicated that the effectiveness of trash fish as a supplemental feed was in doubt. During 1971-1973, the firm had its own formula feed produced by an unaffiliated local concern. In 1974, Marifarms used a Ralston-Purina proprietary diet with good success, feeding 2.2 million pounds during that year.

Harvesting is done with a standard shrimp trawler operated in the 2,500-acre embayment. In the marine lakes, harvesting is accomplished by trawling from a special skiff--storing the harvest in a transfer net which is unloaded into icewater at dockside every hour or so. In both cases, the shrimp undergo preliminary processing and are delivered to nearby processing plants within 24 hours of their time of harvest. According to Marifarms, this results in a product far superior to that obtained from traditional shrimping operations wherein it may be more than a week on ice for the shrimp between the time of harvest and the time of delivery to the processing plant.

Marifarms also employs multiple cropping. Brown shrimp are hatched in February and March and harvested in June through August. White shrimp are hatched in May and June and are moved progressively through the same areas employed for the browns to be harvested during the October-December period. Thus, Marifarms makes maximum use of its capital facilities and maintains an essentially year-round operation.

In the summer of 1975, Hurricane Eloise inflicted a reported level of damage amounting to $770,050 on Marifarms capital improvements. The hurricane also resulted in the loss of virtually all the stock in the 2,500-acre embayment. After considerable effort, the firm obtained adequate loans from public and private sources and operations have now resumed. The story of Marifarms' courageous attempts to achieve economic success with shrimp farming (which is told all too inadequately here) is a tribute to American determination and tenacity, as well as a rather clearly marked trail for those who choose to follow. May they show a healthy profit in 1977.

F. Discussion Period

In the Galveston workshop, the four brief presentations summarized under C and D above were followed by a discussion period during which additional information surfaced.

In response to questions from the floor, Dr. MacGrath mentioned that the total bottom slope of Ralston-Purina's ponds is 1 to 1 1/2 feet. Stocking densities of 1-gram juveniles employed by Ralston-Purina are about 40,000/acre. No gut analyses have been done to determine what the shrimp are eating or if the diets

Photo by W. D. Madden

Figure V-4
A portion of AQUACOP's extensive PVC piping system.

consumed vary from pond to pond. Dr. MacGrath guestimated that, from Ralston-Purina's experience, optimum production pond size lies between 10 and 20 acres. Dr. Webber guessed it was closer to 10 acres. To simplify construction, ponds typically are rectangular, the trick being to have the berm material just equal the volume of material scraped out to dig the pond, thereby leaving no waste material. This tends to create rather long and narrow ponds and these also act to facilitate harvesting, feeding, biomass sampling, and so forth.

Mr. Mock's earlier statement that the NMFS, Galveston, closed system was not yet economically viable raised a question of just what the major cost factors were. Mr. Mock responded that maintaining optimum grow-out temperature (either through insulation or active heating) combined with feed costs were the overriding concerns.

Dr. Holcomb responded to floor queries by noting that two of the Galveston ponds he mentioned earlier are being employed as static ponds rather than flow-through. One is stocked at 25, 000 PL's/acre and the other at 60, 000/acre. Shrimp in both ponds are growing acceptably but the water takes on ambient temperatures and usually remains down around 22° C. Consequently, growth rates may be expected to be somewhat depressed. By and large, feeding rates are not calculated on the basis of biomass samples in today's practices. Rather, there are sliding scales, based upon growth and survival assumptions, which result in increased feeding rates as the shrimp gain weight. These scales may be adjusted when biomass samples indicate significant variance from the assumptions. None of the pond system panelists mentioned in the workshop that excessive rainfall had caused them severe problems yet and no active measures to deal with the eventuality were mentioned. Yet others have found intense rains to impose severe salinity and water stratification problems. Although by and large, penaeids are quite tolerant of both low and high salinities, the danger of both freshwater intrusions and overly high salinities resulting from evaporation must always be considered and, where real, must be handled in the design of the culture system.

In closing, Dr. Jack Parker of Texas A&M University observed that intensive and extensive penaeid grow-out systems need not be viewed as mutually exclusive competitors. Instead, the two have features which can be complementary (using extensive systems for overwintering of pond-reared animals, for example), and it would be a mistake to ignore the possibility of the two approaches operating cooperatively.

Dr. Parker also noted that the Texas A&M experience leads to a conclusion that different penaeid species perform differently in different locations and environments. It is, therefore, dangerous to select a given species for culture in one region on the basis of its performance in another. In the long run, the shrimp farmer may be better off by investing time and money in species comparisons at his own location.

Moreover, two additional factors cause present grow-out technology to remain an art rather than a science. First, there are no valid rules for feeding rates, per biomass, by species, for any of the various feed formulae. Second, optimum water depth by species and location remains a mystery solved only by guess and it is frequently more a function of site topography and hydrology than a function of penaeid physiological requirements.

Add these final observations to the uncertainties which emerged throughout this chapter and it becomes abundantly clear that the state of the art of penaeid grow-out today remains very much just that--an art.

VI. DISEASES OF PENAEID SHRIMP

Opinions concerning the importance of disease in penaeid shrimp farming vary. Some workers consider it a major roadblock to successful culture; others consider the disease threat to be serious but probably not overwhelming; and a few seem relatively unconcerned. Rick Paquette's poll (see Chapter XII) indicates that about 90% of shrimp culturists consider disease control to be a serious problem. This chapter sketches the problems and some of the solutions that have emerged so far. But before looking at specific pathologies, it is appropriate to mention several broader ideas.

First, indications from the workshop left little doubt that virtually everyone actively engaged in shrimp farming has encountered some disease problems; not always devastating, but the pathogens definitely are there. Dr. Carl Sindermann, in an address before the 1971 gathering of the World Mariculture Society, noted a propensity for aquaculturists to look away from the threat of disease problems rather than face the subject squarely. He emphasized that making believe would not make such problems disappear and that, in fact, looking intensely for pathogens in aquaculture, spotting them early, and then developing effective preventions and cures seemed to make better sense. It seems worthwhile to restate that message here as was done in the workshop.

Second, even though many disease organisms may be present in a culture, epizootics frequently can be avoided so long as the shrimp are not otherwise stressed. Physical injury from inexpert handling or containment, extremes in oxygen levels, temperature shock, inadequate nutrition, and a variety of toxins in the water all can stress animals and leave them vulnerable. When populations of stressed animals are dense, even low levels of stress may give pathogens the toehold they require to enter upon the exponential segment of their growth curve and cause devastating epizootics. Thus, <u>the</u> primary disease preventatives are a healthy environment and an adequate diet.

Third, it is appropriate to place the emphasis on prevention. Procedurally it is easier and perhaps it is personally more gratifying for pathologists to diagnose and cure epizootics already underway. But the shrimp farmer is done a far greater service if routine prevention measures result in his animals never getting sick in the first place.

Lastly, judged by the weight of opinion, penaeid shrimp pathology deserves far more intense and nationally coordinated effort than it is receiving today. And this statement in no way detracts from the excellent work already being conducted by many individuals and institutions.

Penaeids are subject to virtually all the basic forms of infection that attack marine and terrestrial creatures including:

Larvae of metazoons

Protozoa

Slime molds

Fungi

Bacteria

Virus

There also appears to be a possibility that penaeids could play the role of intermediate host for some human pathogens, although there is to date no direct evidence that this is so.

At this time, the information available on shrimp diseases is incomplete and not without contradictions. With this in mind, we have chosen to draw directly from selected literature and from the reports given in the workshop. For a more complete review of shrimp diseases, the bibliography contains over 30 references.

A. Dr. C. J. Sindermann

Without doubt, Dr. Sindermann's benchmark book "Principal Diseases of Marine Fish and Shellfish" (Ref. 479) provides the most complete summary of shrimp pathology up to the time of its publication in 1970. Accordingly, information from that publication initiates this part of this chapter.

Sindermann mentions only one unsubstantiated report by Hall (Ref. 483) of a nonpathogenic virus infesting *Penaeus indicus* from the Rufiji Delta. Its presence is indicated by a "whitish blotchy disfigurement dorsally under the integument."

As of 1970, this was the only information relating to viral infections of shrimp and one of the very few relating to viral infections of decapods at large.[a]

Sindermann cites Anderson & Conroy (Ref. 347) briefly describing a "brown spot disease" of *Palaemon serratus* which was characterized by erosion and destruction of the exoskeleton and was particularly evident in crowded culture tanks. A myxobacterium was thought to be the causative agent but this was only by association. The etiology was not determined.

As for the fungi, Sindermann cites Anderson & Conroy again mentioning an organism tentatively identified as *Pythium* sp. which was associated with exoskeleton damage and characterized by massive invasions of hemocoel, muscles, and sometimes the gonads. Infection was rapid and caused mortalities in culture tanks. Sindermann also mentions a chytrid-like parasite in the gills of *Dichelopandalus leptocerus* taken off the northeast coast of the U.S. The infection was associated with a dark brown discoloration of the gills and it was suggested that the infection impaired respiration and could be an important mortality factor in shrimp.

According to Sindermann, protozoan infections of shrimp are all too common. Microsporidians (*Nosema* sp. and *Thelohania* sp.) have been reported from all three of the major commercial species of Gulf of Mexico shrimp and from *Penaeus braziliensis* as well. These infections typically are associated with a white opaque coloration of the body muscles as well as external black spots. Infected individuals do not survive well in tanks and are discarded in processing plants, a potentially serious loss to the industry.

Gregarine protozoa of the genus *Nematopsis* are reported to infect Gulf of Mexico shrimp. Infections occur in the digestive tract and are demonstrably a result of shrimp eating mucous strings from bivalve molluscs where the spore stages of the parasite occur. Another gregarine, of the genus *Cephalolobus* (Kruse) was found in the intestine of pink and brown shrimp from Florida and in *Solenocera membranacea* from the French Mediterranean. Sindermann does not go into the pathology of these infections.

Ciliate protozoa of the genera *Lagenophrys* and *Terebrospira* have been reported as ectoparasites on shrimp from Japan (*Leander squilla*) and on *Palaemon varians*. The organisms may penetrate the exoskeleton and, with the latter genus at least, a complex life cycle intimately related to molting of the host is capable of producing heavy infestations in restricted environments.

---

[a]Dr. William MacGrath, in a personal communication, mentioned viral problems being encountered by Marifarms, Inc., in Florida between 1972 and 1976. Principals of Marifarms have not yet confirmed this to the authors of this report.

According to Sindermann, larval helminths occasionally infect all three major commercial species of Gulf of Mexico shrimp as well as species from other parts of the world, although he does not mention the symptoms associated with the infestations. Adult endoparasitic females of the tubellarian, Kronborgia caridicola have been reported in shrimp from the North Atlantic and may be fatal to the host. Larval nematodes of the genus Contracoecum have been reported in Florida shrimp as have adults in one specimen of Pandalus borealis (Krøyer) from British Columbia.

Finally, Sindermann cites numerous reports of epicardian isopods infesting shrimp from far-flung locations and usually causing suppression of breeding characters sometimes associated with unilateral protuberances of the exoskeleton.

B. Dr. John Barkate, Microbiological Laboratories of Ralston-Purina, Inc.

An excellent recent report of diseases affecting shrimp of the Gulf of Mexico was done by Barkate et al in 1974 (Ref. 480). We extract here from that report.

Barkate and his associates have identified a very large filamentous bacterium which can cause death in postlarval P. setiferus and P. aztecus within 48 hours. In rearing tanks, the organism can be widespread among the bottom detritus without becoming epizootic, so long as water quality remains good. Mortalities of 30-100% are associated with fouling of the water and the report does not state which event comes first. Superficially, the bacterial colonies resemble Sphaerotilus or Caryophanon but the very large size and other characteristics suggest no direct relationship to these two genera. It may also be related to Bacillus cereus var. mycoides; but it did not clearly fit any description obtained from the literature. No prophylaxes was described.

Barkate reports repeated infections of very young shrimp by a marine fungus phycomycete of the genus Lagenidium. The species had not been determined. Infected PL shrimp tanks experienced complete mortality in as little as two days. At the time of the report, attempts to isolate the organism on culture media had been unsuccessful and the life cycle has not been described. No curatives were offered.

One case of heavy mortality of PL's was associated with heavy infestations of the stalked protozoan Glossatella in the water of the rearing tanks and concentrated around the gill regions of the hosts in sufficient density to interfere with respiration.

Barkate also reported that a slime mold of the genus Acrasis had been identified in grape-like clusters in the bodies of shrimp dead for several hours. At the time of the report, this protist was thought to be a secondary condition probably feeding on heavy concentrations of bacteria in the dead hosts. If this is true, the so-called "Grape Disease" has no pathological significance.[a]

C. Dr. S. K. Johnson, Texas A&M University

Dr. Johnson's 1974 paper (Ref. 481) brings in some new and some old ectoparasites and extends the species range of a few previously reported infections. In Texas, the native shrimp species are *P. aztecus*, *P. setiferus*, and *P. duorarum* and the "exotic" species Johnson mentions are *P. vannamei*, *P. stylerostris*, and *P. occidentalis*.

Ectocommensals found with some frequency by Johnson in penaeids were the protozoa *Zoothamnium* sp., *Epistylis* sp., and *Acineta* sp.; the former seeming to prefer the gills and the two latter rarely found on the gills. *Zoothamnium* probably causes shrimp mortality in low-oxygen environments and the pathogenic importance of the other two ectocommensals remains uncertain. Formalin at 25 ppm appears to control *Zoothamnium*, but oxygen concentrations should be kept high during treatment.

The ciliate *Lagenophrys*, not previously reported on penaeids from the Americas, was found on *P. setiferus* and *P. vannamei*. This organism, as does *Acineta*, seems to prefer exterior body surfaces rather than the gills. Johnson also confirms the occurrence of the gregarine protozoan *Nematopsis penaeus* in the intestines of the Gulf Coast animals he examined; the life cycle stage found usually is the trophozoite.

A filamentous bacterium which may have been *Leucothrix* was observed frequently and appeared to prefer the body areas beset with setules unless the infection was very heavy, in which case the gills were involved.

In 1973 a brownish discoloration associated with atrophy of the gill filament tips was observed but no biological agent was isolated.

The parasitic isopod *Probopyrus pandalicola* was seen on the small shrimp *Palemonetes pugio* living in some of the penaeid ponds but was not recorded on the penaeids themselves.

D. Dr. Johnson in the Workshop

Dr. S. K. Johnson was the first speaker in the workshop as he was the last writer cited above. In the workshop he underlined a high-quality rearing environment and careful handling of penaeids as the primary disease deterrents. Any

---

[a]Dr. Charles E. Bland (personal communication) questions this diagnosis and suggests that the organism may have been a member of the genus *Thraustochytriu[m]* a member of the Labyrinthulales.

stress, be it physical, chemical, or biological, will lower disease resistance and open the way for epizootic explosions of some of the multitude of potential pathogens always present in any shrimp culture.

It also seems possible that some pathogens could employ marine shrimp as intermediate hosts on their way to infecting humans. There is no direct evidence of this as yet and no human pathologies have been traced to raw *Penaeus japonicus*--a delicacy in Japan. Nevertheless, those whose business it is to worry about such matters do not dismiss the possibility that some trematodes and nematodes could be involved in this sort of a cycle as could, perhaps, some viruses. Since no one has yet done anything empirical in this regard, all is conjecture at this point in time.

E. Dr. Donald Lightner - Disease Problems of ERL

The University of Arizona/University of Sonora shrimp culture research facility at Puerto Peñasco, Sonora, Mexico, is an excellent facility for the study of shrimp diseases. Shallow, clear waters in the tanks and raceways allow easy visual observation of abnormal conditions, and the high-density rearing concept being developed there virtually assures that disease conditions will occur from time to time.

The marine fungus *Lagenidium* sp. has infected the larvae of *P. stylirostris* and *P. californiensis* on several occasions, and produced mortalities that ranged from less than 10% to nearly 100% of affected hatchery tank populations. Malachite green therapy as described by Bland (in this chapter) has been successfully used to combat epizootics due to *Lagenidium* sp. at Puerto Peñasco.

The development of preventative measures and/or therapy for filamentous gill disease, which is due to a *Leucothrix*-like filamentous epiphyte, has received considerable emphasis in the research program at Puerto Peñasco. Of the chemicals tested to date, potassium permanganate and various copper compounds have shown possible beneficial effects in therapeutic use. Cutrine Plus, a commercial algacide, appears to be particularly effective.

Fusarium solani infections have resulted in substantial losses in juvenile and adult *P. californiensis*. The first epizootic due to *F. solani* observed at Puerto Peñasco followed a 10 ppm potassium permanganate treatment given for one hour to 10- to 12-g juvenile *P. californiensis* for filamentous gill disease. Gill damage resulting from the treatment provided sites for establishment of *F. solani* infections. No effective therapeutic methods for *Fusarium* infections are known for shrimp.

Epizootics due to Vibrio spp. have on occasion been damaging to ERL experiments. Vibrio epizootics typically have followed some sort of stress situation, and have been successfully treated by the use of feeds medicated with broad-spectrum antibiotics such as Furanace, Furacin, and Terramycin. Methylene blue and malachite green have been tried as prophylactics with varying degrees of success.

Among the protozoan pests, the peritroch Zoothamnium and an unidentified hypotrich have on occasion become abundant on the gills of tank- and raceway-reared P. californiensis and P. stylirostris and resulted in some mortalities. Successful therapy has been accomplished with Formalin at 25 ppm for 24 hours or higher concentrations for shorter periods of time. No microsporidia have been observed in tank- or raceway-reared penaeids at Puerto Peñasco, although wild shrimp infected with a Nosema sp. have been observed in commercial landings from Gulf of California waters off Puerto Peñasco.

A new disease of shrimp, given the name "black death" by ERL workers, has caused significant losses during several epizootics. Affected P. californiensis postlarvae or juveniles displayed characteristic blackened areas under the cuticle of the general body surface, in the stomach and hindgut walls, and in the gills and gill cavity. P. stylirostris displayed blackened gills. The black pigmentation has been demonstrated to be melanin deposited by agglutinated and necrotizing hemocytes that were present in dense masses or nodules. The etiology of the disease remains unknown, but dietary deficiencies of L-ascorbic acid have been used to induce the disease in P. californiensis.

F. Dr. Donald H. Lewis of Texas A&M

Dr. Lewis stated that bacteria of the genus Aeromonas occur frequently in marine crustacea and in shrimp. Vibrio are also seen frequently and the only gram-positive forms are Gaffkya which have been infectious agents in other decapods but have not yet been identified as pathogens of shrimp. Luminescent bacteria (probably Photobacterium--ed.) are found on dead penaeids, but pathogenic actions have not yet been associated with this form.

Dr. Lewis stated that the antibiotic susceptability of Vibrio was high and that "Gentomyocin" was the best all-around antibiotic he had found, not only in terms of its bacteriocidal effect but also in terms of the degree to which it is incorporated into shrimp tissue.

According to Dr. Lewis, shrimp definitely do produce an immune response to bacterial infection but it is not yet clear that this response is effective in combating infections that have taken a firm toehold.

G. Dr. Charles Bland, East Carolina University

Dr. Bland is a mycologist presently working on controls for Lagenidium. He has found the most promising fungicides to be:

- "Treflan" (trade name) at 3-5 ppm
- "Surflan" (trade name) a possibility
- DS 90-73 (trade name) produced by ICI-America, used at 1.3 to 3.2 ppm, is very effective but its toxicity to shrimp has not yet been tested (no longer being manufactured)
- Malachite green is still the most promising of all used at 0.006 to 0.01 ppm
- "Terrachlor" (trade name) at 3-5 ppm is probably somewhat more toxic to shrimp than malachite green.

Fungicides are first tested in the laboratory on pure cultures and then those which appear effective in the laboratory are tested on infected penaeids. Reference 514 contains the details of Dr. Bland's recent work with Lagenidium.

H. Conclusion

There is little doubt that this brief chapter is not complete. Nevertheless, the patterns that emerge from it probably are valid. Penaeid pathology is still a long way from achieving the status of a unified subdiscipline. It lacks an adequate base of organized literature and it lacks an adequate pool of dedicated talent and institutional momentum. Possibly it is inappropriate to apportion heavy resources to shrimp diseases until other serious barriers are well overcome. But it will be very frustrating to spend years overcoming the other roadblocks only to find the industry stymied for another decade by repeated epizootics unrestrained by the talent and research base needed to control them.

VII. PENAEID SHRIMP NUTRITION AND FEEDS

Frieda Taub, in opening the nutrition workshop of the 1972 World Mariculture Society meeting, characterized approaches to aquacultural nutrition as tending toward three categories: 1) imitation of natural diets, 2) trial and error experimentation with formula feeds intended for other organisms, and 3) controlled (formula) feeding with chemically defined diets. All three of these trends are discussed to some extent in this chapter, although the search for a breakthrough in formula feeds receives the weight of attention.

One who has not yet fallen into the pandora's box of penaeid nutrition might think that all there is to it is to find a feed that works and then use it. This is true as far as it goes; the problems lie in finding a feed that works. The crux of this problem is that one finds one's self dealing with a multivariate phenomenon of imposing dimensions. There are at least eleven major variables, all of which interact with one another. They are:

- Stage of growth
- Species of shrimp
- Water quality and temperature
- Feed stability (binding)
- Presentation
- Percentage and derivation of lipids
- Percentage and derivation of carbohydrates
- Percentage and amino acid composition of protein
- Health of shrimp
- Effect of feeds which occur naturally in the rearing environment
- Feeding rates

The formula for calculating the possible permutations in this list is $n^2 - 1$ which equals 120. If we now recognize that this list of eleven contains only major headings and that usually many more than ten significant species can easily be listed under each heading, we can see that the actual number of potential permutations facing the few hard-core researchers in this specialty is more like 120^2 to 200^2 or between fifteen and forty thousand! Nutrition experiments may last from a month up to six months and ideally should be replicated at least once to be scientifically acceptable. If we take a mean of three months for each experiment and another three months for its replication and then agree that the number of experiments demanded by a systematic approach to the problem is somewhere in the neighborhood of 25,000, then we have agreed that 12,500 experiment-years are required for a systematic attack on the nutrition problem. Of course this is ridiculous and of course we must rely on creative scientific insight to throw out *a priori* the majority of possible permutations. Nevertheless, the permutations, whatever their real number, are truly many and the "*a priori*" decisions must have a sound foundation in prior empiricism. So this quick logical detour does allow one to stand away a little and glimpse the real magnitude of the shrimp nutrition problem.

Since larval nutrition is so intimately bound up in larval rearing, that subject was covered about as well as it can be now in Chapter II. Therefore, after a quick look at natural feeds, this chapter devotes itself, as did the workshop, primarily to the problems of developing formula feeds for growout.

A. Natural Feeds

Except possibly for a few highly controlled rearing experiments, all cultured shrimp are taking advantage of some natural feeds. When one recognizes that dissolved organics, bacteria, plankton, and detritus as well as lab-lab are all natural feeds, it becomes clear that keeping even the most highly controlled culture entirely free from natural feeds is as difficult as it is inadvisable. According to informed views expressed in the workshop, the practical shrimp farmer is well advised to encourage the growth of all varieties of natural feeds in his cultures and then to supplement as experience proves appropriate with formula feeds and/or "natural" feeds in the form of seafood wastes. This approach, of course, cannot be overdone without impairing water quality and courting disastrous epizootics.

B. Formula Feeds

Lipids from animals and vegetables, carbohydrates from terrestrial crops of all sorts, protein from marine and terrestrial animals as well as terrestrial plants, binders from a plethora of common and exotic organic sources, all are being tried as shrimp formula feed constituents. Dr. Sam Meyers mentioned in the workshop that, so far, researchers know vaguely the sorts of lipids, carbohydrates, and proteins that seem important in penaeid nutrition; however, they don't know the relative proportions by species, let alone by growth stage within species. This will become clearer as this all too brief chapter unfolds.

Constituents of Formula Feeds

First of all, the amino acid demands of penaeids by species (if there are significant species differences) are virtually unknown. The amino acid composition of shrimp protein is known but which ones the animals synthesize is not. Moreover, these demands are likely to vary by growth stage and such variances have not yet been studied. The literature indicates that soybean meal, shrimp and fish meal, and squid meal all are relatively acceptable sources of protein for penaeids. In the workshop, Ms. Zoula Zein-Eldin expressed the opinion that there are wide species differences in demands for protein percentages in formula feeds, as well as differences in required protein levels as a function of the formula used. Her *P. aztecus* showed depressed growth rates when protein in one experimental formula dropped below 53% while 32% was adequate in another formula. Dr. McGrath reported good growth of *P. setiferus* in Florida on feeds of another formula containing only 20-25% protein.

The efficiency with which protein is assimilated by penaeids very likely is affected by the relative proportions of lipids and carbohydrates in the formula as well as the amino acid composition of whatever protein source is employed. Additionally, trace constituents may be active determinants of protein assimilation but have received no research attention whatever as yet. Ms. Zein-Eldin noted that the 53% threshold mentioned above was a threshold for one formula and gave growth rates comparable to 32% protein contained in another formula. In comparison, she noted, two Japanese formulas containing 57% and 58% protein, gave, respectively, growth rates equal to and *twice as great as* the growth rates obtained with her 32% protein control formula. The protein source was not given. Sick and Andrews (Ref. 431), using *P. duorarum*, obtained relatively high growth rates employing soybean meal as a single protein source in formula feeds; in fact, in their experiments, growth rates with soybean meal were higher than the growth rates obtained with shrimp meal or with menhadden meal laced with *Artemia*. Clearly, the protein issue remains confused and the effect of various protein sources (and their percentages) in formula diets must be intimately bound up with lipid and carbohydrate compositions and fractions.

The workshop speakers attended primarily to protein but other workers have devoted some attention to the other constituents. Sick and Andrews (ibid.), for example, found that the type and percentage of both lipids and carbohydrates affected growth rates and produced differences in the composition of the tail meat. To sum it up, the source and the fraction of proteins, carbohydrates, lipids, and probably fatty acids, as well as their chemical form in formula feeds, is critical to the growth rates and final biochemical composition of penaeid shrimp. In addition, there very likely are differences among species and between growth stages of each species. So far, it is rare to find a formula feed producing growth comparable to combinations of natural feeds such as clam, shrimp, or squid meat coupled with natural pond bioproductivity. The research remains to be done. Notice that there probably are several thousand meaningful permutations waiting to be tested in this realm alone. Unless some lucky investigator experiences a revelation from above, the road to truly effective formula feeds is likely to be long and expensive. By truly effective, of course, we mean formula feeds that are inexpensive, reliable, nonpolluting, and which result in growth rates at least comparable to those obtained with natural feeds in extensive grow-out systems.

Binders and Forms of Formula Feeds

According to Dr. Meyers and Ms. Zein-Eldin (Ref. 421), a variety of seaweed extracts, land plant extracts, and animal extracts all are candidate binders for shrimp formula feeds. In the workshop, Ms. Zein-Eldin mentioned that some work was being done in the Galveston laboratory on the use of chitosan as a binder and that the early results were encouraging. Dr. Meyers mentioned that, while shrimp do produce chitenase, it has not been shown yet that they can digest and assimilate chitosan. This leads to a consideration of the nutritive value of the binder itself and beyond that to the ability of shrimp digestive chemistry to break down the binder so as to release the other nutritive elements in the feed. The trick is to find a binder that works with the other constituents, that is water stable for at least 24 hours (and preferably several days), that does at least not reduce (and preferably enhances) the attractiveness of the feed to shrimp, and that is fully digestible by the animals. If the binder contains critical nutritional elements, too, that is fringe benefit. In addition, acceptable binders should be low in cost, adaptable to economical large-scale feed manufacture, storable, selective in leaching (so that chemical shrimp attractants are released while nutrients are not), and themselves noninjurious to water quality in direct chemical terms and in terms of their susceptibility to attack by water-fouling marine microorganisms.

Meyers and Zein-Eldin list 15 possible source categories of binding agents and many of the categories contain a variety of specific sources. Moreover, specific binders work differently at varying concentrations and with differences in feed composition. If there are differences among species and between growth stages in penaeid responses to binders, then there may be a few thousand additional permutations to be investigated in this realm, too. To compound the issue a bit further, there is little doubt that the chemistry of the binder affects the shrimp's responses to the other components of the feed. Also, the effectiveness of the binder as a binding agent can be enhanced under some circumstances by sequestering agents which themselves may affect the shrimp's ability to utilize the feed components.

The full breadth of forms in which feeds may be presented to cultured aquatic animals, according to Meyers and Zein-Eldin, are:

Hard pellets

Expanded pellets

Crumbles

Blocks

Agglomerates

Blower feeds

Autolysed and hydrolized feeds

Oregon moist pellet

Abernathy pellet

Freeze-dried pellet

Not all of these forms may be attractive to shrimp and those which are will vary to some extent in their attractiveness according to species and to growth stage within species. They also will vary in their appropriateness to different culturing methods. Moreover, the effectiveness of the binders varies with the particle size of their nutritional constituents. Lastly, the forms of feeds listed above clearly affect the feed characteristics mentioned previously. Thus, the number of permutations to be investigated under the heading of feed composition and form is further expanded.

Dr. Meyers has suggested (personal communication) that the binder stability problem could be reduced if shrimp could be induced to feed more often. More frequent feeding would reduce the amount of feed fed each time as well as allowing it to dissolve more rapidly. This is an interesting idea which might be put into effect through manipulating the grow-out environment, selective breeding, or both.

C. Measuring Nutritional Effectiveness

It was pointed out in the workshop that present measures of dietary effectiveness are limited to survival rate, increase in overall length, and weight increase. It was then suggested that these measures should be supplemented with such measures as:

- Tail weight
- Digestive gland weight
- Gonad weight
- Gut weight

It was also suggested that biochemical analysis of tail meat should be done routinely in order that we ultimately may learn to feed so as to maximize the protein content of shrimp and shrimp products.

D. Additional Points

Dr. Meyers mentioned that yeasts are a protein source potential which could stand more exploration. Some bacterial cells do contain chiten too.

Mr. Harold Goodwin brought up the question of bacterial flock as a feed potential but the workshop response indicated that the Americas have paid little attention to this Japanese research avenue, and it probably is much more appropriate to penaeid larvae than to juveniles. With respect to this subject, Dr. Meyers has written (personal communication) that more and more workers are employing single-cell protein in shrimp diets and that the Japanese have found that "marine" yeast protein is quite effective as a proteinaceous substrate addition, even when used at concentrations as high as 40%. Dr. Meyers also points out that microbial protein is readily and inexpensively available in the form of brewery (or any fermentation process) wastes in most regions in which shrimp farming is underway now.

Dr. Meyers also offered the opinion that lipids and fatty acids as well as chiten were undoubtedly important in penaeid nutrition. He felt there very likely was a strong interaction between lipids and fatty acids in shrimp diets and it is the types of lipids and fatty acids employed that is of primary significance. He also suggested that water temperature and pH were important in determining rates of chitenase production in shrimp. Penaeids, as a matter of interest, appear to have two chitenase systems, one largely in the hepatopancreas and the other centered around chitenoclastic bacteria in the digestive gland. Still the significance of chiten is largely presumptive since definitive experiments have not yet been done. Nevertheless, all shrimps and prawns are known to feed upon molted exoskeletons and the entire chiten cycle very likely is critical to their growth and health; however, recent indications suggest that it may not be chiten itself but rather the intermediary products from chiten breakdown that are important to exoskeleton formation.

It also appears, from energy balance investigations into pond rearing, that the formula feeds given to pond-reared shrimp do not come close to accounting for the energy demands of the shrimp biomass in ponds. Thus, natural bioproductivity in the ponds as well as cast exoskeletons probably are contributing heavily to penaeid nutrition in the ponds studied. This brings us back to the idea mentioned earlier that commercial growers are well advised to encourage natural productivity in their ponds and to use formula feeds as supplemental diets. It also suggests that research efforts aimed at thorough understanding of shrimp culture pond energy dynamics is in order since this may be a valuable, if not critical, factor in increasing grow-out densities which reduce feed costs.

Returning to the business realities of formula feeds, Dr. McGrath of Ralston-Purina emphasized the lack of leverage aquafeeds have in the formula feed industry. They are, in fact, a tiny fraction of the total formula feed production in the U.S. today and therefore the feed industry is unlikely to devote many of its resources to their development. The industry is also unlikely to get terribly excited about new wonder formulas developed in nonprofit laboratories unless these formulas are inexpensive to produce, not only in terms of their ingredients, but also in terms of the capital investment required for facilities and machinery. To the extent that aquaculture becomes a large formula feed market, this will turn around gradually. But, for the immediate future, the commercial shrimp grower faces a situation similar to that of the original buyers of model T Fords: he can have any formula he wants so long as it is one of those already in production--or he can build his own.

E. Conclusion

We have not here gone into specific formulae for feeds. In some cases, they are proprietary and in others they are purely experimental. Nor has this brief chapter even begun to delve into the details of the penaeid nutrition problem. The bibliography contains no less than 95 references dealing directly with penaeid shrimp nutrition and it would be impractical here to do adequate service to each of these. Moreover, the information in many of these references invites individual interpretation; little of it is conclusive and some of it is contradictory. As Ms. Zein-Eldin said in the workshop, the only dietary demand of penaeids we know with certainty today is that for cholesterol. Nonetheless, there are commercial marine rations that do work in adequately bioproductive shrimp grow-out ponds. The challenge is to make them less expensive and more effective as complete diets.

VIII. PRODUCTION ECONOMICS

In point of fact, this whole report is about the economics of producing penaeid shrimp by culturing; consequently, this chapter necessarily mentions points already made earlier while bringing in additional factors. The intent of the chapter is to characterize the problem as a whole, to cite the major cost factors in shrimp farming, to illuminate the major tradeoffs to be considered in the design of an operation, and to review the points made in the workshop in order to provide an idea of where analysis of production economics lies today. The acute reader will be aware of some redundancy as we bring up points already mentioned. He will also become aware that the structure used in this chapter is only one of many possible structures of this subject. We hope it will prove to be a worthwhile one. First, the major cost variables are discussed. These are followed by examinations of other significant tradeoffs faced by the designer of a new shrimp farm. Then, any points made in the workshop and not already covered in the two major sections of this chapter are reviewed.

A. Major Cost Factors in Shrimp Farming

The way the major cost factors involved in shrimp farms are categorized here, they number twelve. We have not here formally distinguished between capital and operating costs since it did not appear particularly pertinent.

Land

Land is a major cost factor in the design of any shrimp farm unless a company or individual already owns appropriate land not in use; even then, using the land for shrimp farming must be weighed against selling or leasing it or using it otherwise. Land near major markets usually is more expensive. Land that is highly appropriate, since it typically is waterfront property, is relatively expensive. Soil type is important. Substrata are important. Water quality is important. Cheaper lands may bring with them increased costs in other factors. See

Chapter V for more on the land issue. Also, the buy or lease issue is important here.

Water

Usually, the water will be free at its source. The costs of water usually are the costs of getting it from its source to the rearing ponds, tanks, or raceways in a condition that is appropriate for the rearing of shrimp. This means that its physical and chemical characteristics must be appropriate and it must be relatively free from predators, competitors, and disease organisms. An additional cost may lie in processing output water so that it meets any environmental quality regulations that apply. Although, in smaller-scale operations, many predators and competitors can be excluded by fine-mesh filters, manual cleaning of filters is a bother in small operations and a significant expense in larger ones. If this is added to the problem of treating water contaminated by undesirable chemicals and/or low temperature, the combined costs can be high enough to warrant a search for water that is acceptable in the first place. Chapter V has more to say on this subject.

Power

Shrimp farming in the Americas typically involves the use of electricity and, in some areas, may demand some heating. The latter can come from natural gas, solar heating, electricity, or biogas generators. Electricity right now can come from power lines, from fueled generators, or from aerogenerators. Also, pumping water is something that windmills do well.[a] With the costs of conventional forms of power rising frighteningly, it may be valuable in the design of shrimp farms to examine solar energy alternatives and, with present solar technology, this may be particularly true for hatchery operations. One who does so will find the initial capital outlay imposing. True, it usually can be amortized in 5 to 15 years, depending upon the escalation curve used in estimating conventional power prices and depending upon the suitability of the selected site for solar energy conversion devices, but the initial capital needed is expensive and difficult to acquire and the wisdom of investing it is closely coupled to the long-term economic viability of the farm.

Labor

Except for the operation of successful hatcheries, the technical skills demanded by shrimp farm labor frequently are not high. People engaged in the

---

[a] It should be noted here that most commercially available windmills are designed for high hydrostatic head use and most do not produce high flow rates. Thus, windmills may have to be modified or water storage tanks employed for aquacultural use.

occupation do, however, need to understand the biology of the animals as well as the design and operating policies of the farm. Frequently, labor coming from non-Western cultures has difficulty understanding the goal orientation of Western aquaculture. Semanticists recognize that these people very likely have different mental constructs of their world than do Western minds. Therefore, communication barriers may exist and be totally perplexing to the profit-oriented westerner. This is the sort of problem the shrimp farmer invites when he initiates operations in so-called lesser developed countries. The approach he usually ends up following is to maintain a nucleus of westerners in charge of the operation and to use local labor primarily for the more routine tasks.

Smaller operations, of course, can be more or less family run and the present U.S. overburden of labor protection regulations can thereby be circumvented to some extent. But larger operations stateside must hire from the general labor pool and thereby deal with the entire overburden.

Of course, to the extent that the farm is automated, labor costs for routine operations can be reduced. And this is the classical capital versus labor intensity tradeoff. Such capital intensiveness, in order to pay off, demands that the design of the farm be rather well fixed and that prospects for financial viability be very good. It also removes operational flexibility and thereby reduces the latitude of the farmer to refine his operations on the basis of experience.

Cost of Postlarvae

Depending on the size at which the shrimp crop is harvested and marketed, the cost of PL's may be of major or minor importance. For example, if the game plan of the farm is to grow out in one month and market 200 count shrimp, then PL stocking densities must be very high or the biomass carrying capacity of the culture medium is not being fully utilized. This means that many more PL's must be purchased or produced by the internal hatchery and, if the price of PL's is high, this can become a significant cost item. Conversely, if the plan for the farm is to grow out to the 10-15 count size, fewer fry per capital facility and per harvest will be required and the cost of fry may relegate itself to a secondary concern. This issue is closely related to the problem of maximizing utilization of biomass carrying capacity of the culture medium. That problem is discussed further on in this chapter.

Cost of Feed

Feed costs can range from zero to as much as 50 to 60% of the total costs of grow-out. In very extensive pond cultures, natural bioproductivity alone can feed the crop. This approach, of course, very likely will demand a much higher investment in land, water, and probably in capital improvements. It also tends to increase the cost of harvesting. To the extent that grow-out is intensive, supplemental feeding must be employed since the natural bioproductivity simply cannot keep up with the demand of high-density crops. In highly intensive culture systems, nearly all the feed must be supplied by the farmer. However, as grow-out becomes more intensive, production estimates may very well become more reliable, growing space per unit of harvest can be reduced; water intake and discharge volumes can be reduced; harvesting becomes less expensive; and such factors as disease control,[a] predator/competitor control, and so forth tend to become easier to manage and less expensive. Presently, formula feeds cost between 15 and 20 cents per pound. They usually are fed in typical ponds at rates which gradually decrease over the grow-out period from more than 15% to as low as 5% of the estimated shrimp biomass per day. Where the average biomass over the grow-out period is several tens of tons, it is easy to see that the monthly feed bill can be imposing.

Cost of Disease Prevention and Cure

The disease spectre represents a cost to shrimp farmers one way or another. If disease is not combated successfully, the cost takes on the form of diminished or nonexistant harvests and thereby lost revenues. It also may result in some added costs involved in cleaning up the mess left by an epizootic. If disease is combated successfully or not, the chemicals and labor to employ them involve additional costs to the farmer. Probably the least expensive approach to effective disease control is to assure high water quality, avoid other stresses on the animals, and maintain a close watch on the condition of the crop. Once an epizootic gains a foothold, the costs represented by curative measures and counted in lost harvests can soar.

Predator and Competitor Control

As Chapter V pointed out, control of predators and competitors consists largely of physical barriers and clever designs which usually represent both capital outlays and maintenance costs. Combating predators and competitors that have gained entry to the culture may take the form of chemical and/or physical measures. These represent additional operating expenses.

Capital Improvements

Ponds, roads, buildings, plumbing, pumps, fences, possibly solar energy conversion devices, utility extensions, rolling stock, laboratory equipment, furnishings and on and on, all fall into the capital improvement category in that they are expenses that can be depreciated for tax purposes. They and their

---

[a] Disease control also becomes much more necessary since high population densities invite epizootics.

maintenance and replacement do, of course, represent real costs that must be calculated into the costs of production on the basis of some annual depreciation/ amortization schedule. The designer of a shrimp farm must consider tradeoffs between initial costs on the one hand and probable maintenance and replacement demands on the other. He must consider the adequacy of such facilities to do the job and realize that, to the extent that he saves money with inadequate capital improvements, he may well spend money on additional labor and reduce income through lower productivity. The buy or lease question also is important to capital improvements.

Taxes

This is a subject sometimes ignored by the technically oriented. Yet taxes can be a significant cost item in an aquaculture operation. There are land taxes, there may be water taxes, there are sales taxes, and taxes on capital improvements. Taxes also show up in the fringe benefits associated with labor costs and in import and export duties. Some lesser developed countries have less complex and more flexible tax structures than do the various states of the U.S. and its Federal Government. It was suggested in the workshop that it may prove worthwhile to seek royalty arrangements with countries hosting a U.S. shrimp farm since they might prove advantageous for both parties.

Transportation and Communications

The farther away a shrimp farm is from its major markets, from its suppliers of technology, and from its corporate headquarters, the more the costs of transportation and communication are likely to become a significant share of the total. When the farm is remote and national borders must be crossed, subtle costs in the form of stalled operations can arise when a critical item scheduled for delivery on May 2 actually arrives on June 23--if at all. These sorts of costs along with the probability of their occurrence must be factored along with routine travel and communications in any cost analysis dealing with the desirability of locating a shrimp farm remotely.

Natural Disaster

It may seem a bit strange at first to include natural disasters as a cost item in the planning analysis for a shrimp farm, but it cannot be denied that they do cost money. Nor can it be denied that such tradeoffs as site selection and increased investment in capital improvements may reduce the probable financial impact of such phenomena. Thus, it is clear that assessing the probability of occurrence and the likely financial impact of natural phenomena such as earthquakes, tsunamis (tidal waves) and hurricanes is a legitimate component of a complete shrimp farm economic analysis.

Most of those who spoke to this subject in the workshop voiced the opinion that the thing to do was to select the site for a very low probability of natural disaster and then to ignore the subject in subsequent production economics calculations. One lone voice, however, stated that any realistic analysis of production economics would assume a total wipeout of the crop at least one year out of every eight. This figure, however, may have included wipeouts from epizootics. On the other hand, one might wonder whether even this approach is always adequately pessimistic. If each eighth-year wipeout occurs on the eighth year after a farm has built up seven years worth of fiscal momentum, that is one thing; but suppose two successive wipeouts occur on years one and two. By the end of the second year, will the prospective shrimp farmer care that there will now not be another such occurrence for another 14 years?

All in all, the planning for a commercial aquaculture venture of any kind should seek to avoid areas where natural disasters could be a significant problem and then weigh carefully the tradeoffs between protective capital improvements and trusting to the whims of nature. Yes, there are insurance policies to be bought; usually, however, they are very expensive for this sort of protection. In fact, the cost of such insurance is a good index of the natural disaster proneness of a region. To insure or not to insure is yet another tradeoff.

B. Other Significant Economic Tradeoffs

The foregoing cost factors appear to be more or less fundamental. Those discussed here in Section B may be somewhat less obvious in terms of their tradeoff potential, yet they are equally significant. There are six which come to mind:

Extensive versus Intensive

The arguments for and against grow-out extensiveness and intensiveness permeate most of this book. The real situation is a continuum in which five animals per hectare are more intensive than three, and 20 animals per liter are more intensive than 15. Pond cultures using modern flow-through approaches to water exchange may harvest up to 40,000 or so 25-30 count tails per acre but this sort of pond intensity is unusual so far.

Closed and high-intensity flow-through culture systems generally are not considered to be economically attractive at this juncture. The more intensive approaches are likely to become competitive when the technology is well enough in hand to be routine and, as Mr. Mock stated, when they can disconnect themselves from commercial power sources after the prices of commercial power and everything else escalate a great deal. This is an oversimplification of a sort because the discovery of a very cheap and effective formula feed and probably a number of other significant perturbations in the way things are now could change the present picture drastically.

Broadly speaking, the more extensive a culture, the more land and water it will require; the less feed per unit production it will demand; and the more sensitive to environmental variations, predation, and competition it will be. It will also very likely be more expensive to harvest. Conversely, the more intensive a culture, the less land and water it will demand; the more capital improvement outlay it may require; the higher its feed costs will be; and the lower its harvesting costs are likely to be. It usually will be far easier to control disease, predation, and competition, but disease control is also more likely to be needed. Finally, intensive systems tend to be far less sensitive to variations in their external environments.

As of today, the penaeid shrimp industry lies toward the center of the intensive/extensive grow-out spectrum; the weight of commercial pilot operations (with the notable exception of Marifarms, Inc.) is being directed toward ponds using flow-through water systems, employing supplemental feeding at between 5% and 15% of assessed biomass per day, and stocking at densities between 25,000 and 75,000 PL's or juveniles per acre. These ponds may infrequently harvest as many as 40,000 animals per acre but harvests of 15,000-20,000 per acre of 20-30 count shrimp is closer to the rule.

Stocking Densities, Harvesting Densities, and Survival Rates

There are some generalities that can be made with respect to this subject, but there are few hard and fast rules. Generally, high stocking densities will result in better feed conversion rates for a pond, lower survival rates, and slower growth rates. Lower stocking densities are more likely to result in poorer utilization of ponds, lower feed conversion rates, higher survival rates, and faster growth. Today very high pilot operation stocking densities are as high as 40,000-50,000 juveniles per acre and this seems to work out best if the animals have enjoyed one to two months nurturing in nursery and intermediate ponds and have reached a size of about 0.8-1.2 grams before being stocked in the final grow-out ponds. This approach seems to result in high survival as well as good growth in the final grow-out ponds while not depressing growth noticeably in the nursery and/or intermediate ponds. It also allows for more complete utilization of ponds, since all ponds are stocked nearer to their carrying capacity. If skillfully managed, this approach can result in two or even three crops per growing season. But it is done at the expense of added handling of the animals, which results not only in added labor but sometimes in some decrement in survival due to handling injuries. However, if the handling is skillful, the survival cost is minimal.

Species Selection

A very few broad generalities can be made about species selection and probably none of them are true enough to carry much weight. First, exotics are usually less predictable than native varieties; they may be more subject to disease and simply may not grow and survive as well in a foreign location. Secondly, P. monodon, because of its rapid growth and large adult size, probably is the species of choice--when it can be obtained in adequate numbers and if it appears to do well in the site selected. This is a futuristic statement since the life cycle of this species has been closed only recently, has not been closed with well-proven reliability, and has been closed only by AQUACOP in Tahiti and SEAFDEC in the Philippines. Thus, it could be some time before PL's of this form are available in commercially important numbers at attractive prices in the Americas--unless a sizable commercial P. monodon hatchery is established in the Americas with reliable life cycle control. And even then, it has not yet been established that this species will grow well this side of the international date line--although we know of no reason why it should not.

Lastly, and this may be the only truly useful statement made under this heading, the prospective shrimp farmer is well advised to try pilot operations with a variety of candidate species before he places his economic future in the pereiopods of any of them. The innate propensity of the selected species (in terms of such factors as disease resistance, feed conversion, growth rate, and tolerance of physical and chemical water conditions) is a fulcrum between economic success and financial disaster (see "Species Selection" in the bibliography).

Obtaining Postlarvae

If a reliable external source of PL's at an attractive price is available, a shrimp farm may decide to buy its fry rather than investing in its own hatchery. If it does so, the purchase price of PL's thereby becomes an externality in that operation's economic analysis. Since it is an externality, the operation thereby is placed to some extent at the mercy of PL prices and the reliability with which they are supplied. The alternative is to include a hatchery operation in the farm and thus gain a measure of internal control over the fry supply. This brings in added capital, maintenance, and labor costs; it makes the water quality, predator/competitor, and disease issues more critical; and it brings with it the question of larval feeds. The question of larval feeds then brings up a decision as to whether one should have his own Artemia (or other animalcule) hatchery as well as his own phytoplankton generators; or whether he should place his shrimp hatchery at the mercy of external supplies of these two critical variables. To be sure, if a farm's hatchery is successful, it can become an income producer as well as being a source of less expensive PL's--if it is more successful than whatever competition there may be. It is not a simple question, but it boils down to a decision to take responsibility for one's own bucket of worms or to avoid that responsibility by placing the shrimp farm at the mercy of other people's worm buckets.

Mr. Hysmith has suggested (personal communication) that one alternative may be to form cooperative hatcheries owned and operated by five to ten shrimp farms. This would tend to dilute capital requirements and some income might be produced through marketing PL's to smaller operators. Mr. Hysmith feels that a hatchery able to produce 250,000 PL's/year could expand its production tenfold with minimum additional capital.

Pond Size

The pond size tradeoff is covered early in Chapter V. Please refer to that discussion.

Size of the Operation

Are there economies of scale in shrimp farming and if so what are they? The answer to the first half of the question appears definitely to be "yes" and the answer to the second half seems to be "it all depends". If an operation is labor intensive, then it should be large enough to utilize fully all of its permanent employees and particularly the higher-priced personnel. To the extent that an operation is capital intensive, the same basic rule applies: high-cost capital improvements should be utilized productively to the maximum. If land can be had cheaper by the square mile than by the acre, this tradeoff must be examined too, and the same idea applies to water supply systems. Feeds and chemicals for disease and predation/competition control may be significantly cheaper by the dozen (tons) and, if so, this factor should be considered. Also, pond construction and maintenance equipment, unless it is the property of the farm, is expensive and charged portal to portal; if the portal is very far away, the more onsite use of it that can be made, the less expensive it is per unit work accomplished. If it is owned, it certainly should be put to productive use a goodly share of the time. If these considerations lead to a fairly large-scale operation, then that operation probably should have its own processing and storage plant. If this is so, then it could be that the operation should be even larger in order to fully employ such a plant of minimum economic size. If this leads to an operation with a very high-volume output at a rate that can be controlled and to a superior product quality, then the operation begins to move from a position of price taking into one of price setting. And this leads to questions that are examined in the next chapter.

Multiple Use of Facilities

One possibility for improving the economic posture of a shrimp farm located where the winters are cool or cold might be to farm some cold-water species during the winter months. Trout and salmon might be likely candidates. Fry of either of these salmonids should reach about 3/4 pound in 3 to 4 winter months. Although trout do grow well in brackish water, if a sufficiently economic freshwater source were available, keeping shrimp ponds in fresh water for a few months each year would serve to eliminate marine predators, competitors, and disease organisms. To culture a different animal during winter would, of course,

Photo courtesy of Centre Oceanologique, Tahiti

Figure VIII-1
Harvesting shrimp in AQUACOP's larger grow-out ponds.

involve some additional capital. And this would have to be weighed against the advantages of keeping major capital facilities and the farm's standing labor force productive all year.

C. More Production Economics Points from the Workshop

Dr. Harold Webber, President of Groton Associates, chaired this session. He introduced the topic briefly by reiterating that the biology of the species selected, the environment that comes with site selection, the state of the art of all applicable technology, and the socio-economic environment from which materials and technology are drawn, and into which the product is marketed, are all major determinants of the economics of shrimp production. To the extent that the biology of the animal is understood, it can be spawned and reared economically. To the extent that applicable technology is "off the shelf", it is relatively inexpensive; conversely, to the extent the applicable technology is experimental or so low in its volume of production that it essentially is custom built, it will be relatively expensive. The economics of siting already have been discussed. Additionally, the socio-economic business environment is a major determinant of how much the things and people needed to run a shrimp farm will cost as well as the determinant of how much of the product can be marketed at a price above the farmer's break-even point. A good point not made until now is that the more intensive a culture system, the more it is dependent on technology and so the more dependent it is on externalities such as energy, capital equipment, and feeds having prices and reliability of supply not under the control of the aquafarm management.

Mr. Robert Bidlingmier and Ralston-Purina's Mathematical Modeling Project

Ralston-Purina has picked up the mathematical decapod production model originally developed by Louis Botsford for lobsters when he was a graduate student at the University of California's Bodega Bay Marine Laboratory. This is probably the most sophisticated computer model of its type in the aquaculture industry. It is of the interactive, on-line type which means that the model user can communicate directly with the computer through a typewriter console and that the model's programs running in the computer will ask questions of the user until the model acquires all the data and assumptions it requires to complete a run. Mr. Bidlingmier mentioned that one of the real values Ralston-Purina has found in employing a model of this sort is that it forces the staff to explicate all the data and assumptions they are employing and to explicate them in direct terms of production costs versus market return.

The model demands that the users feed it information such as: expected stocking density, survival rates, feed conversion ratios, expected harvest per unit area, acres of harvest area and so forth. Among its many outputs are such figures as tail count and pounds of harvest per acre, grow-out times, required working capital, costs as a percentage of sales, cash flow projections, and incremental balance sheets.

Ralston-Purina finds the model highly valuable for such objectives as determining the tail count at harvest that will maximize return on investment, determining those tradeoffs to which the operation is particularly sensitive under a given set of assumptions, and for focusing empirical research on the most critical facets of their shrimp-farming operations. Experience with the model provides a rational underscore for the intuitively obvious: production economics are highly sensitive to sale price. A more recent addition to the modeling system provides it with a capability to simulate shrimp growth curves as a function of a variety of simultaneous variables including environmental parameters and feeding rates.

Ralston-Purina also is using computing machinery to keep books on daily operations at Crystal River, Florida, and in Panama. The computer follows survival and weight gain on a weekly basis, indicates when to harvest each pond, provides comparisons of actual performance and stated goals, summarizes harvest data, and compares the production performance of the several species being tested. Ralston-Purina's experience with the modeling approach so far indicates that it can be a valuable tool in optimizing farm designs and operations so as to achieve a predictable profit posture in the farming of penaeid shrimp.

In response to questions from the floor, Mr. Bidlingmier affirmed that the model did take taxes into account and could give cost breakdowns by production element (see Mr. Hodges' report). In response to concerns over the cost of this modeling approach being prohibitive for the smaller growers, he stated that adapting Botsford's model for Ralston-Purina's uses had cost the company only about $5,000. Mr. Goodwin then suggested that perhaps the aquaculture community together could develop a generalized model along the Botsford line; a model that could be used effectively with any species at all reasonable scales and with respect to any locale. This might be a possibility so long as the model is restricted in scope to monocultures. An effective generalized polyculture model might "take a little longer" (see Aquatic Ecosystem Analysis and Modeling 1975 — A Sea Grant Perspective by Joe A. Hanson, The Oceanic Institute).

At the conclusion of Mr. Bidlingmier's talk, Dr. Webber mentioned that, conceivably, aquacultural economics can be viewed from at least two perspectives. The first is the perspective being employed here, the perspective of the farmer buying what he needs and offering his produce to an unpredictable marketplace beyond his control. The second perspective is that of vertically integrated industry. This sort of approach would have the aquaculture business rooted in developing its own biology and technology, not to mention its seedstock; its branches and leaves would be represented in the retail food trade right up to and including fast food and retail market packaged products outlets and the public relations that go with them. This cleaves closely to concepts of modern business enterprise developed by J. K. Galbraith in his book "The New Industrial State". Perhaps it is an approach that should be given more attention by aquabusiness.

James Herrin and the Production Economics Analysis done by Sea Farms, Inc., Honduras

In searching for an optimum location for large-scale operations, Sea Farms, Inc., took into account a wide variety of factors. Among those determined to be most critical were land and labor costs, feed costs, and taxes and duties. This culminated in a decision to lease acreage in Honduras where land, labor, and feed costs could presumably be minimized and the tax/duty structure appeared attractive.

Using the figures employed in the Honduras decision, Sea Farms then projected feed costs at 15¢/lb. and a 3:1 feed conversion ratio along with other data to arrive at a conclusion that the optimum size for their operation lay between a 1,000-acre minimum and a 2,000-acre maximum. Their experience to date shows the actual price of feed produced in Honduras to be a little more than 15¢/lb. and the actual feed conversion rate to be a little better than the 3:1 assumed originally. Consequently, Sea Farms' projections appear rather encouraging at this point. The return rate projected by Sea Farms is 33-50¢/lb. of 30-50 count (tails) shrimp and their production costs work out to about $1.20/lb. not including amortization of capital improvements.

The Sea Farms projections exclude any expectations of major natural disaster. The site was chosen on the basis of historical freedom from such events. The calculations do, however, include probabilities of certain operating assumptions being true and they adjust calculated profitability according to these probabilities.

Mr. Carl Hodges on Intensive Culture Economics

In Chapter V we mentioned that Mr. Hodges' report in the workshop grow-out session on the University of Arizona Environmental Research Laboratory's intensive culture experiments contained information on production economics which would be covered in this chapter. This is that report.

ERL, too, employs a computer program to study the economics of its approach to the intensive culture of penaeid shrimp (described in Chapter V). There are fifteen variables as follows:

1. Size at stocking
2. Size at harvest
3. Growth rate
4. Stocking density
5. Feed cost
6. Feed conversion rate
7. Blowdown rate (rate at which output water can be cleansed and returned to the external environment)
8. Unit cost of the aquacells
9. Amortization rate
10. Labor cost
11. Costs of electrical power
12. Fuel costs
13. Mortality rates
14. Price per pound of product marketed
15. Administration costs

ERL's empirical experience and that with the computer model results in feed being the single highest cost factor in that laboratory's intensive rearing equation. Feed costs range from 30-35% of the total of annual operating expenses and this percentage appears relatively insensitive to variations in the other parameters. Capital is a surprisingly small 16% while labor runs 8-10%. These figures hold true with the presently good growth rates of about 2.5 g/month being obtained in the aquacells presently. Mr. Hodges predicts that growth rates of 4.5 g/m are not beyond reasonable expectations with improved feeds and possibly some selective breeding. This may modify the percentage ratios and lower production costs--if the improved feeds are also not too expensive.

D. Conclusions

A priori assessments of production economics are no more cut and dried than are the other aspects of penaeid shrimp culture. In point of fact, it could hardly be otherwise since production economics virtually are the "bottom line" toward which all the other subjects of concern lead. However, despite the fact that the data to insert into mental models of production economics remain uncertain, the aquaculture community nonetheless has produced such models. Some are manifested in computer programs and some in pencil and paper form. The validity of any and all of them, like truth and beauty, probably exists more in the eye of the beholder than anywhere else; such models must be thoroughly understood in order to be used effectively. And still they all appear to be in general agreement that, the way things are today, feed, labor, land, and capital improvements are the big four cost items. We say, "the way things are today", because a major breakthrough in biology or technology could change the picture virtually overnight.

IX. THE FUTURE MARKET FOR CULTURED SHRIMP

The original title of this session of the Galveston workshop was "Processing and Marketing". As a matter of fact, the session turned out to be a fascinating and highly provocative dialogue between Dr. Ian Chaston of Booth Fisheries and Dr. Samuel Gillespie of Texas A&M University's Marketing Department with questions and comments from the floor. The thrust of that dialogue was an inquiry into the _future_ potential of cultured shrimp in the American marketplace. Processing _per se_ was ignored as were the nuts and bolts of selling cultured shrimp in _today's_ marketplace. To our way of thinking, the tack taken by this session was perfectly appropriate. Cultured American shrimp today, what there are of them, can easily be sold iced, frozen, or even live in local markets; their volume is low enough that, with possible rare exceptions, local markets can absorb the cultured product easily at going prices. Cultured shrimp may have a noticeable impact on national and international markets within the next ten years, and it may be longer. This is the era to which Messrs. Chaston and Gillespie spoke. For those interested in the details of shrimp processing, the book "Open Sea Mariculture" (Ref. 482) contains an excellent summary and bibliography by Jeanne M. Collier. This chapter will limit itself to a summary of the Chaston-Gillespie dialogue.

Dr. Chaston began by noting that most shrimp farmers and prospective shrimp farmers today are thinking in terms of marketing 26-35 count (tails) shrimp. Growing them larger is costly and marketing them much smaller brings a significant reduction in unit price. However, it might be that the magnitude of demand for larger shrimp already is changing. What seems to be happening is that consumers, bedeviled on all sides by inflation, are turning away from shrimp with its ever-rising prices and turning toward lower-cost sources of protein. The same shrimp that retailed for $2.00/lb. a few years ago now cost $5-6.00/lb. and this is a rate of inflation that very much exceeds the general trend. It seems, too, that the market for very-high-priced jumbo shrimp is contracting faster than that for the smaller, lower-cost shrimp. The underlying message here is that there is no doubt that shrimp must compete with other sources of protein for an expansive market. High-priced shrimp is a "Tiffany food" and its market elasticity will diminish with mounting general inflation.

For example, the retail frozen, breaded shrimp market today has a volume of about $75 million. Already about $10 million of this volume lies in the sale of cheaper 200 count Indian shrimp extended with soy or rice fillers. It may well be, then, that the size of the traditional shrimp market is declining and that it is being replaced by a potential market for extended shrimp products that can compete price-wise with other sources of protein. This is not idle conjecture; the shrimp industry as a whole is very actively looking into processes and products that will extend shrimp by combining shrimp meat and flavor with vegetable proteins and fillers to form a new product that can be competitive with other staple proteins. If this trend expands, shrimp farmers who are utilizing a 60 count product for $1.95/lb. in their economic models (1975 dollars) may find that they are unable to obtain this price in actual practice. The large buyers may not pay such a price because the consumer demand for shrimp at such high wholesale prices simply will not be there.

If this indeed is the trend of the future, American shrimp farmers should also be looking at selling 200 count shrimp at more like $1.00/lb. (1975 dollars). This might be an economically viable possibility if farmers can design their production in such a way that they can accept a smaller unit gross profit margin per crop and increase the number of crops per year significantly. This may not be too unrealistic if we recognize that culturing risks are reduced by the shorter grow-out times and that culturing facilities can be more densely stocked and more completely utilized. However, as mentioned in the previous chapter, this approach will make the price of PL's critical. If a farmer pays $5.00/1,000 for PL's and then sells 200 count shrimp for, say, $1.20/lb., that leaves him less than 20¢/lb. for all grow-out costs and for his profit. If the 200 count approach is to be taken, PL's will have to be available for $2.00/1,000 or less.

Dr. Gillespie discussed the general problems shrimp farmers face with respect to their market. The basic formula is:

$$P = TR - TC$$

where:

P = Profit

TR = Unit price x quantity sold

TC = PC + MC

PC = Product cost

MC = Marketing cost

Most of the shrimp farming R&D effort today is being directed toward lowering product cost (PC above). The cost of marketing (MC) as well as the unit price obtained and quantity sold (TR) are sensitive to a market over which the farmers may or may not have any degree of control. Prevalent cultural attitudes relating to the desirability and even the status value of whole shrimp and the extended shrimp products mentioned by Dr. Chaston are active determinants of market size and elasticity at comparatively high prices. And desirability is a relative thing that is closely coupled to the price as it compares with the price of alternative protein sources including other forms of seafood, nonseafoods, and wild shrimp. In other words, to the extent that shrimp farmers are unable to assess the impact of competitive protein sources on their markets, they will find their markets unpredictable and very possibly unstable. A major potential source of market instability for the immediate future lies in the activities of foreign shrimp fisheries. These activities enjoy cheaper labor and apparently are prepared to operate at lower unit gross profit margins. Moreover, they are less hindered by FDA restrictions than are domestic growers or fisheries. Not only do they have the capacity to glut American markets--bringing sudden price dives--but they may glut it with inferior products which so offend consumers that years will be needed for the market to regain its original size. To the extent that this phenomenon is a real danger, American shrimp farmers are at its mercy. There is nothing they can do about it short of creating their own private contract markets.

And private contract markets for cultured shrimp definitely appear to be the ray of hope and promise in an otherwise tumultuous market perspective. If shrimp farmers can indeed produce a uniformly superior product with demonstrated reliability of supply at a relatively stable price, then they are in an advantageous position to negotiate long-term contracts with the wholesale buyers. If this occurs, then American shrimp farming will begin to move from a position of price taking toward a position of price setting. At the present time, wholesale buyers and processors must live in a world of seasonal and unreliable supply. Since demand is comparatively constant, these firms must maintain large reserve inventories. Not only are these inventories themselves very expensive (12% money, as Dr. Chaston puts it), but they must be maintained with facilities that involve large capital investments and major fixed overheads. There seems to be little doubt that the big buyers would be willing to enter into contracts and to pay premium prices if they were assured of uniformly superior quality and a reliable year-round supply; the premium paid could be made up readily in reduced inventory and inventory maintenance costs.

If shrimp farming grows to become a major industry, it will be able to afford promotion of cultured shrimp to the level that not just any old shrimp, but cultured shrimp, becomes an item of cultural food preference. To the extent that this type of public relations campaign can be carried out successfully, American shrimp farmers can move even further toward a position of price setting. They will be in a position to create demand through their promotional activities and to satisfy it with flexibility of production. Conceivably, large markets for low-cost 200 count shrimp can be created and filled as can smaller markets for 26-35 count (tails) "superior" shrimp and smaller markets still for very expensive 10-15 count (tails) "miniature lobsters" sold live or frozen and possibly with heads on to further the lobster image.

All this will take time, planning, and a good deal of effort and money. Moreover, it is an approach that can be successful when, and only when, reliable and flexible production of uniformly high-quality cultured shrimp in very large volumes has been achieved or is within reach beyond a reasonable doubt. Only then will industrial capital in adequate amounts be attracted to the new industry.

X. THE LEGAL AND REGULATORY WEB

The legal and regulatory session of the workshop was chaired by Mr. Terry Leary of the Texas Parks and Wildlife Department. There were two panelists: Dr. Earl Mills of the U.S. Army Corps of Engineers and Capt. Paul Mulvihill (USN Ret.) of Aquaculture Research/Environmental Associates, Inc. (AREA). Mr. Mock and Dr. Webber were among the several who joined enthusiastically into a freewheeling discussion following the opening statements by the chairman and panelists.

The pattern that emerged from this stimulating if not encouraging session is shown in Figure X-1. Any aquaculture enterprise within the U.S., it seems, can only exist legally within a network of laws and regulations which stem from local, state, and Federal legislation. Seven categories of legislation were mentioned and not one of these was aquaculture. The truth of the matter is that aquaculture is virtually a nonentity in the legislative world. There are no institutional elements within government yet that are chartered to regulate and promote the interests of aquaculture; the structures of state and Federal governments ignore aquaculture and it is a rare employee of Federal and state agencies having jurisdiction over aquaculture who is informed with regard to aquaculture. So far, aquaculture in the U.S. has been little more than cottage industry. It has no effective lobby and very seldom any effective voice at the state and local levels either.

Aquaculture takes up land, uses water, deals with animals, uses chemicals, imports and exports, and sells food in the marketplace. Thereby it finds itself legally under the jurisdiction of a variety of government agencies intended to protect the public welfare. As we've just said, these agencies are structurally and, for the most part, intellectually ignorant of aquaculture. Moreover, their jurisdictions may overlap while being unintentionally damaging to aquaculture interests. Very few regulations in existance today are purposely designed to protect

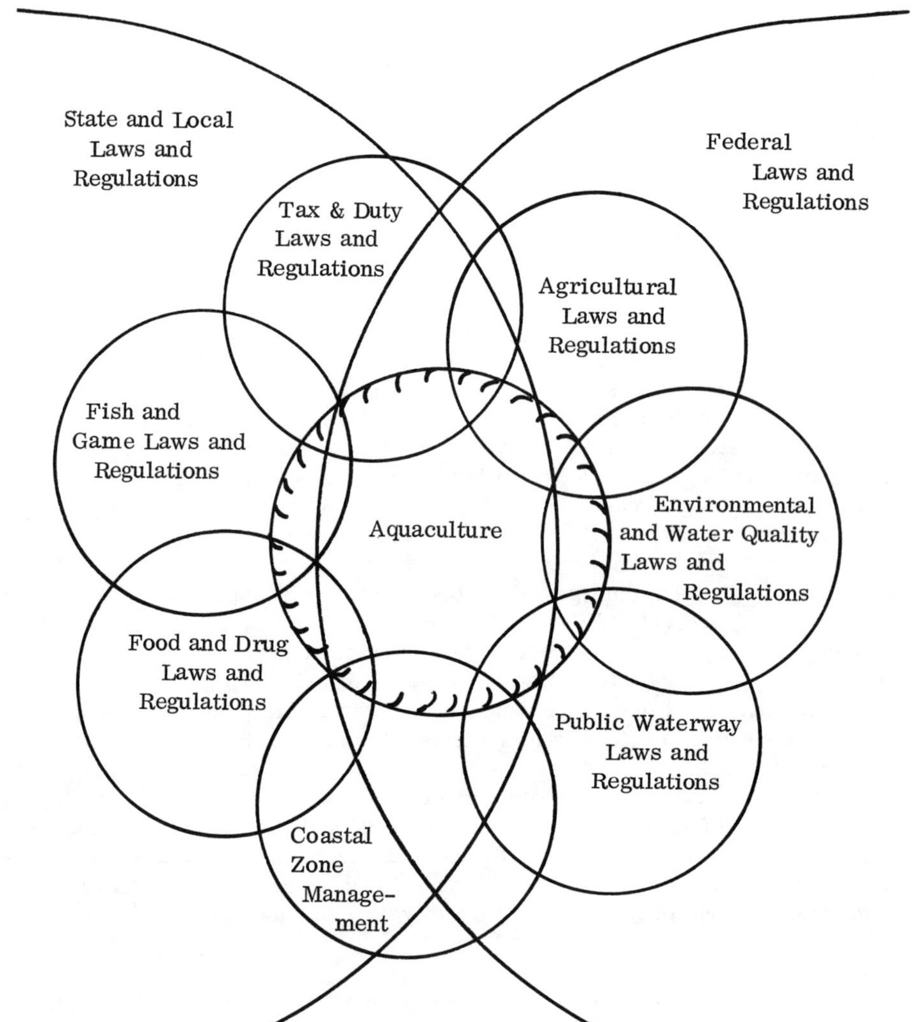

Figure X-1
The legal and regulatory web.

or enable aquaculture.[a] Neither are there any guidelines which tell the prospective aquaculturist how he might best go about hanging his operation viably in the legal and regulatory web. And we underscore here that, even though this chapter appears in a book devoted to penaeid shrimp, it applies to aquaculture in general.

As an example of what aquaculturists face, Capt. Mulvihill has written (personal communication) from his extensive experience as follows:

> "The Federal, state, and local agencies are somewhat confused when the term "aquaculture" is used. In the case of EPA, there is one regulation entitled "Aquaculture Projects, Requirements for Approval of Discharges," published in the Federal Register, Volume 39, Number 115, Part II, June 13, 1974. This regulation was designed primarily to govern utilization of heat effluents from power plant discharges which would be used for aquaculture purposes. If you write the agency and request information on aquaculture, the likelihood is great that this is the regulation you will receive, along with Department of Defense, Corps of Engineers, Permits for Activities in Navigable Waters or Ocean Waters, Volume 40, Number 144, Part IV, July 25, 1975; and Environmental Protection Agency, Navigable Waters, Discharge of Dredged or Fill Material, Part II, September 5, 1975.
>
> "On the other hand, if you write and address fish-farming regulations, you will receive a copy of National Pollutant Discharge Elimination System, Application for Permit to Discharge, Short Form B--Agriculture, which contains in Part III a fish and other aquatic animal production facilities form. It appears EPA defines fish farming and aquaculture as two separate subjects.
>
> "In Dade County, Florida, when applying for a permit to construct and operate a freshwater shrimp hatchery and growout, friends informed us to be sure to request the permit under "fish hatchery" because "aquaculture facility" or "shrimp hatchery" per se does not appear in those terms in laws or rules. (There was no trouble in obtaining a permit.) For example, under the Florida State statutes, under Chapter 72-18, Taxation, Ad Valorem, Taxes, Classification of Rates, there is a provision for providing assessment factors to be used on qualified lands. In definition for aquaculture purposes, such pursuits are defined as "horticulture, floriculture, viticulture, forestry, dairy, livestock, poultry, bee, pisciculture". Webster's Dictionary describes pisciculture as fish culture. You can see if you request information from the State on aquaculture laws and regulations, how the average State employee would fail to refer to this particular law.

---

[a]The states of Texas and Florida do have some such statutes.

"Under the State of Florida Statutes, Chapter 12A-1.39, Sales and Use Tax Law, provides for a purchaser's blanket resale and exemption certificate. A number of items under the certificate have definite application, such as "Portable containers used for processing farm products, use as cloth, plastic or similar material used for shade, mulch, protection from frost or insects on a farm, use as nets by commercial fisheries and is exempt under the provisions of Rule 12A-1.01(8), and feeds for raising poultry and livestock on farms and for feeding dairy cows".

"Chapter 12A-1.87, Paragraph (6) says "Nurserymen, dairymen, poultrymen, cattle ranchers, apiarists (beekeepers), and breeders and producers of fish are in the same category as farmers and are entitled to the same limitation or exemption on their purchases of tangible personal property.

"The State of Florida Statute #253 authorized the Trustees of the Internal Improvement Fund to lease submerged lands owned by the State in its sovereign capacity and the water above to persons desiring to engage in aquaculture activities; prescribing procedures for making application for such leases and the granting of same; prescribing the essential features of such lease contracts; providing penalties for persons violating the exclusive rights of lessees and for lessees violating the terms of their leases; authorizing the trustees to adopt rules and regulations.

"The State of Florida Statute 618.01 (1) of 1971 was amended in 1975 to include aquatic products within the definition of agricultural products. State of Florida Statute 619.01 was also amended permitting those engaged in the production or marketing of aquatic products to form a nonprofit cooperative association.

"Last but most important is a Federal regulation which is referred to under Short Title as the "Consolidated Farmers Home Administration Act of 1961." Within this Act they define Section 401. As used in this title (1) the term "farmers" shall be deemed to include persons who are engaged in, or who, with assistance afforded under this title, intend to engage in, fish farming, and (2) the term "farming" shall be deemed to include fish farming. (Pub. L. 87-128, title III, s 343, as added Pub. L. 87-703, title IV s 401 (5), Sept. 27, 1962, 76 Stat. 632.)"

This chapter condenses the remarks offered in the workshop under the seven headings shown in the rosette around aquaculture in Figure 14. There is no particular attempt to distinguish between local, state, and Federal jurisdictions even though these distinctions do emerge now and then.

A. Fish and Game Laws

Fish and game laws usually are state centered and are intended to preserve wild populations in the public interest. In this case, the public interest is held to be served by sport and commercial fishing regulated to preserve the wild stock and by the preservation of any aesthetic qualities the stock may be considered to add to the environment.

These laws, more frequently than not, are inimical to aquaculture interests. For example, they tend to set seasons which may mean limiting the times during which aquaculturists can legally "possess" their cultured stock. "Possess" is usually interpreted to mean harvest, fortunately. They may also set lower and/or upper limits on the size of animals harvested and sometimes on the sex. For example, they may restrict the collection of spawners without special permits and frequently close the season during spawning months--making it illegal to possess shrimp without a special permit. State fish and game laws may also restrict harvesting methods and gear. If anyone cultured grunion in California, for example, he could harvest them legally today only with his bare hands. Similarly, Texas laws intentionally restrict the efficiency of shrimp harvesting gear in public waters in order to protect wild stocks from decimation by ambitious technology.

In addition, the fish farmer who would like to culture non-native species usually will find that he needs special permission to import them. This sort of thing usually is handled on a case-by-case basis and opposition from environmentalists may appear in the form of public hearings, injunctions, and other delays and even prohibitions.

B. The Coastal Zone Management Act

The Federal Coastal Zone Management Act (CZM) provides for Federal support to each state in developing a CZM master plan and in then enacting appropriate legislation and management structures to carry out the plan. The purpose of CZM plans is to resolve conflicts and assure that the highest and best use of the coastal zone is made in terms of the long-range public interest. Therefore, they are inherently limiting with respect to new activities in the coastal zone. As the effects of the CZM act gain momentum, aquaculturists more and more will be required to demonstrate that their proposed projects are indeed the highest and best use of the area they propose to employ; Mr. Mills feels that aquaculture is doing this inadequately now. Even if aquaculture is successful at this, there still may be setback laws to be dealt with, laws governing the right of public access, and inhibitions on the intake and discharge of water.

CZM planning is relatively new. In many states, it is one area in which aquaculture interests can have a voice if they are organized and act immediately.

C. Environmental Protection and Water Quality Laws

Aquaculture seems to face a perplexing paradox with respect to water quality laws. Waters that are sufficiently uncontaminated to provide feedwater for aquaculture frequently have been assigned A or AA water quality classifications. If they are classified AA, any discharge whatever into them is prohibited. Even Class A waters have very stringent discharge limitations. Thus, the aquaculturist may well be required to clean his discharge water to very high standards before re-releasing it to its source. At the moment, it does not seem to matter that water discharged from aquaculture operations will not necessarily be detrimental to the aquatic ecosystem into which it flows. The laws in some states are very clear on how much of what classes of materials such discharge water may contain. And, when it comes to heads and wash water from shrimp-processing plants, there are few states which will permit direct discharge into very many of their contiguous waters.

In this regard, Capt. Mulvihill writes again (personal communication) as follows:

> "It is fully realized that we must maintain under EPA regulations certain standards of water quality from our discharges. What I was trying to stress (in the workshop) was that "one man's trash is another man's treasure". In south Florida, adjacent to the Everglades National Park, a catfish raceway operation was conducted for years. The discharge from this operation was as great in volume as the tons of feed fed to these catfish. The discharge spread into low-lying areas and dead-end canals; it was like a never-ending spring. The population of game fish exploded in the canals and low areas. It became a feeding station for the wildlife of the park. It was inspected by County, State, and Parks officials and all agreed that it was not detrimental but, rather, that it enhanced the environment.
>
> "We must admit this is a unique case. Dade County is primarily composed of limerock formations with little or no nutrients so a discharge of this nature enriches the water and the substrate acts as a nitrifying bacteria filter. It is an exception to the rule.
>
> "In addressing Dr. Ling and the Japanese representatives (in the workshop), my intent was to point out that for centuries, with their very large (human) populations discharging wastes into rivers and lakes, the fish populations did not appreciably suffer. With the introduction of our new-found technologies, our rivers, oceans, and lakes become polluted by pesticides, PBC's, nonorganic fertilizers, etc. What portion of the pollutants in these waterways would necessarily be attributed to human and animal wastes?"

Other environmental protection laws may impose restrictions on pond designs, on grading, and on the chemicals which a fish farmer can employ. They also may demand the filing and approval of an environmental impact statement (EIS), and this may lead to public hearings and a variety of other costly delays.

D. State and Federal Agricultural Regulations

Agricultural regulations tend to regulate the use of herbicides, pesticides, fertilizers and other chemicals and may in some cases impose grading restrictions on lands zoned for agriculture.

On the other side of the coin, they may permit agriculture practices that are damaging to nearby aquacultural interests. For example, a shrimp farm wiped out by unintentional fallout from nearby crop dusting probably would be required to sue in civil court to recover damages; usually, no state laws would have been broken. Usually, the deleterious effects of agriculture on aquaculture are less dramatic, consisting of lowered culture water quality resulting from percolation or runoff of chemical fertilizers, pesticides, and herbicides into aquaculture ponds directly or into feedwaters that supply them. There are no regulations which purposely protect aquaculture from agriculture or from other forms of industry for that matter.

An interesting point that Capt. Mulvihill brought out with respect to this subject is that restrictions on the use of human and animal excreta as fertilizer for crops intended for human consumption are severe in the U.S. The FDA, for example, defines such resources as "filth" and regulates their use accordingly. Comments by Dr. Ling and Japanese participants confirmed Capt. Mulvihill's point that organic fertilizers in appropriate concentrations by and large were beneficial to aquatic life whereas inorganic chemical fertilizers must be employed with a great deal of discretion. It is true that no food grower wants to cooperate with the life cycle of any disease organism and this will demand some treatment of human and animal wastes before they are used in intelligent culture. But the main barricade now is that their use is not legal under any circumstances; instead, they remain a disposal problem.

E. Food and Drug Regulations

The Federal Food and Drug Administration (FDA) regulates what can and cannot cross states' boundaries commercially for human consumption, and counterparts in state governments usually protect the eater and drinker from inappropriate food practices internally.

Regulations within this category very definitely set upper limits on the concentrations of a wide variety of chemicals and organic and inorganic substances that can occur in foodstuffs intended for human consumption. This almost certainly will have some impact on the chemicals and antibiotics that can be used by aquaculturists to treat or prevent disease. It may even inhibit aquacultural employment of some feed waters. Some contaminated water may not harm the cultured animals but the animals' biology tends to concentrate contaminants in the meat beyond acceptable limits. Dr. Webber pointed out that the contaminant concentration phenomena probably would be particularly irksome where aquaculture was carried out in the heated discharge water of nuclear power plants. This water always contains low levels of radioisotopes which tend to concentrate through marine food chains. Allowable levels of radioactivity in human foods are so low now that it would be quite possible for stock reared in such waters to exceed the maxima.

F. Laws Governing Public Waterways

The designers of some aquaculture projects may wish to erect structures in waters that have been designated as public waterways. Thus, they may be subject to Coast Guard, Corps of Engineers, and state transportation department regulations. Even if no such structures are planned, they may still face the U.S. Army Corps of Engineers.

Dr. Mills stated that the Rivers and Harbors Act of 1899 (Section 10) requires permits for all structures and work in or affecting navigable waters of the U.S. "Navigable waters", defined as waters of the U.S. in the Federal Water Pollution Control Act (FWPCA) as amended, according to a recent Federal court interpretation, include not only the waters themselves but also contiguous and adjacent wetlands; the FWPCA concerns dredge and fill activities but not most structures. As it stands now, this means that aquaculture activities in any region that can be interpreted to be navigable water or contiguous and adjacent wetlands very likely will require a Corps of Engineers permit and this will demand that that agency's criteria, as well as those of the Environmental Protection Agency (EPA), be met. At the moment, these criteria are not designed with an eye toward assisting aquaculture in coastal areas. In fact, the court ruling is so recent that the Corps is now gearing up to handle its expanded responsibility effectively. There will be public hearings and there will be an opportunity for aquaculture to raise its voice if it so chooses. Probably, there will be a matter of some time before the Corps is in a position to handle its expanded responsibility with great expediency. However, Dr. Mills writes (personal communication) that, in the public hearings held so far, the voice of aquaculture was silent.

G. Taxation

Taxes and duties are the final category in the breakdown used in this chapter. Aquaculture conducted within the U.S. will be subject to local, state, and Federal taxes and, except when it imports foreign goods or exports its products, it will be relatively free from duty regulations. However, the regulatory web may be at least one factor stimulating serious fish farmers to look toward foreign countries as sites for their operations. In this case, duty regulations could become important. The workshop participants mentioned this as a factor but carried the subject no further.

H. Conclusion

We have provided here little more than a flimsy sketch of the legal and regulatory problems faced. No solutions have been proposed and, in fact, the problems have been defined with inadequate specificity. The National Sea Grant Office has sponsored and is sponsoring a number of studies which delve more deeply into various aspects of the problem we have brought up here. Readers interested in pursuing the subject further should contact that office to obtain up-to-date information on available reports and studies in progress.

But the big issue seems to be what aquaculturists might do to improve the legal and regulatory situation in which they find themselves. One alternative already has been mentioned: they can take their operations outside the U.S. And some already are doing so (although it is not clear that they are in fact being driven out by the legal and regulatory web alone). For those which remain, it must by now be clear that aquaculture must gain a voice or remain unnoticed and unserved in the legislative process. In our opinion, a truly thorough survey of all laws and regulations affecting aquaculture of all types is warranted. The study would focus upon the Federal level and, underneath that, on a state-by-state basis. It would be an invaluable first step in gearing up to cause the legislative process to serve aquaculture on a nondiscriminatory basis. Such a work also could be used by the aquaculture community to assess realistically the problems new projects would be likely to face. Sea Grant has supported research projects of this type on a regional basis and perhaps now should consider this national-level proposal. But with or without this project, it is clear that aquaculture must organize to some degree and then develop lobbying power at both the state and Federal level. If it does not, the time at which legislation will enable rather than inhibit aquaculture may be rather far away and it may be never. Capt. Mulvihill has suggested that perhaps aquaculturists should first organize themselves and then join the already powerful agriculture lobby.

XI. STATE-OF-THE-ART SUMMARY

This session of the workshop recalled the principals in earlier discussions to the podium to give summaries of their topics. In some cases, just summaries resulted and in others, the summaries stimulated further discussion. It warrants recognition that the summaries and resulting discussions occurred in the context of the wealth of information that surfaced in the workshop. Everyone benefited from new information and no one was immune from having his thinking about shrimp culture grow in the process of the workshop. After all, that was one of the purposes. Therefore, while most of the summaries sketched here conform rather well to their corresponding earlier chapter, others offer bits of new information and somewhat changed perspectives. The previous chapter headings are the headings of the summaries given here.

A. Hatchery Systems

The basic question here, as posed by Mr. Goodwin, is, "Is penaeid shrimp hatchery technology now far enough along that it is not a limiting factor in shrimp culture?" Mr. Mock replied that the answer very definitely was "yes". However, a substitute for *Artemia* (nauplii-eggs) is sorely needed. *Artemia* are becoming scarce and extremely expensive. It also would be nice to have a good substitute for the early-stage diatoms that are being produced now with some trouble and expense. Other participants enthusiastically underscored the need to find a replacement for *Artemia*. Note here that successful large-scale commercial culture of *Artemia* almost anywhere could obviate the need for a replacement and that, if replacements are demanded, they may come in the form of alternative animalcules (such as rotifers, *Brachionis* sp., copepods, etc.) or in the form of flaked formula feeds.

Dr. Parker did not entirely agree with Mr. Mock, pointing out that the presumed cost for PL's is still far too high if shrimp farmers are to grow out for the 200 count market. This is primarily because hatchery survival is not yet high enough and the costs of feeds and operations are still far too high.

Dr. Parker's remarks stimulated several remarks about costs of larvae. In essence, all larval cost estimates appear to be little more than informed estimates because, as of today, no one in the Americas is actually operating a commercial penaeid hatchery. The estimates offered ranged from about \$5.00/1,000 through Dr. Webber's rather firm assertion that he could produce PL's for \$4.00/1,000 down to remarks by others that \$1.00/1,000 appeared possible under the right circumstances. And the right circumstances today appear to demand as a minimum that the hatchery be large in size and market volume, be located near excellent shrimping grounds, and own its own trawler. Thus, the boat would be employed in regular shrimping (making a profit at that) when not busy collecting spawners for the hatchery. Mr. Mock chose this opportunity to point out that the real cost is not the cost at the front door of the hatchery, but at the back door of the grow-out farm. He noted that it costs him approximately \$35-40 to airship 10,000 PL's and the shipment alone thereby works out to \$3.50 to \$4.00/1,000.

It should also be kept in mind that hatcheries today are based on catching wild spawners and the preceding discussion reflected this. When life cycles are closed on commercial scales, the cost of PL's probably will be affected favorably.

B. Obtaining Spawners

Hopefully, this is an issue that is on its way out as a major concern within a few years. It certainly is an issue upon which there are almost as many opinions as there are opinionees. The majority, however, voted for owning one's own boat as the nearest thing to the high road to sanity in this activity. Mr. Sweat mentioned that, with the shrimping business the way it is today, one can purchase a good trawler for around \$40,000 and operate it himself. Yet Mr. Mock mentioned that one culturing company sets aside as much as \$75,000 a year for spawner capture alone. It seems to depend on who is doing it, how, and where.

There was a good deal of speculation on the subject of spawning aboard the boat and this even extended into speculation on developing a barge-mounted hatchery which would go to the spawning grounds for eggs and then deliver PL's to the farms. There was some discussion of the government role in all this and Mr. Goodwin suggested that providing PL's to the fledgling shrimp industry free or at less than cost was an appropriate role for government.

At the moment, given current hatchery production volumes, it appears that spawners can be captured economically if one owns his own boat, is in a good shrimping location, and operates wisely.

C. Closing the Life Cycle

Dr. Clark spoke to this subject offering that the shrimp community has learned enough in the past two years to assure itself that the life cycle of most species can be closed in captivity. Very possibly there will be fairly reliable egg production for several species within the next two years.

But the real goal is to reproduce all attractive species in quantity and with high reliability leading to a high degree of genetic control--getting the right sperm into the right eggs so as to improve dramatically the strains being cultured. This is going to be a long-term process and it is going to be expensive. Right now, that work is being done more by the seat of the pants than anything else. Teams of biochemists, endocrinologists, nutritionists, and geneticists are required. This probably will be accomplished only if there is a firm long-term commitment on the part of government--an occurrence that many consider to be unlikely.

D. Grow-out Systems

Dr. Jack Parker of Texas A&M University spoke to the subject of extensive (pond) grow-out systems noting first that research in this area has gone about as far as it can go until more and better seed stock is available. Feeding rates have been pretty well established--even though it takes an experienced hand to estimate pond biomass, water quality tolerances are pretty well known, predation is rather well understood if not always under control, and few serious disease problems are cropping up. Right now, good grow-out survival rates are about 60% for *P. setiferus*, 75% for *P. vannamei*, and 50% for *P. stylirostris*. These figures are for good, healthy ponds where there are no dead shrimp noted and water quality stays high. With more reliable supplies of more PL's, a great number of experiments that would result in further refinements of pond grow-out technology could be done.

E. Disease Prevention and Control

Mr. Goodwin, in what probably was a straw man play, challenged the assertion that disease indeed is a major roadblock to economic shrimp culture. Dr. Lightner and the majority stood firm in the opinion that if it is not now, disease will be a major obstacle when other roadblocks are removed. Dr. Lightner noted particularly that disease already is a major problem in intensive cultures and that it may be an unrecognized factor in lowering survival rates in pond grow-out--even though it does not usually reach epizootic proportions.

So far the shrimp culture community probably has not yet identified all the pathologies to which shrimp are subject, nor has it identified causative agents for the recognized pathologies nor defined effective treatments for the agents which are known. This makes it hazardous to introduce exotics since a microorganism that might be almost benign in one species might be highly virulent in another.

It was noted also that, although salmonella has not yet been identified in penaeids, it could occur and could infect humans through this vector if the shrimp meat is not adequately processed and prepared. All in all, the disease issue appears to remain a critical one in the minds of the penaeid culture community.

F. Nutrition and Feeds

Dr. Meyers summed up this issue and the summary is a potpourri of interrelated points which reflect the magnitude of the whole nutrition issue. First, he pointed out that the workshop discussion had in fact not dealt with nutrition but rather with diets. The nutritional requirements by species and growth stage are not now known; in fact the penaeid community is a long way from knowing them. For example, there appears to be an "*Artemia* factor" in larval nutrition; later larval stages which get *Artemia* nauplii in their diet do well and those which don't, don't. Nutritionists don't know why. Neither do they have any good, scientifically based rules for mixing proteins, carbohydrates, lipids, fatty acids, binders, and trace factors in formula feeds to achieve optimum diets. There are rules of thumb that have grown out of experience but they are not based on adequate knowledge of the animal's nutritional requirements. Moreover, as Dr. Clark pointed out, there very likely are some special nutritional requirements related to the process of female sexual maturation and researchers have not yet begun to look into that question.

In order to plan for adequate shrimp formula feed production, the U.S. feed industry needs to be able to predict both the qualitative and quantitative aspects of feed demand some years in advance. Moreover, the feed constituents that are available in quantity and inexpensively today are not those which necessarily will be available a few years hence. So the feed producers face gross uncertainties on both ends: they cannot predict their raw material supply reliably and the penaeid community cannot give them valid formula feed demand forecasts. It is unlikely, therefore, that the major feed producers are going to become very excited about shrimp feeds until the situation develops a good deal more certainty.

Returning to the subject of larval diets, Dr. Meyers mentioned that experimental flake feeds in his laboratory already were supplementing larval diets heavily and successfully and that it appeared entirely possible that a formula flake feed could replace natural feeds altogether if this thrust was made a focus of priority research. Dr. Webber pointed out that there were vast amounts of agricultural wastes to be had very inexpensively in almost all countries in the Americas and that most of them would make excellent substrates for culturing microorganisms to be used as larval feeds and also for grow-out.

In conclusion, it was agreed that nutritional research is not adequate at present and that it may be a long time before dietary formulation is transformed from an art into a science.

G. Production Economics and Processing and Marketing

Dr. Webber spoke to both the production economics and processing and marketing issues. There are a number of mathematical models for analyzing the economics of shrimp production. By and large, according to Dr. Webber, they are all excellent tools for defining our ignorance. Whether they operate with hand calculators or sophisticated computers, they all demand that their users formalize their data and their assumptions. Models thereby make it very difficult to ignore the unknowns. And of these, there are plenty. The shrimp-culture community doesn't yet have the hard cost and production figures and so the answers provided by the computers are very truly no better than the assumptions, guesses, and estimates that went into them, he said. The answers may even be worse than the inputs since human nature tends to recognize the frailties of its own guesses and yet to deify computer-generated answers.

Dr. Webber, drawing from his own extensive experience, offered that a vertically integrated approach, stretching from the capture of gravid females to final marketing of the cultured product, appeared to hold the highest economic potential. Dr. Webber said that Groton Associates' operations are headed in just this direction. So far, Groton has found its trawler costs to be quite high and its original estimates of front-end capital investment demands to be alarmingly low. However, these capital costs can be amortized over a number of years and Dr. Webber believes still that shrimp farming can be a competitive attraction for venture capital. He also reiterated that his experience with hatchery operations was by now considerable and that a price of $4.00/1,000 postlarvae looked very realistic for the time being.

Dr. Webber feels that, for the time being, power, labor, and related costs can be predicted fairly accurately although power prices could become unstable within a few years. The major unpredictable element in the production economics equation is feed. With this fact in mind, improved overall grow-out ecosystem management--management that will maximize natural feed productivity and thereby minimize supplemental feeding costs--seems to be the best insurance against the instability of feed prices. Pasture managers do this type of management very successfully and shrimp farmers should be able to learn the necessary tricks, too.

In closing, Dr. Webber reiterated the often-stated truism that, in the final analysis, production economics are entirely at the mercy of market prices. Thus, again, the name of the game for shrimp farmers is to produce for the market rather than attempting to market whatever the shrimp farms produce.

The high-return market products must be correctly identified and cultured shrimp production then directed accordingly. Only with superior quality shaped to the market demand can shrimp farmers obtain the premium prices they desire and deserve.

H. Legal and Regulatory Issues

These questions were summed up very quickly in the closing minutes of the workshop with a single blanket and very valid statement; to wit, "aquaculturists at large will enjoy favorable legislation only after they have come together effectively and developed some lobbying power." At this point in history, the development of Coastal Zone Management regulations offers an excellent opportunity for aquafarmers to make their voices heard and effective at the Federal and state levels; the extent to which aquaculture participates in coastal zone management planning may be a revealing barometer of its future economic potential. In addition, since all Federal legislation appears in draft form in the "Federal Register" long before it is voted into law, aquafarmers need to read the Register religiously and then focus whatever lobbying power they can muster on issues which will affect them.

XII. RESEARCH PRIORITIES IN PENAEID SHRIMP CULTURE

Mr. Robert Wildman of the National Sea Grant Office attended and participated in every one of the workshop's sessions as well as talking extensively with the participants between sessions. As befits his position in government, he engaged in these activities with an eye toward deriving his own views of the research priorities of penaeid shrimp culture. The final session of the workshop concluded with Mr. Wildman's results. In addition, sometime before the workshop, Mr. Rick Pacquette of the University of Washington had sent questionnaires on research priorities to most of the workshop participants. This work was part of a Sea Grant-sponsored project at the University of Washington. Mr. Pacquette's results precede Mr. Wildman's remarks in this final chapter.

A. The University of Washington Poll by Mr. Rick Pacquette

The results of Mr. Pacquette's survey are reproduced here with only superficial editing.

> "Ninety questionnaires were distributed. Sixty were returned, although nine of these were not completed. The response rate was about 57%.
>
> "The results of the survey expressed as a percentage of the returns are presented in Table XII-1. The general consensus was that further research is important to the development of a commercially viable penaeid shrimp industry. If all the research areas could be lumped together to form a general pattern, 23.1% of the respondents felt research was urgently needed (category A), 51.4% felt research necessary (category B), 14.3% felt research not necessary (category C), and 11.2% had no opinion (category D).
>
> "Since most of the attendees of the workshop represented activities oriented around research of one kind or another, it is not surprising that categories A and B were checked as often as they were. The analysis below will pinpoint those responses that did not fit this general pattern.

"There was a close relationship between categories C and D. I think many of the respondents felt that the categories represented a continuum of research importance from most (A) to least (D). Actually, a discontinuity between C and D should have been inferred. If this was not the case, then perhaps categories C and D should be pooled, indicating that 25.5% or less felt that research was not necessary.

"In the meetings, certain areas were emphasized as requiring immediate attention. These were:

- Completing the life cycle,
- Nutritional research to determine the larval, juvenile, and adult requirements,
- Developing economical commercial feed.

These were emphasized on the questionnaire as well. The questionnaire also suggested that other areas could be described as "urgent" (receiving greater than 30% in category A):

- Disease control,
- Analyses of costs for growing shrimp under various conditions, and
- The development of intensive culture systems in middle America.

"In general, categories A and B together were chosen by 75% of the respondents. However, three areas stood out by being chosen by more than 85%:

- Disease identification,
- Growth and survival studies with respect to food consumption,
- Techniques for water reuse.

We might label these areas as "important for research" to distinguish them from the "urgent" areas mentioned before.

"Areas which did not fit the general pattern mentioned above and differed in having significantly higher percentages in the <u>no opinion</u> and <u>research not necessary</u> categories were:

- Nutrient physical form,
- Analyses of holding containers and methods for transferring stock between them, and
- Instrumentation.

Evidently, either technology is presently well ahead of culture methodology or the problems in methodology are so severe that we don't know the kind of technology we need.

"Comments were scattered throughout the questionnaire and touched on a variety of subjects. Several mentioned food and disease problems. For instance, it was suggested that genetic research is necessary to develop more food-efficient animals, and that we need disease research to understand the basic disease mechanisms and the immune systems of shrimp. It was also pointed out that little effort is oriented toward research on the normal microflora of the shrimp gut, although this most likely influences the nutritional and health status of the animal. Relationships between growth/survival and photoperiod (in larval stage), pH, and prolonged pond stratification also need study.

"There were many comments in the areas concerning law and culture techniques for the small farmer. Several people suggested that legislation was needed to:

- Provide subsidies for private pilot projects,
- Encourage commercial efforts, and
- Provide low-interest loans or tax concessions until commercial status was attained. Also, EPA regulations regarding shrimp mariculture need clarification.

"According to the respondents, we need culture techniques that are low energy, closed, and extensive, although they may be labor intensive. Also, they should require low entry cost and small working capital for culturing shrimp at the small-farmer level. One person conceded that this was probably "too far in the future" and that small farm efficiency would "probably never be as high as large farms."

Table XII-1
Marine shrimp questionnaire results.

Area		Results (%)			
		A	B	C	D
BIOLOGY					
1. Life cycle--maturation		78	22	-	-
2. Hatchery development		23	47	12	18
-government operated	14%				
-industry operated	86%				

Table XII-1 (cont.)

Area	Results (%)			
	A	B	C	D
3. Genetic research--growth	23	55	4	18
-survival	16	58	8	18
- hybrids	21	61	12	6
4. Disease--identification	28	58	2	12
-diagnosis	29	53	4	14
-control	39	51	2	8
5. Nutrition--larval requirements	47	33	10	10
-juv-adult requirements	41	47	4	8
-commercial feed	39	47	12	2
-regimes	25	57	10	8
-physical form	8	25	29	18
6. Growth/survival curves				
-temperature	10	45	33	12
-salinity	10	45	31	14
-oxygen	8	53	29	10
-food	27	63	6	2
CULTURE SYSTEMS				
7. Intensive systems	30	30	20	20
8. Multi-species techniques				
-complementary species	8	72	14	6
-physical plants	8	60	20	12
9. Input water treatment	23	51	18	8
10. EPA standards	20	52	16	12
11. Water reuse	23	63	10	4
12. Improved materials	4	51	10	15
13. Size and shape	4	56	24	16
14. Handling stock	4	55	21	12
15. Instrumentation	18	35	37	10

Table XII-1 (cont.)

Area	Results (%)			
	A	B	C	D
LEGAL FACTORS, MARKETING				
1. Legislation to improve availability	25	43	12	22
2. Cost analyses	35	49	10	6
3. Byproduct market analyses	18	63	6	10
4. Cost profile	29	57	6	8
5. Small-farmer techniques	17	57	10	16
ALL AREAS AVERAGED TOGETHER	23.1	51.4	14.3	11.2

B. Robert Wildman's Observations

Mr. Wildman stated his conclusions succintly after emphasizing that he was assigning priorities for government-sponsored research for the immediate future rather than ranking the importance of problems in shrimp farming. There is a real difference between the two and this distinction is important. The priorities as Mr. Wildman stated them are:

1. High:
   a. Life cycle control
2. Medium to high:
   a. Disease prevention and control
   b. Nutrition and feeds
3. Low to medium:
   a. Development of improved grow-out systems
   b. Processing and marketing
   c. Production economics
   d. Legal and regulatory problems

APPENDIX A

WORKSHOP ON THE CULTURE OF PENAEID SHRIMP

October 8-11, 1975

Galveston, Texas

Sponsored by:

The Oceanic Institute
The National Sea Grant Program
The National Marine Fisheries Service Gulf Coastal Fisheries Center
The Texas Agricultural Extension Service
Texas A&M University, a Sea Grant College

Location: Holiday Inn of Galveston
600 Strand
Galveston, Texas

General Chairman: Harold L. Goodwin, for The Oceanic Institute

Registration and Arrangements: Libby Goodwin, Goodwin & Goodwin; and Zoula Zein-Eldin, Gulf Coastal Fisheries Center

October 8

3:00 p.m. to 5:00 p.m. Registration and Tour, Gulf Coastal Fisheries Center, National Marine Fisheries Service

8:00 p.m. to 10:00 p.m. Registration continues, Holiday Inn

October 9

8:30 a.m. Welcome and Introduction to the Workshop: Goodwin

8:45 a.m. SESSION ONE

Chairman: Richard Neal, Gulf Coastal Fisheries Center

1. Hatchery Systems:

   Discussion Leader: Mr. Mock

2. Obtaining Spawners:

   Discussion Leader: Mr. Sweat

3. Life Cycle Control:

   Discussion Leader: Dr. Clark

12:00 Lunch, Holiday Inn. Speaker: Dr. Alain Michel, Centre Oceanologique du Pacifique, Tahiti

SESSION TWO

Chairman: Dr. Jack Parker, Center for Marine Resources, Texas A&M University

1. Grow-out Systems:

   Discussion Leader: Mr. Holcomb

2. Disease Prevention and Control:

   Discussion Leader: Dr. Barkate

3. Nutrition and Feeds:

   Discussion Leader: Dr. MacGrath

5:00 p.m. Adjourn

6:30 p.m. Attitude Readjustment, Cash Bar, Holiday Inn

7:00 p.m. Dinner, Holiday Inn. Speaker: John Glude

October 10

8:30 a.m. SESSION THREE

Chairman: Joseph Angelovic, Gulf Coastal Fisheries Center

1. Production Economics:

   Discussion Leader: Dr. Webber

2. Processing and Marketing:

   Discussion Leaders: Dr. Chaston and Dr. Gillespie

3. Legal and Regulation Problems:

   Discussion Leader: Terry Leary

12:00 Lunch

1:30 p.m. SESSION FOUR

Chairman: John Glude, NOAA Aquaculture Coordinator

1. Summary of the State of the Art:

   Rapporteur: Hal Goodwin

2. Priorities for Penaeid Shrimp Research:

   Rapporteur: Robert Wildman

Participants in Session Four were the Session Chairman and Discussion Leaders.

4:30 p.m. Adjourn

October 11

Tour, Texas A&M and Texas Parks and Wildlife Shrimp Facilities.

APPENDIX B

AN INCOMPLETE COMPILATION OF PEOPLE AND ORGANIZATIONS INVOLVED IN SHRIMP CULTURE IN THE AMERICAS

Booth Fisheries (Division of Consolidated Foods)

General Offices
2 North Riverside Plaza
Chicago, Illinois 60606

Dr. Ian Chaston
Aquaculture Manager

Shrimp Procurement and
Processing Division
P.O. Box 592
Brownsville, Texas 78520

Mr. L. Hodgson, General Manager

Booth is one of the largest shrimp companies in the world. The procurement/processing operation is based in Brownsville, Texas. This office also directs overseas operations in Central America, the Middle and Far East, and Mexico.

Following withdrawal from active research on economic feasibility studies on shrimp culture systems, Booth is inactive in the field of aquaculture other than monitoring ongoing shrimp research by other organizations.

Centre National pour l'Exploitation des Oceans (CNEXO)
Centre Oceanologique du Pacifique (COP)

Vairao, Tahiti

Telephone: 4 Vairao

Mr. J. de Chazeaux, Director

AQUACOP: COP Aquaculture Team

J. Calvas, Environmental conditions
G. Cuzon, Nutritionist
J. M. Griessinger, Crustacean and fish culture
J. F. le Bitoux, Biopathologist
J. L. Martin, Unicellular algal culture
A. Michel, Coordinator aquaculture project

The Pacific Oceanologic Center (COP) began, in 1972, research experiments in the culture of marine shrimps in a coralien tropical environment.

The following fields are under investigation:

Maturation and spawning in captivity

Larval rearing

Disease prevention and control

Nutrition and feeds

Grow-out systems

University of Arizona Environmental Research Laboratory (ERL)

Tucson International Airport
Tucson, Arizona 85706 Telephone: (602) 884-2931

Mr. Carl N. Hodges, Director

Associates:
Dr. Donald V. Lightner, Biopathologist
Dr. L. Bernard Colvin, Biochemist/Nutritionist
Mr. Lee E. Mahler, Engineer (Coordinator)
Mr. David Moore, Facilities Manager
Mr. Christopher W. Brand, Nutritionist
Mr. Billy R. Salser, Production Manager
Mr. Don Donald, Microbiologist
Mr. James Ure, Bioengineering Technician

(All of the above are stationed full time or half time at the ERL coastal shrimp research station at Puerto Peñasco, Sonora, Mexico.)

University of Sonora
Centro de Investigaciones Cietificas y Technologicas
Hermosillo, Sonora, Mexico

Xico Murrieta, Director

Associates: Jose Ramon Infante, Armando Perez, Juan Reprieta, Panchita Reprieta, Carlos Villavcencio, Eneyda Villavcencio, Oscar Villavcencio

Shrimp culture project: Unidad Experimental
Puerto Peñasco
Sonora, Mexico

The Environmental Research Laboratory began, in 1973, an extensive production research experiment in the intensive culture of marine shrimp in controlled environments. The coastal research station on the Gulf of California, which ERL shares with the University of Sonora, was converted into a large shrimp hatchery with a variety of controlled environments, including 30-meter raceways within air-inflated structures. Pathology and nutrition laboratories are on-site. A dozen of the ERL staff are assigned full time to the project; half of these are in residence in Puerto Peñasco (160 air miles southwest of Tucson) and the remainder, senior staff members, commute weekly via regular ERL light-plane flights. Three or more technicians from the University of Sonora work cooperatively in the project and another 12 to 15 Mexican nationals are employed at the station.

The objective of the continuing, long-term project is to develop economically viable systems of intensive marine shrimp culture in controlled environments; to identify and control diseases associated with intensive culture; and to establish optimum nutritional requirements and associated systems of formulating, compounding, and delivering artificial feeds.

Groton Associates, Inc.

P.O. Box 517
Groton, Massachusetts 01450 Telephone: (617) 448-5916

Dr. Harold H. Webber, President

Divisions:

Groton Bioindustries Development Co.
39 Hollis Street
Groton, Massachusetts 01450

Tropical Bioindustries Development Co.
9000 S.W. 87th Court, Suite 104
South Miami, Florida 33176
Telephone: (305) 279-7026

Commercial application (shrimp farm):

Maricultura, S.A.
Apartado 5255
San Jose, Costa Rica
Telephone: 23-50-17

Massachusetts

Ms. Pauline F. Riordan, Vice President, Food Technology - Marketing
Mr. John E. Huguenin, Senior Project Engineer

Mexico

Dr. Simon Williams, Vice President, Economic Development
Mr. James Miller, Economist

Florida

Dr. Durbin C. Tabb, General Manager
Dr. Eric J. Heald, Program Director, Ecology
Dr. Martin A. Roessler, Program Director, Ecology Biometrics
Mr. Gary Beardsley, Biologist

Costa Rica

Mr. Robert E. Smith, President
Dr. Harold H. Webber, Vice President, Technical Director
Mr. James G. Broom, Operations Manager
Mr. William Drummond, Hatchery Manager

Groton Associates are consultants in research, development, and commercial applications in marine shrimp farming.

Gulf Coastal Fisheries Center
Aquaculture Division
National Marine Fisheries Service

4700 Avenue U
Galveston, Texas 77550

East Lagoon
Galveston, Texas

Dr. Richard A. Neal, Division Chief

Telephone: (713) 763-1211

Associates:

Dr. Wallis H. Clark, Jr., Subtask Leader, Maturation Subtask
Mr. Ausbon Brown, Jr., Fishery Biologist
Mr. Richard A. Diener, Fishery Biologist
Mr. Marcel Duronslet, Fishery Biologist
Mr. Ray Wheeler, Fishery Biologist
Mr. Ashley Yudin, Biological Aid
Mr. Kenneth Marvin, Biological Aid

Mrs. Zoula Zein-Eldin, Subtask Leader, Nutrition Subtask
Mrs. Jane McGaffey, Fishery Biologist

Dr. Jorge Leong, Subtask Leader, Disease Subtask
Mr. Clark T. Fontaine, Fishery Biologist
Mrs. Peggy Kenney, Fishery Biologist
Mrs. Imogene Sanderson, Biological Technician

Mr. Cornelius R. Mock, Subtask Leader, Intensive Culture Subtask
Mrs. Loretta Ross, Fishery Biologist
Mr. Kenneth Holden, Biological Aid

The Aquaculture Division is conducting research on basic biological problems which are presently restricting the development of shrimp farming. This research includes studies of penaeid shrimp reproduction and sexual maturation, studies of shrimp nutrition including the formulation of economical diets and studies of shrimp diseases, their recognition, prevention, and treatment. In addition, a part of the staff is developing closed raceway systems for high-density culture of shrimp on a year-round basis. Opportunities are available for trainees to work with the research team in Galveston to learn the techniques being used there.

University of Houston
Department of Biology

4800 Calhoun Boulevard
Houston, Texas 77004

Telephone: (713) 749-2674

Dr. Addison L. Lawrence, Professor

Dr. Lawrence is involved in shrimp nutrition and mariculture work.

Louisiana Wild Life and Fisheries Commission
Marine Biological Laboratory

P.O. Box 37
Grand Isle, Louisiana 70358

Facility located on Grand Terre Island
Telephone: (504) 787-2163

Ralph Latapie, Marine Laboratory Supervisor

Associates:

Stephen Sackett, Marine Biologist
Jack Hotstream, Biological Aide

The agency is primarily concerned with the marine commercial seafoods of Louisiana. When salinities and temperatures in the Louisiana estuaries are optimum for penaeid shrimp, Louisiana has had commercial landings of up to nearly 100 million pounds (heads on) annually with a dockside value approaching $50 million. In addition to monitoring the hydrological parameters which influence shrimp production and year-round field sampling of natural populations, the Marine Laboratory has conducted pond studies of penaeid shrimp since 1962. In addition to laboratory facilities, there are twenty-three 1/4-acre ponds used in various studies on shrimp, oysters, and finfish.

Most of our previous work has been monoculture of shrimp, testing various stocking rates, feeds and feeding rates, and evaluating and developing various culture and handling techniques. Most of the work has been done on *Penaeus setiferus* and *P. aztecus*, with some work done on *P. occidentalis*. Current work is being done on polyculture with penaeid shrimp and *Macrobrachium* studies.

Marifarms, Incorporated

3830 Frankford Avenue
Panama City, Florida 32401
Telephone: (904) 763-5011
and 769-2411

John R. Cheshire, President
Holton R. Harders, Vice President and Treasurer
Joseph Ikeguchi, Vice President and Secretary-Treasurer

Commercial shrimp culture in two 300-acre marine lakes and one 2,500-acre embayment pen. Extensive hatcheries and nursery ponds are included in the facilities.

Marine Nutritional Systems, Inc.

1050 Yuma Street, Suite 101
Denver, Colorado 80204
Telephone: (303) 629-0200

Mr. James Kitchel, President
Mr. Carl Kallansrud, General Manager

Associates:

Mrs. M. A. Murphy Kenslow, Director Algal Research
Mr. Roger Rotvold, Production Supervisor

Marine Nutritional Systems provides natural foods for larval shrimp. Frozen concentrates of the algae *Tetraselmis* sp. and *Skeletonema* sp. are currently available. Other algae species will be cultured upon request.

Shrimp Culture, Inc.

P.O. Box 1123
Key West, Florida 33040
Telephone: (305) 294-9561

James E. Herrin, Jr., President
Dale S. Hunter, Administrative Assistant

Shrimp Culture, Inc. (cont.)

Associates (Republic of Honduras):

Charles S. Hamlin, General Manager
James P. Norris, Hatchery Manager
William N. Rudd, Pond Manager

Facilities:

813 Caroline Street	Choluteca
Key West, Florida 33040	Choluteca, Honduras

Shrimp Culture, Inc., through its subsidiary, Sea Farms de Honduras, S.A. de C.V., operates a marine shrimp culture facility on the Pacific coast of the Republic of Honduras. Shrimp Culture, Inc., and its predecessor, Sea Farms, Inc., have been actively engaged in research and development in marine shrimp culture since 1969 in Key West and since 1973 in the Republic of Honduras.

Texas Agricultural Extension Service
Department of Wildlife and Fisheries Science
Texas A&M University

Dr. Wallace G. Klussmann, Extension Sea Grant Program Leader, College Station, Texas

Dr. Jack Parker, Sea Grant, Mariculture Specialist and Program Manager, College Station, Texas

Dr. Fred S. Conte, Area Mariculture Specialist and Project Leader, Corpus Christi, Texas

Mr. Hoyt W. Holcomb, Area Mariculture Specialist and Project Leader, Angleton, Texas

Mr. David B. Ramsey, Sea Grant Program Aide, Angleton, Texas

Mr. Jim L. Harrell, Sea Grant Program Aide, Angleton, Texas

Mr. David L. Hutchins, Sea Grant Program Aide, Corpus Christi, Texas

----

APPENDIX C

A MARINE SHRIMP CULTURE BIBLIOGRAPHY[a]

John E. Huguenin
and
S. Suzanne Huguenin

Groton BioIndustries
Division of Groton Associates

October 1976

Introduction

This bibliography has been compiled through a series of iterations, each one more comprehensive and better organized than its predecessors. It is in part intended to provide as complete a listing as possible of pertinent published materials, which can be used to clarify the state of the art. Shrimp culture is presently a very dynamic field and such a bibliography, dependent on published information, will not completely reflect current capabilities existing in laboratories or in the field. In addition, since many aspects of importance for commercial application do not get discussed extensively in scientific journals, there are unfortunately data voids in the literature and some of the references may be more general in content than might be desired.

It is hoped that this bibliography will provide an organized method for the serious investigator, whether previously professionally involved in shrimp culture or not, to acquire in-depth scientific and technical information. For these purposes a subject index of selected references and an author index have been provided.

Readers are encouraged to provide additions and corrections to this bibliography and any such information will be very much appreciated.

---

[a]Preparation partially supported through the Massachusetts Institute of Technology Sea Grant Program and the Woods Hole Oceanographic Institution Sea Grant Program.

Subject Index of Selected Shrimp References

References

1. Williamson, D.I. 1968. The type of development of prawn as a factor determining suitability for farming. FAO Fisheries Report, No. 57, Vol. II: 77-84.

2. Hildebrand, H.H. 1954. A study of the brown shrimp (Penaeus aztecus Ives) grounds in the western Gulf of Mexico. Publs. Inst. Mar. Sci. 3(2): 233-266.

3. Hildebrand, H.H. 1955. A study of the fauna of the pink shrimp (Penaeus duorarum Burkenroad) grounds in the Gulf of Campeche. Publs. Inst. Mar. Sci., 4(1):169-232.

4. Burkenroad, M.D. 1934. The penaeidae of Louisiana. Bull. Amer. Mus. Nat. Hist., 68:61-143.

5. Burkenroad, M.D. 1939. Further observations on penaeidae of the northern Gulf of Mexico. Bull. Bingham Oceanog. Coll., Vol. 6, Art. 6:1-62.

6. Lunz, G.R. 1958. Pond cultivation of shrimp in South Carolina. Proc. Gulf and Carib. Fish. Inst., 10:44-48.

7. Pearson, J.C. 1939. The early life histories of some American penaeidae, chiefly the commercial shrimp, Penaeus setiferus (Linn.). U.S. Bureau of Commercial Fisheries, Bulletin No. 30, Vol. 49:1-73.

8. Anderson, W.W., J.E. Wing and M.J. Lindner. 1949. Early stages in the life history of the common marine shrimp, Penaeus setiferus (Linnaeus). Biol. Bull. 96:168-172.

9. Heegaard, P. 1953. Observations on spawning and larval history of shrimp, Penaeus setiferus (L.). Pub. Inst. Mar. Sci. 3(1):75-105.

10. Anderson, W.W. 1956. Observations upon the biology, ecology and life history of the common shrimp, _Penaeus setiferus_ (Linnaeus), along the South Atlantic and Gulf coasts of the United States. _Proc. Indo-Pacific Fish. Council,_ 6(III): 399-403.

11. Lindner, M.J. and W.W. Anderson. 1956. Growth, migration, spawning and size distribution of shrimp, _Penaeus setiferus_. U.S. Fish and Wildlife Service, _Fishery Bulletin 106_, Vol. 56:555-645.

12. Dobkin, S. 1961. Early developmental stages of pink shrimp, _Penaeus duorarum_, from Florida waters. Bureau of Commercial Fisheries, _Fishery Bulletin 190_, Vol. 61:321-349.

13. Cummings, W.C. 1961. Maturation and spawning of pink shrimp, _Penaeus duorarum_ Burkenroad. _Trans. Amer. Fish. Soc._ 90(4):462-468.

14. Tabb, D.C., D.L. Dubrow, and A.E. Jones. 1962. Studies of the biology of the pink shrimp, _Penaeus duorarum_ Burkenroad, in Everglades National Park, Florida. _Fla. Bd. Conserv., Tech. Ser. No. 37_:1-31.

15. Cook, H.L. and M.A. Murphy. 1971. Early developmental stages of the brown shrimp, _Penaeus aztecus_ Ives, reared in the laboratory. _Fish. Bull._ 69(1):223-239.

16. St. Amant, L.S., K.C. Corkum, and J.G. Broom. 1963. Studies on growth dynamics of the brown shrimp, _Penaeus aztecus_, in Louisiana waters. _Proc. Gulf & Carib. Fish. Inst._ 15:14-26.

17. St. Amant, L.S., J.G. Broom, and T.B. Ford. 1966. Studies of the brown shrimp, _Penaeus aztecus_, in Barataria Bay, Louisiana 1962-1965. _Proc. Gulf & Carib. Fish. Inst._ 18th Ann. Sess. 1-17.

18. Burkenroad, M.D. 1949. Occurrence and life histories of commercial shrimp. _Science_ 110:688-689.

19. Williams, A.B. 1955. A contribution to the life histories of commercial shrimps (Penaeidae) in North Carolina. _Bull. Mar. Sci. Gulf & Carib._ 5(2):116-146.

20. Costello, T.J. and D.M. Allen. 1964. Pink shrimp life history. _U.S. Fish and Wildlife Service Circular No. 183_: 30-31.

21. Pérez-Farfante, I. 1969. Western Atlantic shrimps of the genus _Penaeus_. _Fishery Bulletin_ 67(3):461-591.

22. Anderson, W.W. 1970. Contributions to the life histories of several penaeid shrimps (penaeidae) along the South Atlantic Coast of the United States. U.S. Fish and Wildlife Service, Special Scientific Report - Fisheries No. 605, 24 pp.

23. Anderson, W.W. 1966. The shrimp and the shrimp fishery of the southern United States. U.S. Fish and Wildlife Service, Fishery Leaflet 589, 8 pp.

24. Pérez-Farfante, I. 1970. Diagnostic characters of juveniles of the shrimps *Penaeus aztecus*, *P. duorarum* duorarum and *P. brasiliensis* (crustacea, decapoda, penaeidae). U.S. Fish and Wildlife Service, Special Scientific Report - Fisheries No. 599, 26 pp.

25. Joyce. E.A., Jr., and B. Eldred. 1966. The Florida shrimping industry. State of Florida, Board of Conservation Educational Series No. 15, 47 pp.

26. Knopf, G.M. 1970. Opportunities in the shrimp fishing industry of the southeastern United States. Sea Grant Information Bulletin No. 3, Univ. of Miami Sea Grant Institutional Program, 28 pp.

27. Hudinaga, M. 1942. Reproduction, development and rearing of *Penaeus japonicus* Bate. Japanese J. Zool., 10(2):305-493, + 31 pp. of plates (in English).

28. Williams, A.B. 1958. Substrates as a factor in shrimp distribution. Limnol. & Oceanog. 3(3):283-290.

29. Gunter, G. 1961. Habitat of juvenile shrimp (family penaeidae). Ecology, 42(3):598-600.

30. Fujinaga, M. and Z. Kittaka. 1966. Studies on food and growth of larval stages of a prawn, *Penaeus japonicus*, with references to the application to practical mass culture. Inform. Bull. Planktol. Jap., 13:83-94. (Only abstract and figures are in English), translation from Div. of Foreign Fisheries (Translations), U.S. National Marine Fisheries Serv.

31. Sastrakusumah, S. 1971. A study of the food of juvenile migrating pink shrimp, *Penaeus duorarum* Burkenroad. Sea Grant Tech. Bull. No. 9, 36 pp., University of Miami, Jan. 1971.

32. Dall, W. 1968. Food and feeding of some Australian penaeid shrimp. FAO Fisheries Report No. 57, Vol. II: 251-258.

33. Flint, L.H. 1956. Notes on the algal food of shrimp and oysters. Proc. Louisiana Acad. Sci., 19:11-14.

34. Tabb, D.C., W.T. Yang, C.P. Idyll, E.S. Iversen. 1969. Research in marine aquaculture at the Institute of Marine Science, University of Miami. Trans. Amer. Fish. Soc., 98(4):738-742.

35. Holthuis, L.B., H. Rosa, Jr. 1965. List of species of shrimps and prawns of economic value. FAO Fish. Tech. Pap. 52, 21 pp.

36. Kruse, D.N. 1959. Parasites of the commercial shrimps, *Penaeus aztecus* Ives, *P. duorarum* Burkenroad, *P. setiferus* (Linnaeus). Tulane Stud. Zool. 7(4):123-144.

37. Hutton, R.F., F. Sogandares-Bernal, B. Eldred, R.M. Ingle, and K.D. Woodburn. 1959. Investigations on the parasites and diseases of salt water shrimp (penaeidae) of sports and commercial importance to Florida. Fla. Bd. Conserv. Mar. Lab., Tech. Ser. No. 26, 1-38.

38. Iversen, E.S., and R.B. Manning. 1959. A new microsporidian parasite from the pink shrimp (*Penaeus duorarum*). Trans. Amer. Fish. Soc., 88:130-132.

39. Villadolid, D.V., and D.K. Villaluz. 1951. The cultivation of sugpo (*Penaeus monodon* Fabricius) in the Philippines. The Philippine J. of Fish., 1(1):55-66.

40. Delmendo, M.N. and H.R. Rabanal. 1956. Cultivation of sugpo, *Penaeus monodon*, in the Philippines. Proc. Indo-Pacific Fish. Council., 6(III): 424-431.

41. Caces-Borja, P. and S.B. Rasalan. 1968. A review of the culture of sugpo, *Penaeus monodon* Fabricius, in the Philippines. FAO Fish. Rept. No. 57, Vol. 2:111-123.

42. Menon, M.K. 1954. On the paddy field prawn fishery of Travancore-Cochin and an experiment in prawn culture. Proc. Indo-Pacific Fish. Council, 5(II):131-135.

43. Gopinath, K. 1956. Prawn culture in the rice fields of Travancore-Cochin, India. Proc. Indo-Pacific Fish Council., 6(III):419-424.

44. Kow, T.A. 1954. The shrimp industry of Singapore. Proc. Indo-Pacific Fish. Council., 5(II):145-156.

45. Kow, T.A.H. 1968. Prawn culture in Singapore. FAO Fish. Rept. No. 57, Vol. 2, 85-93.

46. Fujinaga, M. 1963. Culture of Kuruma shrimp (Penaeus japonicus). Curr. Aff. Bull. Indo-Pac. Fish. Council, (36):10-11.

47. Kesteven, G. L. and T. J. Job. 1958. Shrimp culture in Asia and the Far East: A preliminary review. Proc. Gulf and Carib. Fish. Inst., 10:49-68.

48. Manning, S. F. 1963 (in two parts). Can you raise shrimp? ... Yes, but the problems of shrimp farming aren't all technical. 4 pp., April, and Many factors figure in building of successful shrimp ponds. 4 pp., May, National Fisherman combined with Maine Coast Fisherman.

49. Rabanal, H. R., H. R. Montalban, and D. K. Villaluz. 1951. The preparation and management of Bangos Fish Pond Nursery in the Philippines. The Philippine Journal of Fisheries, 1(1):3-34.

50. Allen, D. M. 1963. Shrimp farming. Fishery leaflet 551, 8 pp. U.S. Bur. of Comm. Fish.

51. Hickling, C. F. 1961. The Farming of Fish. Pergamon Press, New York, 88 pp.

52. Tang, Y. 1961. The use of Saponin to control predaceous fishes in shrimp ponds. The Progressive Fish-Culturist 23(1):43-45.

53. Lunz, G. R. and C. M. Bearden. 1963. Control of predaceous fish in shrimp farming in South Carolina. Bears Bluff Laboratories, Wadmalaw Island, S. C., No. 36, 9 pp.

54. Lunz, G. R. 1966. Annual Report 1964-1965. Contribution from Bears Bluff Laboratories No. 43, Wadmalaw Island, S. C., 12 pp.

55. Manning, S. F. 1964. Trends in American shrimp cultivation. World Fishing, 13(12):34-36.

56. Zein-Eldin, Z. P. 1961. Shrimp physiology. Circular 129:44-48, U.S. Fish and Wildlife Service, Galveston Biological Laboratory, Fiscal 1961.

57. Hudinaga, M. 1935. Studies on the development of Penaeus japonicus Bate. Rep. Hayatomo Fish. Inst., 1(1):1-51.

58. Iversen, E.S. and C.P. Idyll. 1960. Aspects of the biology of the Tortugas pink shrimp, Penaeus duorarum. Trans Amer. Fish. Soc., 89(1):1-8.

59. Webber, H.H. 1970. The development of a maricultural technology for the penaeid shrimps of the Gulf and Caribbean region. Helgolander wissenschaftliche Meersuntersuchungen, Vol. 20, pp. 455-463.

60. Idyll, C.P. 1970. Status of commercial culture of crustaceans. Food-Drugs from the Sea Proceedings 1969, Marine Tech. Soc., pp. 55-64.

61. Richards, W.L. 1971. Studies of the use of vertical substrates for improving production in the culture of pink shrimp, *Penaeus duorarum* Burkenroad. Sea Grant Tech. Bull. 10, Univ. of Miami, 150 pp.

62. Idyll, C.P. 1950. The commercial shrimp industry of Florida. Fla. St. Bd. Conserv., Education Series No. 6:1-33, (reissued 1957).

63. Fujinaga, M. 1969. Kuruma shrimp, *Penaeus japonicus*, cultivation in Japan. FAO Fish. Rept. No. 57, 3:811-832.

64. Idyll, C.P. 1965. Shrimp nursery-science explores new ways to farm the sea. National Geographic 127 (5):636-659, (May).

65. Hudinaga, M. and M. Miyamura. 1962. Breeding of the 'Kuruma' prawn (*Penaeus japonicus* Bate). J. Oceanographic Soc. of Japan, 20th Anniversary Volume, 1962:694-706, (only abstract, tables and figures in English).

66. Cook, H.L. and M.A. Murphy. 1969. The culture of larval penaeid shrimp. Trans. Amer. Fish. Soc. 98 (4):751-754.

67. Cook, H.L. and M.A. Murphy. 1965. Rearing penaeid shrimp from eggs to post larvae. Paper presented at the meeting of the Southern Division, American Fisheries Society, Oct. 10-13, 1965, Tulsa, Oklahoma, 10 pp., Contrib. U.S. Bur. Comm. Fish. Biol. Lab. No. 209.

68. Hudinaga, M. and J. Kittaka. 1967. The large scale production of the young Kuruma prawn *Penaeus japonicus* Bate. Information Bull. on Planktonology in Japan. Commemorative number of Dr. Y. Matsue, 1967, 34-46 (in English).

69. Hancock, D.A., editor. 1968. Culture of shrimps and prawns. FAO Fish. Rept. No. 57, Vol. 1, Section 4.6, pp. 29-34, June.

70. Yee, W.C. 1970. Seawater utilization in an energy center. Presented at the National Meeting of American Institute of Chemical Engineers, Atlanta, Georgia, Feb. 15-18, 25 pp.

71. Cook, H. L. and M. A. Murphy. 1966. Rearing penaeid shrimp from eggs to post-larvae. Proc. Conf. Southeastern Assoc. Game Commis. 19: 384-388.

72. Cook, H. L. 1969. A method of rearing penaeid shrimp larvae for experimental studies. FAO Fish. Rept. No. 57, Vol. 3:709-715.

73. Kutkuhn, J. H. 1966. Dynamics of a penaeid shrimp population and management implications. Fishery Bull., 65(2):313-338.

74. Cook. H. L. 1970. Larval culture. U.S. Bureau of Commercial Fisheries, Circ. 343, 5-6.

75. Cook, H. L. 1969. Larval culture. U.S. Bureau of Commercial Fisheries, Circ. 325, 5-6.

76. Cook, H. L. 1968. Taxonomy and culture of shrimp larvae. U.S. Bureau of Commercial Fisheries, Circular 295, 7-8.

77. Ewald, J. J. 1965. The laboratory rearing of pink shrimp, *Penaeus duorarum* Burkenroad. Bull. Mar. Sci., 15(2):436-449.

78. Cook, H. L. 1966. Identification and culture of shrimp larvae. U.S. Bur. of Comm. Fish., Circ. 246, 12-13.

79. Aldrich, D. V. and Z. P. Zein-Eldin. 1963. Laboratory results relating to shrimp culture - laboratory studies of shrimp tolerances to salinity and temperature. Bur. of Comm. Fish., Galveston, Texas, 7 pp., manuscript.

80. Zein-Eldin, Z. P. and D. V. Aldrich. 1965. Growth and survival of post larval *Penaeus aztecus* under controlled conditions of temperature and salinity. The Biological Bulletin, 129 (1):199-216, (August).

81. Zein-Eldin, Z. P. and G. W. Griffith. 1966. The effect of temperature upon the growth of laboratory-held post larval *Penaeus aztecus*. Biol. Bull., 131(1):186-196, (Aug.).

82. Zein-Eldin, Z. P. and G. W. Griffith. 1969. An appraisal of the effects of salinity and temperature on growth and survival of post larval penaeids. FAO Fish. Rept. No. 57, Vol. 3, pp. 1015-1026.

83. Zein-Eldin, Z. P. 1963. Effect of salinity on growth of post larval penaeid shrimp. Biol. Bull., 125(1):188-196, (August).

84. Zein-Eldin, Z.P. 1969. Food and experimental environments. U.S. Bureau of Commercial Fisheries, Circular 325, 7-9.

85. Fujinaga, M. 1969. Methods of cultivation of penaeid shrimp. United States Patent Office, Patent No. 3, 477, 406, Nov. 11, 14 pp.

86. Zein-Eldin, Z.P. 1970. Food and experimental environments. U.S. Bureau of Commercial Fisheries, Circular 343, 6-7.

87. Wheeler, R.S. 1970. Culture of juvenile and adult shrimp. U.S. Bureau of Commercial Fisheries, Circular 343, 9-10.

88. Wheeler, R.S. 1966. Cultivation of shrimp in artificial ponds. U.S. Bureau of Commercial Fisheries, Circular 246, 14-15.

89. Wheeler, R.S. 1967. Experimental rearing of post larval brown shrimp to marketable size in ponds. Commercial Fish. Review, 29(3):49-52, (March).

90. Wheeler, R.S. 1969. Culture of juvenile and adult shrimp. U.S. Bureau of Commercial Fisheries, Circular 325, 6-7.

91. Wheeler, R.S. 1968. Cultivation of shrimp in artificial ponds. U.S. Bureau of Commercial Fisheries, Circular 295, 8-9.

92. Costello, T.J. 1963. Pink shrimp life history. U.S. Fish and Wildlife Service, Circular 161, 35-37, Fishery Research Biological Laboratory, Galveston, Fiscal year 1962.

93. Johnson, M.C. and J.R. Fielding. 1956. Propagation of the white shrimp *Penaeus setiferus* (Linn.), in captivity. Tulane Stud. Zool., 4(6): 175-190.

94. Broom, J.G. 1966. Shrimp culture in ponds. Louisiana Conservationist, 18(11-12):7, (Nov.-Dec.).

95. Broom, J.G. 1969. Pond culture of shrimp on Grand Terre Island, Louisiana, 1962-1968. Proc. Gulf & Carib. Fish Inst., 21:137-151.

96. Hull, E.W.S. 1965. Vision-ingenuity-sweat. Geo-Marine Tech., pp. 20-23, (April).

97. Lunz, G.R. 1968. Annual Report 1966-1967, Contributions from Bears Bluff Laboratories No. 46, Wadmalaw Island, S.C., 9 pp. (Jan.).

98. Cook, H. L. 1966. A generic key to the protozoean, mysis, and post larval stages of the littoral penaeidae of the northwestern Gulf of Mexico. U.S. Fish Wildl. Serv., Fishery Bull., 65(2):437-447.

99. Richards, W. L. 1974. The feasibility of penaeid shrimp culture in brackish ponds receiving treated sewage effluent. Wastewater Use in the Production of Food and Fiber - Proceedings, U.S. Envir. Protect. Agency, EPA-660/2-74-041, pp. 504-512, June 1974 (available U.S. Gov. Print. Off., $5.35).

100. Miyamura, M. 1969. Method for the artificial culture of shrimp. U.S. Patent Office Patent No. 3, 473, 509, 11 pp. (October).

101. Cook, H. L. and M. A. Murphy. 1969. The culture of larval penaeid shrimp. Trans. Amer. Fish. Soc. 98(4):751-754.

102. Cook, H. L. and M. A. Murphy. 1971. Early developmental stages of the brown shrimp *Penaeus aztecus* Ives, reared in the laboratory. Fishery Bull., 69(1):223-240.

103. Robinson, P. 1969. Florida shrimp farm nears commercial stage. Nat. Fisherman 49(12):6-C, (March).

104. Stoertz, C. R. Editor. 1969. Florida gets shrimp farm. Marine Resources Digest, Marine Biology Digest, 1(7):4, (April). Newsletter published monthly by Girard Associates, Inc., Mt. Arlington, N. J.

105. Borden, H. 1969. Marine experts endorse Akima farming project. National Fisherman, 49(12): 6-C (March).

106. Wiesepape, L. M. and D. V. Aldrich. 1970. Effects of temperature and salinity on thermal death in post larval brown shrimp, *Penaeus aztecus*. Sea Grant Publication TAMU-SG-71-201, Texas A&M Univ., (October).

107. Karim, M. and D. V. Aldrich. 1970. Influence of diet on the feeding behavior, growth, and thermal resistance of post larval *Penaeus aztecus* and *P. setiferus*. Sea Grant Publication TAMU-SG-70-226, Texas A&M Univ., (September).

108. Beard, T. W. 1972. A preliminary report on the growth and survival of *Macrobrachium rosenbergii* de Man, *Penaeus aztecus* Ives, *P. indicus* H. Milne Edwards and *P. monodon* Fabricius in laboratory recirculation systems. Intern. Council for the Exploration of the Sea (ICES), Shellfish and Benthos Committee, C. M. 1972/K:26.

109. de la Bretonne, L.W., Jr., and J.W. Avault, Jr. 1970. Shrimp mariculture methods tested. The American Fish Farmer 1(12):8-11 (November).

110. Hull, E.W.S., Editor. 1968. A visit to the mariculture state of Florida. Attachment to Ocean Science News 10(36) 6-7 (September).

111. Subrahmanyan, C.B. and C.H. Oppenheimer. 1970. Food preferences and growth of grooved penaeid shrimp. Food-Drugs from the Sea, Proceedings 1969., H.W. Youngher, Jr., Editor, Marine Technology Society, 1970, pp. 65-75.

112. Cleary, D.P. 1969. Demand and price structure of shrimp. Working Paper No. 15, Division of Economic Research, U.S. Bureau of Comm. Fish., 48 pp. (June).

113. Osborn, K.W., B.W. Magnan, and S.B. Drummond. 1969. Gulf of Mexico shrimp atlas. U.S. Bur. of Comm. Fish., Circ. 312, 20 pp.

114. Anderson, W.W., M.J. Lindner, and J.E. King. 1949. The shrimp fishery of the southern United States. Commercial Fisheries Review, 11(2):1-17.

115. Clifford, D.M. 1956. Marketing and utilization of shrimp in the United States. Proc. Indo-Pacific Fish. Council. 6(III):438-443.

116. Longnecker, O.M., Jr. 1968. The place of the shrimp industry in the United States fisheries. The Future of the Fishing Industry of the United States, Univ. of Wash. Publications in Fisheries, New Series, Vol. 4, 111-117.

117. Inglis, A. and E. Chin. 1966. The bait shrimp industry of the Gulf of Mexico. Fishery Leaflet 582, U.S. Bur. of Comm. Fish., 10 pp. (May).

118. Love, T.D. 1967. The sun-dried-shrimp industry of Mexico's north central Gulf. Comm. Fisheries Review, 29(4):58-61, (April).

119. Foo, Y. 1880. U.S. Patent 310-811, Jan. 13, 1880 to Yee Foo.

120. Kane, T.E. 1970. Aquaculture and the law. University of Miami Sea Grant Program, Sea Grant Techn. Bulletin No. 2, 98 pp. (November).

121. Shields, H.W. 1970. Aquaculture and the law. Proc. Gulf and Caribbean Fish. Inst., 22:61-64, 90-96.

122. Foster, W.C. 1968. Fishery regulation by state or Federal government. The Future of the Fishing Industry, Univ. of Wash. Publication in Fisheries, New Series, Vol. 4, pp. 321-325.

123. Prewitt, R. 1971. Rambling along. The American Fish Farmer, 2(2): 22-24, (January).

124. Yee, W.C. 1972. The potential of aquaculture at nuclear energy centers - A systems study. Oak Ridge National Laboratory, ORNL-4488, 66 pp.

125. Weymouth, F.W., M.J. Lindner, and W.W. Anderson. 1933. Preliminary report on the life history of the common shrimp Penaeus setiferus (Linn.). Bull. U.S. Bur. Fish., 48(1940):1-26.

126. Lindner, M.J. 1966. What we know about shrimp size and the Tortugas Fishery. Proc. Gulf & Carib. Fish. Inst., 18th Ann. Sess., 18-26.

127. Mistakidis, M.N., Editor. 1968-1970. FAO world scientific conference on the biology and culture of shrimp and prawns, Mexico City, Mexico, 12-24 June 1967. Proc. in FAO Fish Rept. No. 57, Vol. 1 (June 1968), Vol. 2 (Oct. 1968), Vol. 3 (Oct. 1969) and Vol. 4 (Oct. 1970), (total of 94 papers).

128. Chin, E. and D.M. Allen. 1959. List of references on the biology of shrimp (family penaeidae). U.S. Bur. of Comm. Fish., Spec. Sci. Rept. - Fish. No. 276, 143 pp.

129. Tamura, T. 1966 (2nd edition). Chapter 15 - Propagation of crustaceans. Sections 15.1-15.8 on various species of shrimp (45 pages), Marine Aquaculture, Trans. from the Japanese by M.I. Watanabe, Nat. Sci. Found., 1970, available from NTIS, U.S. Dept. of Comm., PB 194 05IT.

130. Shigueno, K. 1970. Problems of prawn culture in Japan. Proc. Indo-Pacific Fish. Council, 14th Session (Nov. 1970): IPFC/C70/Sym(8).

131. Tabb, D.C., W.T. Wang, Y. Hirono, J. Heinen. 1972. A manual for culture of pink shrimp, Penaeus duorarum, from eggs to post larvae suitable for stocking. Sea Grant Special Bulletin No. 7, Univ. of Miami, 59 pp.

132. Fujinaga, G. 1961. Technics of shrimp farming during zoea, mysis and post larval stages. (In Japanese) Tokkyo Koho (Patent Agency Bulletin), 11 May 1961, Patent Application No. 34-27796:1-3, Translation from Division of Foreign Fisheries (translations), Nat. Mar. Fish. Serv.

133. Lindner, M.J. and H.L. Cook. 1971. Progress in shrimp mariculture in the United States. Proc. Symp. Invest. Resour. Caribb. Sea Adj. Regions, FAO Fish. Rept. 71.2, 157-162.

134. Costello, T.J. and D.M. Allen. 1970. Synopsis of biological data on the pink shrimp Penaeus duorarum Burkenroad, 1939. FAO Fisheries Reports No. 57, Vol. 4, 1499-1537.

135. Lindner, M.J. and H.L. Cook. 1970. Synopsis of biological data on the white shrimp Penaeus setiferus (Linnaeus) 1767. FAO Fisheries Reports No. 57, Vol. 4, 1439-1469.

136. Cook, H.L. and M.J. Lindner. 1970. Synopsis of biological data on the brown shrimp Penaeus aztecus Ives, 1891. FAO Fisheries Reports No. 57, Vol. 4, 1471-1497.

137. Kim, K.D. 1967. Studies on the artificial culture of Penaeus orientalis Kishinouye. Proc. Indo-Pacific Fish. Council, 12(II):253-257.

138. Meyers, S.P., D.P. Butter, and G.F. Sirine. 1971. Encapsulation - a new approach to larval feeding. The American Fish Farmer, 2(8):15-16, 18,20 (July).

139. Sirine, G. 1967. Microencapsulation. Stanford Research Inst. Journal 15:1-6 (June).

140. Meyers, S.P. 1971. Crustacean ration formulation research. Feedstuffs, 43(28):27 (July 10).

141. Ronquillo, I.A. and A. de Jesus. 1959. Notes on growing of lab-lab in Bango nursery ponds. Philipp. Journal Fish., 5(2):99-102.

142. Trust, T.J. and A.J. Wood. 1973. An initial evaluation of ethylene oxide for the sterilization of formulated and pelleted fish feeds. J. Fish Res. Board of Can., 30:269-274.

143. Deshimaru, O. and K. Shigueno. 1972. Introduction to the artificial diet for prawn Penaeus japonicus. Aquaculture 1(1):115-133, May.

144. Meyers, S.P., J. Avault, D.P. Butler, J.S. Rhee. 1970. Development of rations for economically important aquatic and marine invertebrates. Coastal Studies Bulletin, 5:157-172, Louisiana State University.

145. Forster, J.R.M. and T.W. Beard. 1969. Some experiments on the preparation of a compound diet for the prawn Palaemon serratus (Pennat). International Council for Exploration of the Sea (ICES), Fisheries Improvement Committee, C.M. 1969/E:6, 9 pp.

146. Condrey, R.E., J.G. Gosselink, H.J. Bennett. 1972. Comparison of the assimilation of different diets by Penaeus setiferus and P. aztecus. Fish. Bull., 70(4):1281-1291.

147. Shewbart, K.L., W.L. Mies, and P.D. Ludwig. 1973. Nutritional requirements of brown shrimp Penaeus aztecus. Texas A&M Univ., TAMU-SG-73-205, 53 pp.

148. Condrey, R.E. 1971. Comparative assimilation efficiencies of penaeids fed on natural and artificial diets. M.S. Thesis, Louisiana State University, 45 pp.

149. Forster, J.R.M. 1972. Some methods of binding prawn diets and their effects on growth and assimilation. J. Intern. Council for Expl. of Sea, 34(2):200-216.

150. Kanazawa, A., M. Shimaya, M. Kawasaki, and K. Kashiwada. 1970. Nutritional requirements of prawn-I Feeding on artificial diets. Bull. of the Japanese Soc. of Sci. Fish., 36(9):949-954 (in English).

151. de Figueiredo, M.J. 1973. Some food studies in the larval rearing of Palaemon serratus (Pennant). ICES, C.M. 1973/K.5, 7 pp.

152. Hastings, W.H., S.P. Meyers, D.P. Butler. 1971. A commercial process for water-stable fish feeds. Feedstuffs 43(47):38-?, (Nov. 13).

153. Cowey, C.B. and J.R.M. Forster. 1971. The essential amino acid requirements of the prawn Palaemon serratus. The growth of prawn on diets containing proteins of different amino-acid composition. Mar. Biol. (Berlin) 10(1):77-81.

154. Hastings, W.H., S.P. Meyers, D.P. Butler. 1973. New process - Feed hangs together. The American Fish Farmer 4(2):13-14 (Jan.).

155. Sick, L.V., J.W. Andrews, and D.B. White. 1972. Preliminary studies of selective environmental and nutritional requirements for the culture of penaeid shrimp. Fishery Bulletin 70(1):101-109 (Jan.).

156. Balazs, G.H., E. Ress, and C.C. Brooks. 1973. Preliminary studies on the preparation and feeding of crustacean diets. Aquaculture 2:369-377.

157. Forster, J.R.M., P.A. Gabbott. 1971. The assimilation of nutrients from compounded diets by the prawns *Palaemon serratus* and *Pandalus platyceros*. J. Mar. Biol. Assoc. U.K., 51(4):943-961.

158. Sick, L.V., D. White, and G. Baptist. 1973. The effect of duration of feeding, amount of food, light intensity and animal size on rate of ingestion of pelleted food by juvenile penaeid shrimp. The Progressive Fish-Culturist, 35(1):22-26 (Jan.)

159. Meixner, R. 1966. The effects of food supply on moulting, growth and spawning *Crangon crangon*. (L.) ICES Paper, Shellfish Comm., C.M. 1966/M, 7 pp.

160. Kay, M.M. 1971. L-isoleucine: an inducer of the feeding response in decapoda crustaceans. Experientia, 27(1):103-104.

161. Brown, A., Jr. 1972. Experimental techniques for preserving diatoms used as food for larval *Penaeus aztecus*. Proc. Natl. Shellfisheries Assoc., 62:21-25 (June).

162. Sulit, J., R.S. Esguerra, and H.R. Rabanal. 1957. Fertilization of Bangos Nursery Ponds with commercial fertilizer. Philipp. J. Fish., 5(2): 125-133.

163. Vanderwalker, J.G. and E. Chin. 1962. A device for feeding brine shrimp to fishes. Trans. Amer. Fish. Soc., 91(2):230-231.

164. Benoit, D., R. Syrett, and J. Hale. 1969. Automatic live brine shrimp feeder. Trans. Amer. Fish Soc., 98(3):532-533.

165. Persoone, G. and P. Bratland. 1972. An improved separator both for *Artemia* nauplii and other phototactic invertebrates. Helgolander Wiss. Meeresunters, 23(2):243-247.

166. Khandker, N.A. 1962. The composition of shrimp meal from fresh and spoiled shrimp heads. Comm. Fish Rev., 24(2):8-10.

167. Meyers, S.P. and J.E. Rutledge. 1971. Shrimp meal - a new look at an old product. Feedstuffs, 43(49):31-? (Nov. 27).

168. Hutton, R. F., F. Sogandares-Bernal, and B. Eldred. 1959. Another species of Microphallus Ward, 1901, from the pink shrimp, Penaeus duorarum Burkenroad. J. Parasitol. 45(5):490.

169. Overstreet, R. M. 1973. Parasites of some penaeid shrimp with emphasis on reared hosts. Aquaculture, 2:105-140.

170. Villella, J. B., E.S. Iversen, and C. J. Sindermann. 1970. Comparison of the parasites of pond-reared and wild pink shrimp (Penaeus duorarum Burkenroad) in South Florida. Trans. Amer. Fish. Soc., 99:789-794.

171. Sindermann, C. J. 1970. The role and control of diseases and parasites in mariculture. Proc. Food-Drugs Sea 1969, pp. 145-173, 47 Ref., Marine Tech. Soc.

172. Hutton, R. F. 1964. A second list of parasites from marine and coastal animals of Florida. Trans. Am. Microsc. Soc., 83(4):439-447.

173. Huner, J. V. 1968. Use of Fintral-5 to control undesirable fishes in shrimp-oyster ponds. Proc. La. Acad. Sci., 31:58-61.

174. Iversen, E.S. and N. N. VanMeter. 1964. A record of the microsporidian, Thelohania duorara, parasitizing the shrimp, Penaeus brasiliensis. Bulletin of Marine Science of the Gulf and Caribbean, 14(4):549-553.

175. Naqvi, S. M. and D. E. Ferguson. 1970. Levels of insecticide resistance in fresh-water shrimp, Palaemonetes kadiakensis. Trans. Amer. Fish. Soc., 994:696-699.

176. Chin, E. and D. M. Allen. 1957. Toxicity of an insecticide to two species of shrimp, Penaeus aztecus and P. setiferus. Texas J. Sci., 93:270-278 (Sept.).

177. Caillouet, C.W. and D.C. Tabb. 1972. Pink shrimp research at Turkey Point. The American Fish Farmer, 3(5):8-11.

178. GATT-ITC. 1967. Major markets for shrimp and prawns in Western Europe. GATT, Intern. Trade Center. Geneve, Switzerland, 149 pp., I-54, (Feb. 1967), $5.00, updated 1969.

179. Smith, M. 1970. Shrimp farming. Compressed Air Magazine, 75(1):18-20 (Jan.).

180. Robinson, P. 1968. Florida aquaculture gets boost. National Fisherman, May 1968, p. 8C.

181. Hull, E.W.S., Editor. 1971. Some 500,000 pounds of farmed shrimp have been harvested. Ocean Science News, 13(50):1-3, (Dec. 10, 1971).

182. Annon. 1972. Shrimp harvest at Marifarms. The American Fish Farmer, 3(2):5-7, (Jan.).

183. Anderson, A. 1972. Florida shrimp farm loses part of crop to high tides. Maine Coast Fisherman, Section A, p. 10 (April).

184. Forster, J.R.M. and J.F. Wickins. 1967. Experiments in the culture of the prawn *Palaemon serratus* (Pennant). ICES CM 1967, Fisheries Improvement Committee, Doc. E-13.

185. Forster, J.R.M. 1970. Further studies on the culture of the prawn, *Palaemon serratus* Pennant, with emphasis on the post larval stages. Fishery Investigations, Series II, 26(6):1-40.

186. Wilkins, J.F. 1972. Experiments on the culture of the spot prawn *Pandalus platyceros* Brandt and the giant freshwater prawn *Macrobrachium rosenbergii* (deMan). Fishery Investigations, Series II, 27(5):23 (London).

187. Wilkins, J.F. 1972. Developments in the laboratory culture of the common prawn, *Palaemon serratus* Pennant. Fishery Investigations, Series II, 27(4):1-23, (Ministry Agric. Fish and Food, London).

188. Wickins, J.F. 1972. The food value of brine shrimp, *Artemia salina* L., to larvae of the prawn, *Palaemon serratus* Pennant. J. Exp. Mar. Biol. Ecol., 10(2):151-170.

189. Forster, J.R.M. and T.W. Beard. 1974. Experiments to assess the suitability of nine species of prawns for intensive cultivation. Aquaculture, 3(4):355-368.

190. Kitabayashi, K., K. Shudo, K. Nakamura, and S. Ishikawa. 1971. Studies on formula feed for Kuruma prawn-I, On the relationship among glucosamine, phosphorus and calcium. Bull. Tokai Reg. Fish. Lab., (Tokyo), 65:91-108, (only the abstract is in English).

191. Kitabayashi, K., K. Shudo, K. Nakamura, and S. Ishikawa. 1971. Studies on the formula feed for Kuruma prawn-II, On the utilization values of glucose. Bull. Tokai Reg. Fish. Lab. (Tokyo), 65:109-115, (Only the abstract is in English).

192. Kitabayashi, K., K. Shudo, K. Nakamura, and S. Ishikawa. 1971. Studies on the formula feed of Kuruma prawn-III, On the growth promoting effects of both Arginine and Methionine. Bull. Tokai Reg. Fish. Lab. (Tokyo), 65:119-128, (only the abstract is in English).

193. Shudo, K., K. Nakamura, S. Ishikawa, and K. Kitabayashi. 1971. Studies on the formula feed for Kuruma prawn-IV, On the growth promoting effects of both squid liver oil and cholesterol. Bull. Tokai Reg. Fish. Lab. (Tokyo), 65:129-138, (only abstract in English).

194. Kitabayashi, K. K. Shudo, K. Nakamura, and S. Ishikawa. 1971. Studies on formula feed for Kuruma prawn-V, On the growth promoting effects of protein levels in the diet and reexamination of ingredients used. Bull. Tokai Reg. Fish. Lab. (Tokyo), 65:139-147, (only abstract is in English)

195. Yee, W.C. 1972. Thermal aquaculture: engineering and economics. Environmental Science & Technology, 6(3):232-237, (March).

196. Yang, W.T. 1970. Mariculture in Japan using heated effluent water. Proceedings of the Conference on the Beneficial Uses of Thermal Discharges, Albany, N.Y., Sept. 17-18, 1970, pp. 29-43, S.P. Mathur and R. Stewart Editors, N.Y. State Dept. of Environmental Conserv.

197. Yee, W.C. 1970. Food values from heated water - an overview. Proceedings of the 32nd Annual Conference of the Chemurgic Council - Chemurgy for Better Environment and Profits, Wash. D.C., Oct. 22-23, 20 pp.

198. Florida State Legislature. 1969. Florida Laws 69-46. Adding Sections 253.67 to 253.75 to Chapter 253, Florida Statutes, June 1969.

199. Reeve, M.R. 1969. The laboratory culture of the prawn *Palaemon serratus*. Fishery Investigations, Series II, 26(1):1-38, Ministry of Agric. (London).

200. Hull, E.W.S. (Editor). 1972. We've taken a look at the Marifarms patents. Ocean Science News 14(4):2-3, 1/28/72.

201. Pariser, E.R. and S. Bock. 1973. Chitin and chitin derivatives: an anotated bibliography of selected publications. MIT Sea Grant Report No. 73-2.

202. Anderson, W.W. and M.J. Lindner. 1943. A provisional key to the shrimps of the family penaeidae with special reference to American forms. Trans. Am. Fish. Soc., 73:284-319.

203. Kubo, I. 1955. A review of the biology and systematics of shrimps and prawns of Japan. Proc. Indo-Pacific Fish. Council, 6(3):387-398.

204. Hall, D. N. F. 1962. Observations on the taxonomy and biology of some Indo-West Pacific penaeidae. Fish. Publ. Colon. Off. London, 17:1-229.

205. Word, J. and D. Charwat. 1974. Key to shrimp common in Southern California trawl catches. TM-211, 41 pages, Southern California Coastal Water Research Project, 1500 E. Imperial Highway, El Segundo, California 90245.

206. Webber, H. 1971. Design of an aquaculture enterprise. Proceedings of the Gulf and Caribbean Fisheries Institute, 24th Annual Session (Nov.) pp. 117-125.

207. Anderson, L. G. 1973. An economist looks at mariculture. Marine Tech. Society Journal, 7(3):9-15 (May-June).

208. Anderson, L. G. and D. C. Tabb. 1971. Some economic aspects of pink shrimp farming in Florida. Proc. of the Gulf and Caribbean Fisheries Institute, 23rd Ann. Session, pp. 113-124, discussions pp. 161-162, held at Willemstad, Curacao, Nov. 1970.

209. Schwarz, H. 1969. Shrimp production system. Marine Tech. Soc. J., 3(4):19-20.

210. Mock, C. R., R. A. Neal, and B. R. Salser. 1973. A closed raceway for the culture of shrimp. Proc. Fourth Ann. Workshop World Mariculture Soc., 4:247-259.

211. Fujinaga, G. 1960. Facilities for shrimp farming (in Japanese), Tokkyo Koho (Patent Agency Bulletin), 20 July 1960, Patent application No. 33-28986:1-2, Translation from Div. of Foreign Fisheries (Translations), National Marine Fisheries Service.

212. Neal, R. A. 1971. Shrimp culture research at the Galveston Biological Labs. The American Fish Farmer, 2(6):10-11, 18, 20 (May).

213. Kutty, M. N. 1969. Oxygen consumption of the prawns *Penaeus indicus* H. Milne Edwards and *Penaeus semisuleatus* de Hann, FAO Fish. Rept. No. 57, 3;957-969.

214. Subrahmanyam, C. B. 1962. Oxygen consumption in relation to body weight and oxygen tension in the prawn, *Penaeus indicus* (Milne Edwards). Proc. India Acad. Sci., (B), 55:152-161.

215. Egusa, S. 1961. Studies on the respiration of the Kuruma prawn, Penaeus japonicus Bate-II Preliminary experiments on its oxygen consumption. Bull. Jap. Soc. Sci. Fisheries #27 (7):650-659 (in English).

216. Egusa, S. and T. Yamamoto. 1961. Studies on the respiration of the Kuruma prawn, Penaeus japonicus Bate-I. Bull. Jap. Soc. Sci. Fisheries #27 (1):22-27 (in English).

217. Panikar, N.K. 1968. Osmotic behavior of shrimp and prawns in relation to their biology and culture. FAO Fish. Rep. No. 57 (2):527-538.

218. Lindner, M.J. 1971. Shrimp resources of the Caribbean Sea and adjacent regions. FAO Fisheries Report No. 71.2, pp. 149-156.

219. Croker, R.S. 1967. The shrimp industry of Central America, the Caribbean Sea and northern South America. Foreign Fisheries Leaflet No. 74, Feb. 1967, 127 pp., Branch of Foreign Fish., Div. of Intern. Affairs, U.S. Nat'l. Marine Fish. Service.

220. Anon. 1968. General information on the shrimp industries of selected countries: fisheries agencies, sanitation agencies, shrimp firms exporting to the United States, fisheries trade journals, and fisheries trade associations. Foreign Fisheries Leaflet No. 2, Feb. 1968, 108 pp., Branch of Foreign Fish., Div. of Intern. Affairs, U.S. Nat'l Marine Fish. Service.

221. Cleary, D.P. 1970. World demand for shrimp and prawns may outstrip supply during next decade. Commercial Fisheries Review #32 (3): 19-22 (March).

222. Reeve, M.R. 1969. The suitability of the English prawn Palaemon serratus (Pennant) for cultivation - a preliminary assessment. FAO Fish Rep. No. 57 (3):1067-1073.

223. Phillips, G. 1971. Incubation of the English prawn Palaemon serratus. J. Mar. Bio. Assoc. U.K., 51(1):43-48.

224. Kawazawa, A., M. Tanaka, S. Teshima, and K. Kashiwada. 1971. Nutritional requirements of prawn II - Requirements of steroids. Bull. of the Jap. Soc. of Sci. Fisheries, 37(3):211-215 (in English).

225. Kanazawa, A., N. Tanaka, S. Teshima, and K. Kashiwada. 1971. Nutritional requirements of prawns III - Utilization of the dietary sterols. Bull. of the Jap. Soc. of Sci. Fish., 37(10):1015-1019 (in English).

226. Kanazawa, A., N. Tanaka, and K. Kashiwada. 1972. Nutritional requirements of prawns IV - The dietary effect of ecdysones. Bull. of the Jap. Soc. of Sci. Fish. 38(9):1067-1071 (in English).

227. Thorhaug, A., T. Devany, and B. Murphy. 1971. Refining shrimp culture methods: the effect of temperature on early stages of the commercial pink shrimp. Proc. of the Gulf and Caribbean Fisheries Institute, 23rd Ann. Session, pp. 125-132, discussion p. 162, held at Willemstad, Curacao, Nov. 1970.

228. Lunz, G.R. 1956. Harvest from an experimental one-acre salt-water pond at Bears Bluff Laboratories, South Carolina. Progressive Fish-Culturist, 18(2):92-94.

229. Harris, A.H. and C.D. Rose. 1970. Penaeid shrimp production in a managed, unfed, natural pond in South Louisiana. (Assoc. of S.E. Biologist) The A.S.B. Bull. 17(2):46 (April).

230. Sick, L.V. and J.W. Andrews. 1971. Shrimp mariculture. The American Fish Farmer, 2(11):14-16 (Oct.).

231. Schafer, H.J. 1971. Advances in Pacific shrimp culture. Proc. of the Gulf and Caribbean Fisheries Institute, 23rd Ann. Sess. pp. 133-138, held at Willemstad, Curacao, Nov. 1970.

232. Cordover, R. 1972. Penaeid prawn production. The American Fish Farmer, 3(11):4-6 (Oct.).

233. Wheeler, R.S. 1969. Culture of penaeid shrimp in brackish-water ponds, 1966-1967. Proc. 22nd Ann. Conf., S.E. Assoc. Game and Fish Comm., Oct. 1968, 22:387-391.

234. Dobkins, S. 1969. Abbreviated larval development in Caridean shrimps and its significance in the artificial culture of these animals. FAO Fish. Rep. No. 57(3):935-946.

235. Campbell, K.C. 1968. The case for fish farming-Part I: Prawn culture in North America. World Fishing, 17(12):36-37 (Dec.).

236. Ling, S.W. 1968. Role and possibilities of brackishwater fish and shrimp farming in Southeast Asia. Proc. of the Intern. Seminar on Possibilities and Problems of Fisheries Development in Southeast Asia, Berlin 10-30 Sept. 1968, organized by the German Foundation for Developing Countries, pp. 351-359.

237. Campbell, K.C. 1969. Prawn farming in the Far East. World Fishing, 18(1):41-43 (Jan.).

238. Parker, J.C. 1970. Shrimp farming - Texas style. Coastal Land Resources - Conf. Proc., June 16-17, 1970, Texas A&M Sea Grant, TAMU-SG-71-101, pp. 22-26.

239. Idyll, C.P. 1971. Induced maturation of ovaries and ova in pink shrimp. Comm. Fish. Review, 33(4):20.

240. Parker, J.C. 1972. Commercial shrimp farming nearing reality. Texas Agricultural Progress, 18(3):13-16.

241. Parker, J.C., H.W. Holcomb, Jr., W.G. Klussmann, J.C. McNeill. 1972. Effect of fish removal on the growth and condition of white shrimp, *Penaeus setiferus* (Linnaeus), in brackish ponds. Texas A&M Univ. Sea Grant Program TAMU-SG-72-701, also Proc. 3rd Ann. Workshop World Mariculture Soc., 3:287-297, 1973.

242. Hughes, D.A. 1968. Factor controlling emergence of pink shrimp (*Penaeus duorarum*) from the substrate. Biol. Bull., 134(1):48-59.

243. Fuss, C.M., Jr. 1964. Observations on burrowing behavior of the pink shrimp, *Penaeus duorarum* Burkenroad. Bull. Mar. Sci. Gulf Caribb., 14(1):62-73.

244. Hughes, D.A. 1969. Factors controlling the time of emergence of pink shrimp, *Penaeus duorarum*. FAO Fisheries Reports #57(3):971-981.

245. Wickham, D.A. 1967. Observations on the activity patterns of juveniles of the pink shrimp, *Penaeus duorarum*. Bull. Mar. Sci., 17(4):769-786.

246. Fuss, C.M., Jr., L.H. Ogren. 1966. Factors affecting activity and burrowing habits of the pink shrimp, *Penaeus duorarum* Burkenroad. Biol. Bull., 130(2):170-191 (April).

247. Jachowski, R.L., A.A. Myrberg, Jr. 1968. Photo-orientation of the pink shrimp, *Penaeus duorarum*. Bull. Mar. Sci., 18(1):191-202.

248. Wickham, D.A. 1967. Observations on the activity patterns in juveniles of the pink shrimp, *Penaeus duorarum*. Bull. of Mar. Sci., 17(4):769-786.

249. Grady, J.R. 1971. The distribution of sediment properties and shrimp catch on two shrimping grounds on the continental shelf of the Gulf of Mexico. Proc. of the Gulf and Caribbean Fisheries Institute, 23rd Ann. Sess. pp. 139-148, held at Willemstad, Curacao, Nov. 1970.

250. King, J.E. 1948. A study of the reproductive organs of the common marine shrimp, Penaeus setiferus (Linnaeus). Biol. Bull., 94(3):244-262.

251. Berry, R.J. 1970. Shrimp mortality rates derived from fishery statistics. Proc. Gulf & Caribb. Fish. Inst., 22:66-78 (22nd Ann. Sess.-Nov. 1969).

252. Kutkuhn, J.H. 1962. Conversion of "whole" and "headless" weights in commercial Gulf of Mexico shrimps. Spec. Scientific Rep. - Fisheries No. 409, Fish & Wildlife Service, 7 pp.

253. Anderson, W.W., M.J. Lindner. 1958. Length-weight relation in the common white shrimp, Penaeus setiferus. Spec. Scientific Rep. - Fisheries No. 256 (May) Fish & Wildlife Service, 13 pp.

254. Fontaine, C.T., R.A. Neal. 1968. Relation between tail length and total length for three commercially important species of penaeid shrimp. Fishery Bull., 67(1):125-126.

255. Saloman, C.H., D.M. Allen, and T.J. Costello. 1968. Distribution of three species of shrimp (genus Penaeus) in waters contiguous to southern Florida. Bull. of Mar. Sci., 18(2):343-350.

256. Aldrich, D.V., C.E. Wood, K.N. Baxter. 1968. An ecological interpretation of low temperature responses in Penaeus aztecus and P. setiferus post larvae. Bulletin of Marine Science, 18(1):61-71.

257. Littlepage, J.L. and M.N. McGinley. 1965. A bibliography of the genus Artemia (Artemia salina) 1812-1962. Spec. Publ. No. 1, San Francisco Aquarium Soc., Inc., Steinhart Aquarium, San Francisco, California.

258. Anderson, W.W., G.R. Lunz. 1965. Southern shrimp - a valuable regional resource. Leaflet No. 4, Oct. 1965, 7 pp., Atlantic States Marine Fisheries Comm., Tallahassee, Florida.

259. Ingle, R.M. 1961. Synoptic rationale of existing Florida shrimp regulations. Proc. Gulf and Caribb. Fish Inst., 13th Ann. Sess. pp. 22-27.

260. Balazs, G.H. 1973. A simplified method for identifying experimental shrimp. The Progressive Fish-Culturist, 35(1):27 (Jan.).

261. Redmann, G. 1973. Studies on the toxicity of Mirex to the estuarine grass shrimp. Palaemontes pugio. Gulf Research Reports, 4(2):272-277.

262. Sorgeloos, P., G. Persoone. 1973. A culture system for Artemia, Daphnia, and other invertebrates, with continuous separation of the larvae. Arch. Hydrobiol., 72(1):133-138.

263. Sorgeloos, P. 1973. High density culturing of the brine shrimp, Artemia salinas L. Aquaculture, 1(1973):385-391.

264. Economic Research Laboratory, NMFS. 1973. Basic economic indicators - shrimp 1947-1972. Current Fisheries Statistics, No. 6131, Nat'l. Mar. Fish. Serv., 55 pp.

265. Menon, M.K. 1952. The life history and bionomics of an Indian penaeid prawn Metapenaeus dobsoni Miers. Proc. Indo-Pacific Fish. Council, 6(3):80-95.

266. Menon, M.K. 1955. Notes on the bionomics and fishery of the prawn Metapenaeus dobsoni Miers on the southwest coast of India. Indian J. Fish., 2:41-56.

267. Menon, M.K. 1957. Contributions to the biology of penaeid prawns of the southwest coast of India - I Sex ratio and movements. Indian J. Fish., 4:62-74.

268. Racek. A.A. 1959. Prawn investigations in eastern Australia. Res. Bull. State Fish. N.S.W., 6:1-57.

269. Racek, A.A. 1970. Indo-west Pacific penaeid prawns species of commercial importance. Proc. Indo-Pacific Fish. Counc., 14th Sess., (Nov. 1970), IPFC/C70/Sym(3).

270. Racek, A.A. and W. Dall. 1965. Littoral penaeidae (crustacea, decapoda) from northern Australia, New Guinea and adjacent waters. Verh. Kon. Akad. Wet. Amsterdam, 56(3):1-116.

271. Racek, A.A. and W. Dall. 1966. The functional anatomy of the digestive tract of a shrimp, Metapenaeus bennettae. Aust. J. Mar. Freshw. Res., 15(3):699-714.

272. Dall, W. 1958. Observations on the biology of the green tail prawn Metapenaeus mastersii (Haswell). Aust. J. Mar. Freshw. Res., 9:111-134.

273. Dall, W. 1964. Studies on the physiology of a shrimp, Metapenaeus mastersii (Haswell) - I Blood constituency. Aust. J. Mar. Freshw. Res. 15(2):145-161.

274. Dall, W. 1965. Studies on the physiology of a shrimp, Metapenaeus mastersii (Haswell) - II Endocrines and control of moulting. Aust. J. Mar. Freshw. Res., 16(1):1-12.

275. Dall, W. 1965. Studies on the physiology of a shrimp, Metapenaeus sp. - III Composition and structure of the integument. Aust. J. Mar. Freshw. Res., 16(1):13-23.

276. Dall, W. 1965. Studies on the physiology of a shrimp, Metapenaeus sp. - IV Carbohydrate metabolism. Aust. J. Mar. Freshw. Res. 16(2): 163-180.

277. Dall, W. 1965. Studies on the physiology of a shrimp, Metapenaeus sp. - V Calcium metabolism. Aust. J. Mar. Freshw. Res., 16(2):181-203.

278. Roedel, P.M. 1973. Shrimp 73 - A billion dollar business. Mar. Fish. Rev. (Spec. Shrimp Issue), 35(3&4):1-2.

279. Manar, T.A. 1973. Who eats shrimp? Mar. Fish. Rev. (Spec. Shrimp Issue), 35(3&4):3-5.

280. Anon. 1973. Shrimp imports in 1972 total 223.2 million pounds, (breakdown by country of origin). Mar. Fish. Rev. (Spec. Shrimp Issue), 35(3&4):6.

281. Whitaker, D.R. 1973. The shrimp dollar, 1972. Mar. Fish Rev. (Spec. Shrimp Issue), 35(3&4):8.

282. Temple, R.F. 1973. Shrimp research at the Galveston Laboratory of the Gulf Coastal Fisheries Center. Mar. Fish. Rev. (Spec. Shrimp Issue), 35(3&4):16-20.

283. Caillouet, C.W., Jr. and K.N. Baxter. 1973. Gulf of Mexico shrimp resource research. Mar. Fish. Rev. (Spec. Shrimp Issue), 35(3&4):21-24.

284. Marullo, F. 1973. An automatic pumping device for sampling post larval shrimp (Penaeus sp.). Mar. Fish. Rev. (Spec. Shrimp Issue), 35 (3&4):24-26.

285. Gross, G.B. 1973. Shrimp industry of Central America, Caribbean Sea, and northern South America. Mar. Fish. Rev. (Spec. Shrimp Issue), 35(3&4):36-55.

286. Neal, R.A. 1973. Progress towards farming shrimp in the United States. Mar. Fish. Rev. (Spec. Shrimp Issue), 35(3&4):67-70.

287. Mock, C.R. 1973. Shrimp culture in Japan. Mar. Fish. Rev. (Spec. Shrimp Issue), 35(3&4):71-74.

288. Kubo, I. 1949. Studies on the penaeids of Japanese and its adjacent waters. J. Tokyo Coll. Fish., 36(1):1-467.

289. Krantz, G.E. and E.S. Iversen. 1974. Husbandry helps harvest of healthy shrimp. Proc. Gulf and Caribbean Fish. Inst., (26):92-98.

290. Pillai, T.G. 1970. Pests and predators in coastal aquaculture systems of the Indo-Pacific region. Proc. Indo-Pacif. Fish. Counc., 14th Sess. (Nov. 1970):IPFC/C70/Sym(22), 26 pp.

291. George, M.J. 1962. On the breeding of penaeids and the recruitment of their post larvae into the backwater of Cochin. Indian J. Fish., 9(1): 110-116.

292. Voss, G.L. 1955. A key to the commercial and potentially commercial shrimp of the family penaeidae of the western North Atlantic and the Gulf of Mexico. Florida Board Conserv., Tech. Ser. #14, pp. 1-23.

293. Rousseau, J.E. 1960. Shrimp-waste meal: effect of storage variables on pigment content. Comm. Fish. Rev., 22(4):6-10.

294. Tang, Y.A. and S.H. Chen. 1967. A survey of the algal pasture soils of milkfish ponds in Taiwan. FAO Fish. Rep. #44(3):198-209.

295. Nose, T. 1964. Protein digestibility of several test diets in cray and prawn fish. Bull. Freshw. Fish. Res. Lab., (Tokyo), 14:23-28, (in Japanese, English summary)

296. Meyers, S.P., D.P. Butler, and W.H. Hastings. 1972. Alginates as binders for crustacean rations. Prog. Fish-Culturist, 34(1):9-12 (Jan.).

297. Hood, M.A., S.P. Meyers, and A.R. Colmer. 1971. Bacteria of the digestive tract of the white shrimp, *Penaeus setiferus*. Bacteriological Proceedings, 71:6-147.

298. Shewbart, K.L., W.L. Mies, and P.D. Ludwig. 1972. Identification and quantitative analysis of the amino acids present in protein of the brown shrimp *Penaeus aztecus*. Mar. Bio., 16:64-68.

299. Sick, L.V. and G.J. Baptist. 1973. The effects of selected physical and nutritional factors on the rate of ingestion of pelleted food by post larval penaeid shrimp. Journal of Elisha Mitchell Sci. Soc. 89(2):75-81.

300. Andrews, J.W., L.V. Sick, and G.J. Baptist. 1972. The influence of dietary protein and energy levels on the growth and survival of penaeid shrimp. Aquaculture 1:341-347.

301. Wiesepape, L.M., D.V. Aldrich, and K. Strawn. 1972. Effects of temperature and salinity on thermal death in post larval brown shrimp, *Penaeus aztecus*. Physiological Zoology, 45:22-33.

302. Broad, A.C. 1951. The shrimp in North Carolina. Survey of Marine Fisheries of North Carolina, pp. 191-204, edited by H.F. Taylor, The Univ. of N. Carolina Press, Chapel Hill, N.C.

303. Broad, A.C. 1965. Environmental requirements of shrimp. Publs. Publ. Hlth. Serv., Wash., 999-WP-25:86-91.

304. Renfro, W.C. 1963. Life history and stages of Gulf of Mexico brown shrimp. U.S. Fish and Wildlife Service, Circular 183:94-98.

305. Renfro, W.C. 1963. Small beam net for sampling post larval shrimp. Galveston Biological Laboratory Fishery Research for the Year Ending June 30, 1962, U.S. Fish & Wildlife Serv., Circular 161:86-87.

306. Webber, H. 1975. Shrimp farming. The Comm. Fish Farmer, 1(4):8-12 (Mar/April).

307. Pérez-Farfante, I. 1971. A key to the American Pacific shrimps of the genus *Trachypenaeus* (decapoda, penaeidae) with the description of a new species. Fishery Bull., 69(3):635-646.

308. Pérez-Farfante, I. 1970. Diagnostic characters of juveniles of the shrimps *Penaeus aztecus aztecus*, *P. duorarum duorarum* and *P. brasiliensis* (crustacea, decapoda, penaeidae). Special Scientific Report - Fisheries No. 599, 26 pp., U.S. Fish and Wildlife Service.

309. Fontaine, C.T. 1971. Conversion tables for commercially important penaeid shrimp of the Gulf of Mexico. U.S. Nat. Mar. Fish. Serv., Data Report 70, 9 pp.

310. Meyers, S.P. and J.E. Rutledge. 1973. Utilization of economically-valuable by-products from the shrimp processing industry. Food-Drugs from the Sea - Proceedings 1972, edited by L.R. Worthen, Mar. Tech. Soc., pp. 75-85.

311. Eldred, B. 1958. Observations on the structural development of the genitalia and the impregnation of the pink shrimp, Penaeus duorarum Burkenroad. Florida State Board of Conservation Marine Laboratory, St. Petersburg, Florida, Technical Series No. 23, 26 pp.

312. Elfred, B. 1958. Meioceras lermondi as food for Penaeus duorarum. Nautilus, 71(4):152.

313. Eldred, B. and R. F. Hutton. 1960. On the grading and identification of domestic commercial shrimp (family penaeidae) with a tentative world list of commercial penaeids. Quart. J. Fla. Acad. Sci., 23(2):89-118.

314. Eldred, B., R.M. Ingle, K.D. Woodburn, R.F. Hutton, and H. Jones. 1961. Biological observations on the commercial shrimp, Penaeus duorarum Burkenroad, in Florida waters. Fla. State Bd. Conserv. Mar. Lab., Professional Papers, Series No. 3, 139 pp.

315. Williams, A. B. 1960. Influence of temperature on osmotic regulation in two species of estuarine shrimps (Penaeus). Biol. Bull., 119:560-571.

316. Loesch, H. 1965. Distribution and growth of penaeid shrimp in Mobile Bay, Alabama. Publ. Inst. Mar. Sci. Texas, 10:41-58.

317. Loesch, H. 1971. Some observations and recommendations on (Penaeus) shrimp management practices and results as related to Honduras. (Spanish-English), Projecte Regional de Desarrollo Pesqueroen Centro America, Bol. Technico, 6(1):1-28.

318. Jacob, J. and H. Loesch. 1971. A preliminary two-year comparison of Penaeus aztecus: growth rate, distribution and biomass in the Barataria Bay area, Louisiana. La. State U. Coastal Studies Bull., 6:45-55.

319. Wiesepape, L.M. 1975. Thermal resistance and acclimation rate in young white and brown shrimp, Penaeus setiferus Linn. and Penaeus aztecus Ives. TAMU-SG-76-202, Texas A&M Univ. Sea Grant Program, Nov.

320. Ford, T.B. and L.S. St. Amant. 1971. Management guidelines for predicting brown shrimp, Penaeus aztecus, production in Louisiana. Proc. of the Gulf and Caribbean Fisheries Institute, 23rd Ann. Sess., held at Willemstad, Curacao, Nov. 1970, pp. 149-161.

321. Kelly, C.E. and A.W. Harmon. 1972. Method of determining carotenoid contents of Alaska pink shrimp and representative values for several shrimp products. Fishery Bull., 70(1):111-113.

322. Munro, J.L., A.C. Jones, and D. Dimitriou. 1968. Abundance and distribution of the pink shrimp (Penaeus duorarum) on the Tortugas Shelf of Florida, Aug. 1967-Oct. 1967. U.S. Fish & Wildlife Service, Fisheries Bulletin, 67(1):165-181.

323. Shealy, M.H., Jr. and P.A. Sandifer. 1975. Effects of mercury on survival and development of the larval grass shrimp Palaemonetes vulgaris. Marine Biology, 33:7-16.

324. Vanderzant, C., B.F. Cobb, and R. Nickelson. 1974. Role of microorganisms in shrimp quality: a research summary. TAMU-SG-74-201, 46 pp., Texas A&M Univ. Sea Grant Program.

325. Lightner, D.V., B.R. Salser, and R.S. Wheeler. 1974. Gas bubble disease in the brown shrimp (Penaeus aztecus). Aquaculture 4(1):81-84.

326. Jones, A.C., D.E. Dimitriou, J.J. Ewald, and J.H. Tweedy. 1970. Distribution of early developmental stages of pink shrimp, Penaeus duorarum, in Florida waters. Bull. Mar. Sci., 20(3):634-661.

327. Venkataramaiah, A., G.J. Lakshmi, and G. Gunter. 1972. The effects of salinity, temperature and feeding levels on the food conversion, growth and survival rate of the shrimp Penaeus aztecus. Food-Drugs from the Sea - Proceedings 1972, edited by L.R. Worthen, Mar. Tech. Soc., pp. 29-42.

328. Delinendo, M.N., P.E. Osborn, R. Halili, E. Martinez. 1970. Construction of ponds for aquaculture. Proc. Indo-Pacific Fisheries Council, Session 14 (Nov. 1970): IPFC/C70/Sym(26).

329. Bursey, C.R. and C.E. Lane. 1971. Ionic and protein concentration changes during the molt cycle of Penaeus duorarum. Comparative Biochemistry and Physiology, 40A:155-162.

330. Band, F.B. 1970. Disease mechanisms in crustacean and marine arthropod. A Symposium on Diseases of Fish and Shellfish, edited by F. Staninlas and S. Snieszko, Amer. Fish Soc. Special Publication 5:383-404.

331. Sutton, A.H., G. Main, and A. Ronald. 1969. An instrument for counting the larvae of the prawn Palaemon serratus and the brine shrimp Artemia salina. Laboratory Practice, 18(4):433 (April).

332. Corbett, M.G. 1970. Machine for separating northern shrimp, Pandalus borealis, from fish and trash in the catch. Fishery Industrial Research, 6(2):53-62.

333. Phillips, G.C. 1971. Marking shrimps and prawns by Latex injections. Crustaceana, 22(1):84-86.

334. Emiliani, A. 1971. Equipment for holding and releasing penaeid shrimp during marking experiments. Fishery Bull., 69(1):247-251.

335. Fontaine, C.T., S.E.P. Gislason, and W.L. Trent. 1972. A system for collecting large numbers of live post larval shrimp. Fishery Bull., 70(4):1298-1302.

336. Klussman, W.G. and J.C. Parker. 1972. Mariculture pond demonstrations. Sea Grant Program Annual Report 1970-71, Texas A&M, TAMU-SG-72-104, pp. 17-21.

337. Gehting, W.R. 1971. Sound preception and production in the pink shrimp, Penaeus duorarum Burkenroad. Sea Grant Technical Bulletin No. 8, (Jan. 1971) Univ. of Miami.

338. Teinsongrusmee, B. 1965. The effect of temperature on growth of post larval pink shrimp Penaeus duorarum Burkenroad. Thesis, Univ. of Miami, Coral Gables, Fla., 66 pp.

339. Vanderzant, C., et al. 1970. Microbial flora of Gulf of Mexico and pond shrimp. Journal of Milk and Food Technology, 33(8):Aug. 1970, 5 pp.

340. Wolvekamp, H.P. and T.H. Waterman. 1960. Respiration. The Physiology of Crustacea, edited by T.H. Waterman, New York, Academic Press, Vol. 1, pp. 35-100.

341. McIlhenny, W.F. 1972. Shrimp culture research. Sea Grant Annual Report 1970-71, Texas A&M, TAMU-SG-72-104, pp. 24-35.

342. Kirkegaard, I. and R.H. Walker. 1969. Synopsis of biological data on the tiger prawn, Penaeus esculentus Haswell 1879. Aust. Commonw. Sci. Ind. Res. Organ. Div. Fish. Oceanogr. Fish Synop. 3, 30 pp.

343. Gillespie, W.C., J.C. Hite, and J.S. Lytle. 1969. An economic analysis of the U.S. shrimp industry. S.C. Agric. Exp. Sta. Clemson AE330, Bibliogr. 78 pp., Dec. 1969.

344. Worth, H. 1952. Shrimp Cookery. Citadel Press, New York, 222 Park Avenue, South, N.Y. First Paper Back Edition 1964, 182 pp., $1.25.

345. Jarvis, N.D. 1943. Canning crabs, lobsters and shrimp. Fishery Leaflet 85, 24 pp., U.S. Fish & Wildlife Serv., Reprint from "Principles and methods in the canning of fishery products", Research Report No. 7, 366 pp. (pp. 259-283).

346. Liu, C-Y. 1971. The experimental monoculture of Penaeus monodon Fabricius. Aquaculture 1(2):25-30 (only abstract and illustrations in English), publication of the Tainan Fish Culture Station and Tung Kang Marine Laboratory, Taiwan.

347. Anderson, J.I.W. and D.A. Conroy. 1968. The significance of disease in preliminary attempts to raise crustacea in sea water. Proc. III Sympos. Comm. Off. Int. Epizoot. Etude Maladies Poissons, Separate No. 3, 8 pp., also Off. Intern des Epizooties Bull. 69:1239-1247.

348. Purvis, C.E. and E.G. McCoy. 1972. Over wintering pink shrimp, Penaeus duorarum, in Core and Pamlico Sound. Special Scientific Report No. 22, Div. of Commercial and Sport Fisheries, North Carolina Dept. of Natural & Econ. Resources.

349. Bardach, J.E., J.H. Ryther, and W.O. McLarney. 1972. Chapter 32 - Shrimp culture. Aquaculture - The Farming and Husbandry of Freshwater and Marine Organisms, pp. 587-632, Wiley - Interscience, N.Y.

350. Shigueno, K. 1975. Shrimp Culture in Japan, 153 pp., available from: Japan Publications Trading Co., Ltd., 1255 Howard St., San Francisco, California 94103, $13.50.

351. Price, V.A. and K.K. Chew. 1969. Post-embryonic development of laboratory-reared spot shrimp. Proc. Natl. Shellfish. Assoc., (1968) 59:8.

352. Price, V.A. and K.K. Chew. 1972. Laboratory rearing of spot shrimp larvae (Pandulus platyceros) and description of stages. J. Fish. Res. Bd. Canada, 29(4):413-422.

353. Barr, L. 1973. Studies of spot shrimp, Pandalus platyceros, at Little Port Walter, Alaska. Mar. Fish. Rev. (Spec. Shrimp Issue), 35(3&4):65-66.

354. Olsen, J.C. 1973. Pandalid shrimp life history research at Kachemak Bay, Alaska. Mar. Fish. Rev. (Spec. Shrimp Issue), 35(3&4):62-64.

355. Barr, L. and R. McBride. 1967. Surface-to-bottom pot fishing for pandalid shrimp. Special Scientific Report - Fisheries No. 560, 7 pp., U.S. Fish & Wildlife Service.

356. Digby, P.S.B. 1961. Sensitivity of the shrimp Crangon vulgaris to hydrostatic pressure. J. Physiol. (London), 158:12-13.

357. Digby, P.S.B. 1961. Mechanism of sensitivity to hydrostatic pressure in the prawn, Palaemonetes varians Leach. Nature 191(4786):366-368.

358. Broad, A.C. 1957. Larval development of Palaemonetes pugio Hothuis. The Biological Bulletin, 112(2):144-161.

359. Broad, A.C. 1957. The relationship between diet and larval development of Palaemonetes. The Biological Bull., 112(2):162-170 (April).

360. Lofts, B. 1956. The effect of salinity changes on the respiration rate of the prawn Palaemonetes varians (Leach). J. Exp. Biol., 33:730-736.

361. Rajyalakshmi, T. 1960. Observation on the embryonic and larval development of some estuarine palaemonid prawns. Proc. Natl. Inst. Sci. India (B) 26:395-408.

362. Rajyalakshmi, T. 1961. Studies on maturation and breeding in some estuarine palaemonid prawns. Proc. Natl. Sci. India (B) 27(4):179-188.

363. Davis, C.C. 1965. A study of the hatching process in aquatic invertebrates - XIV An examination of hatching in Palaemonetes vulgaris (Say). Crustaceana, 8(3):233-238.

364. Scattergood, L.W. 1952. The northern shrimp fishery of Maine. Comm. Fish. Rev., 14(1):1-16.

365. Andersen, W.W. and M.J. Lindner. 1971. Contributions to the biology of the royal red shrimp Hymenopenaeus robustus Smith. Fishery Bulletin 69(2):313-336 (April).

366. Nomura, H. and J.F. Filhe. 1968. A shrimp exploratory survey in northeastern and northern Brazil, with some biological observations on Penaeus aztecus. FAO Fish. Report No. 57, Vol. 2:219-232.

367. Squires, H.J. 1968. Relation of temperature to growth and self-propagation of Pandalus borealis in Newfoundland. FAO Fish. Report No. 57, Vol. 2:243-250.

368. Meixner, R. 1968. Reproduction of the sand shrimp, Crangon crangon (L.). FAO Fish. Rep. 57, Vol. 2:259-264.

369. Rao, P.V. 1968. Maturation and spawning of the penaeid prawns of the southwest coast of India. FAO Fish Report No. 57, Vol. 2:285-302.

370. Schafer, H.J. 1968. Storage materials utilized by starving pink shrimp, Penaeus duorarum Burkenroad. FAO Fish. Report No. 57, Vol. 2: 393-404.

371. George, M.J., K.H. Mohamed, and N.N. Pillai. 1968. Observations on the paddy-field prawn filtration of Kerala, India. FAO Fish. Report No. 57, Vol. 2:427-442.

372. Mohamed, K.H., P.V. Rao, and M.J. George. 1968. Post larvae of penaeid prawns of southwest coast of India with a key to their identification. FAO Fish. Report No. 57, Vol. 2:487-504.

373. Khandler, N.A. 1968. Some aspects of the biology of white shrimp Penaeus schmitti Burkenroad, in Lake Unare, Venezuela. FAO Fish. Report No. 57, Vol. 2:505-512.

374. Pease, N. L. 1968. The design and field testing of an electro-shrimp trawling system. FAO Fish. Report No. 57, Vol. 2:513-520.

375. Pannikar, N.K. 1968. Osmotic behavior of shrimps and prawns in relation to their biology and culture. FAO Fish. Report No. 57, Vol. 2:527-538.

376. Jones. S. 1969. The prawn fishery resources of India. FAO Fisheries Report No. 57, Vol. 3:735-748.

377. Tirmizi, N.M. 1969. Commercial prawns of West Pakistan. FAO Fish. Report No. 57, Vol. 3:749-764.

378. Ewald, J.J. 1969. The Venezuelan shrimp industry. FAO Fish. Report No. 57, Vol. 3, pp. 765-774.

379. Berry, R. J. and K.N. Baxter. 1969. Predicting brown shrimp abundance in the northern Gulf of Mexico. FAO Fish. Report No. 57, Vol. 3: 775-798.

380. Ivanov, B. 1969. The biology and distribution of the northern shrimp (Pandalus borealis Kr.) in the Bering Sea and the Gulf of Alaska. FAO Fish. Report No. 57, Vol. 3:799-810.

381. Roessler, M.A., A.C. Jones, and J.L. Munro. 1969. Larval and post larval pink shrimp Penaeus duorarum in south Florida. FAO Fish. Report No. 57, Vol. 3:859-866.

382. Hughes, D.A. 1969. On the mechanisms underlying tide-associated movements of Penaeus duorarum Burkenroad. FAO Fish. Report No. 57, Vol. 3:867-874.

383. Gunter, G. and J.C. Edwards. 1969. The relationship of rainfall and fresh-water drainage to the production of the penaeid shrimps (Penaeus fluviatilis Say and Penaeus aztecus Ives) in Texas and Louisiana waters. FAO Fish. Report No. 57, Vol. 3:875-892.

384. Smidt, E. 1969. Pandalus borealis in Greenland waters: its fishery and biology. FAO Fish. Report No. 57, Vol. 3:893-902.

385. Kutty, M.N. 1969. Oxygen consumption of the prawn Penaeus indicus H. Milne Edwards and Penaeus semisulcatus de Haan. FAO Fish. Report No. 57, Vol. 3, 957-970.

386. Pullen, E.J. and W.L. Trent. 1969. White shrimp emigration in relation to size, sex, temperature and salinity. FAO Fish Report No. 57, Vol. 3:1001-1014.

387. Tabb, D.C. and N. Kenny. 1969. A brief history of Florida's live bait shrimp fishery with description of fishing gear and methods. FAO Fish. Report No. 57, Vol. 3, pp. 1119-1134.

388. Iversen, E.S. 1969. Microsporidosis in commercial penaeid shrimp. FAO Fish Report No. 57, Vol. 3:1135-1140.

389. Neal, R.A. 1969. Methods of marking shrimp. FAO Fish. Report No. 57, Vol. 3:1149-1165.

390. Tiews, K. 1970. Synopsis of biological data on the common shrimp Crangon crangon (Linnaeus, 1758). FAO Fish. Report No. 57, Vol. 4:1167-1224.

391. Simpson, A.C., B.R. Howell, and P.J. Warren. 1970. Synopsis of biological data on the shrimp Pandalus montagui Leach, 1814. FAO Fish. Report No. 57, Vol. 4:1225-1250.

392. Mohamed, K.H. 1970. Synopsis of biological data on the jumbo tiger prawn Penaeus monodon Fabricius, 1798. FAO Fish. Report No. 57, Vol. 4: 1251-1266.

393. Mohamed, K.H. 1970. Synopsis of biological data on the Indian prawn Penaeus indicus H. Milne Edwards, 1837. FAO Fish. Report No. 57, Vol. 4:1267-1288.

394. Butler, T.H. 1970 Synopsis of biological data on the prawn Pandalus platyceros Brandt, 1851. FAO Fish. Report No. 57, Vol. 4:1289-1316.

395. Kunju, M.M. 1970. Synopsis of biological data on the penaeid prawn Solenocera indica Nataraj, 1845. FAO Fish. Report No. 57, Vol. 4:1317-1334.

396. George, M.J. 1970. Synopsis of biological data on the penaeid prawn Metapenaeus dobsoni (Miers, 1878). FAO Fish. Report No. 57, Vol. 4: 1335-1358.

397. George, M.J. 1970. Synopsis of biological data on the penaeid prawn Metapenaeus affinis (H. Milne Edwards, 1837). FAO Fish. Report No. 57, Vol. 4, pp. 1359-1376.

398. Dahlstrom, W.A. 1970. Synopsis of biological data on the ocean shrimp Pandalus jordani Rathbun, 1902. FAO Fish. Report No. 57, Vol. 4: 1377-1416.

399. Pérez-Farfante, I. 1970. Sinopsis de datos biologicos sobre el camaron blanco Penaeus schmitti Burkenroad, 1936. FAO Fish. Report No. 57, Vol. 4:1417-1438.

400. George, M.J. 1970. Synopsis of biological data on the penaeid prawn Metapenaeus monoceros (Fabricius, 1798). FAO Fish. Report No. 57, Vol. 4:1539-1558.

401. George, M.J. 1970. Synopsis of biological data on the penaeid prawn Metapenaeus brevicornis (H. Milne Edwards, 1837). FAO Fish. Report No. 57, Vol. 4:1559-1574.

402. Rao, P.V. 1970. Synopsis of the biological data on the penaeid prawn Parapenaeopsis stylifera (H. Milne Edwards, 1837). FAO Fish. Report No. 57, Vol. 4, pp. 1575-1606.

403. Hendry, R.R. 1971. The Florida mariculture law. Proc. of the First Ann. Workshop, World Mariculture Soc., 1:47-49.

404. Bente, P.F., Jr. 1971. Application of first mariculture law, operation under the law. Proc. of the First Ann. Workshop, World Mariculture Soc., 1:50-54.

405. Davies, C.B. and H.W. Shields. 1971. Mariculture and the law. Proc. of the First Ann. Workshop, World Mariculture Soc., 1:55-59.

406. Broom, J.G. 1971. Shrimp culture. Proc. First Ann. Workshop World Mariculture Society, 1:63-68.

407. Subrahmanyam, C.B. and C.H. Oppenheimer. 1971. The influence of feed level on the growth of grooved penaeid shrimp in mariculture. Proc. of the First Ann. Workshop World Mariculture Society, 1:91-100.

408. Mock, C.R. and M.A. Murphy. 1971. Techniques for raising penaeid shrimp from egg to post larvae. Proc. of the First Ann. Workshop World Mariculture Soc., 1:143-156.

409. Yee, W.C. 1971. Thermal aquaculture design. Proc. of the Second Ann. Workshop World Mariculture Soc., 2:55-65.

410. Parker, J.D., W.G. Klussman, H.W. Holcomb, Jr., and J.C. McNeill, IV. 1971. Distribution of aquatic macrofauna in a marsh on West Galveston Bay, Texas and possible effects thereon resulting from impoundments for shrimp culture. Proc. Second Ann. Workshop World Mariculture Soc. 2:97-110, also Texas A&M, TAMU-SG-71-208.

411. Kane, T.E. 1973. Mariculture: raising legal problems. Proc. of the Third Ann. Workshop World Mariculture Soc., 3:39-44.

412. Baumeister, J.W. 1973. The Refuse Act Permit Program and its application to mariculture. Proc. of the Third Ann. Workshop World Mariculture Soc., 3:45-49.

413. Perry, W.G., A.B. Ensminger, and W.R. Latapie. 1973. Marsh pond construction. Proc. of the Third Ann. Workshop World Mariculture Soc., 3:149-165.

414. Broom, J.G. 1973. Shrimp culture studies in Honduras, 1969-1971. Proc. Third Ann. Workshop World Mariculture Society, 3:193-204.

415. Caillouet, C.W., Jr. 1973. Ovarian maturation induced by eyestalk ablation in pink shrimp, *Penaeus duorarum* Burkenroad. Proc. of the Third Ann. Workshop World Mariculture Soc., 3:205-225.

416. Neal, D.A. and W.R. Latapie, Jr. 1972. Pond culture on Grand Terre Island Louisiana, 1969-1971. Proc. of the Third Ann. Workshop World Mariculture Soc., 3:227-240.

417. Latapie, W.R., Jr., J. G. Broom, and D.A. Neal. 1973. Growth rates of *Penaeus aztecus* and *P. setiferus* in artificial ponds under varying conditions. Proc. Third Ann. Workshop World Mariculture Soc. 3:241-254.

418. Venkataramaiah, A. and G. J. Lakshmi. 1973. Plastic wading pools for holding and raising penaeid shrimp under laboratory conditions. Proc. of the Third Ann. Workshop World Mariculture Soc., 3:255-265.

419. Venkataramaiah, A., C. J. Lakshmi, and G. Gunter. 1973. The effects of salinity and feeding levels on the growth rates and food conversion efficiency of the shrimp Penaeus aztecus. Proc. of the Third Ann. Workshop World Mariculture Soc., 3:267-284.

420. Mock, C. R. 1973. The culture of Penaeus japonicus in Japan. Proc. of the Third Ann. Workshop World Mariculture Soc., 3:285-286 (abstract only)

421. Meyers, S. P. and Z. P. Zein-Eldin. 1973. Binders aid pellet stability in development of crustacean diets. Proc. of the Third Annual Workshop World Mariculture Soc., 3:351-364.

422. Hysmith, B. T., J. R. Booth, H. L. Cook, and W. L. Mies. 1972. A study of the effects of feeding synthetic diets to brown shrimp (Penaeus aztecus). Proc. of the Third Annual Workshop World Mariculture Soc. 3:365-388.

423. Forster, J. R. M. 1973. Studies on compounded diets for prawns. Proc. of the Third Ann. Workshop World Mariculture Soc., 3:389-402.

424. Andrews, J. W. and L. V. Sick. 1973. Studies on the nutritional requirements of penaeid shrimps. Proc. of the Third Ann. Workshop World Mariculture Soc., 3:403-414.

425. Caillouet, C. W., Jr., J. M. Heinen, and D. C. Tabb. 1972. Pink shrimp feeding experiments with wheat bran. Proc. of the Third Annual Workshop of the World Mariculture Soc., 3:415-459.

426. Grajcer, D. and R. Neal. 1972. Growth of hatchery-reared Penaeus aztecus on experimental diets. Proc. of the Third Ann. Workshop World Mariculture Soc., 3:461-470.

427. Jannke, T. E. 1973. An improved method for detecting ovarian maturation of pink shrimp. Proc. of the Fourth Ann. Workshop World Mariculture Soc., 4:185-194.

428. Gould, R. A., D. V. Aldrich, and C. R. Mock. 1973. Experimental pond culture of brown shrimp (Penaeus aztecus) in power plant effluent water. Proc. of the Fourth Ann. Workshop World Mariculture Soc., 4:195-213.

429. Parker, J.C. and H.W. Holcomb, Jr. 1973. Growth and production of brown and white shrimp (Penaeus aztecus and P. setiferus) from experimental ponds in Brazoria and Orange Counties, Texas. Proc. of the Fourth Ann. Workshop World Mariculture Soc., 4:215-234.

430. Holcomb, H.W., Jr. and J.C. Parker. 1973. Efficiency of seine and drain harvesting techniques in experimental penaeid shrimp culture ponds. Proc. of the Fourth Ann. Workshop World Mariculture Soc., 4:235-245.

431. Sick, L.V. and J.W. Andrews. 1973. The effect of selected dietary lipids, carbohydrates and protein in the growth, survival and body composition of Penaeus duorarum. Proc. of the Fourth Ann. Workshop World Mariculture Soc., 4:263-276.

432. Shewbart, K.L. and W.L. Mies. 1973. Studies on nutritional requirements of brown shrimp - The effect of linolenic acid on growth of Penaeus aztecus. Proc. of the Fourth Ann. Workshop World Mariculture Soc., 4:277-287.

433. Griffith, G.W., M.A. Murphy Kenslow, and L.A. Ross. 1973. A mass culture method for Tetraselmis sp. - A promising food for larval crustaceans. Proc. of the Fourth Ann. Workshop World Mariculture Soc., 4:289-294.

434. Zein-Eldin, Z.P. and S.P. Meyers. 1973. General consideration of problems in shrimp nutrition. Proc. of the Fourth Ann. Workshop World Mariculture Soc., 4:299-317.

435. Johnson, S.K., J.C. Parker, and H.W. Holcomb. 1973. Control of Zoothamnium sp. on penaeid shrimp. Proc. of the Fourth Ann. Workshop World Mariculture Soc., 4:321-331.

436. Lewis, D.H. 1973. Response of brown shrimp to infection with Vibrio sp. Proc. of the Fourth Ann. Workshop World Mariculture Soc., 4:333-338.

437. Parker, J.C., F.S. Conte, and W.S. MacGrath. 1974. An intensive culture system for penaeid shrimp. Proc. of the Fifth Annual Workshop World Mariculture Soc., 5:65-80.

438. Harris, A.H. and R.H. Kilgren. 1974. Overwintering of white shrimp in estuarine impoundments in Louisiana. Proc. of the Fifth Ann. Workshop World Mariculture Soc., 5:81-86.

439. Chan, E.S. and R.L. Lawrence. 1974. The effect of antibiotics on the respiration of brown shrimp larvae and post larvae (Penaeus aztecus Ives) and the bacterial populations associated with the shrimp. Proc. of the Fifth Annual Workshop World Mariculture Soc., 5:99-124.

440. Elam, L.L. and A.W. Green. 1974. Culture of white shrimp (Penaeus setiferus Linnaeus) in static water ponds. Proc. Fifth Ann. Workshop World Mariculture Soc., 5:87-97.

441. Caillouet, C.W., J.P. Norris, E.J. Herald, and D.C. Tabb. 1974. Growth and yield of pink shrimp (Penaeus duorarum Burkenroad) in a feeding experiment in ponds. Proc. of the Fifth Ann. Workshop World Mariculture Soc., 5:125-135.

442. Riley, J.R., K.T. Dorsey, and V.C. Supplee. 1974. The shrimp culture program at the Encironmental Research Laboratory, University of Arizona. Proc. of the Fifth Ann. Workshop World Mariculture Soc., 5:421-430.

443. Moore, D.W., Jr., R.W. Sherry, and F. Montanez. 1974. Maturation of Penaeus californiensis in captivity. Proc. of the Fifth Ann. Workshop World Mariculture Soc., 5:445-449.

444. Boschi, E.E. and M.A. Scelzo. 1974. Rearing the penaeid shrimp, Artemesia longinaris, from egg to juvenile in the laboratory. Proc. of the Fifth Ann. Workshop World Mariculture Soc., 5:443-444.

445. MacKay, R.D. 1974. A note on minimal levels of oxygen required to maintain life in Penaeus schmitti. Proc. of the Fifth Ann. Workshop World Mariculture Society, 5:451-452.

446. Milne, P.H. 1972. Section 3 - Sea farm construction techniques. Fish and Shellfish Farming in Coastal Waters, Fishing News (Books) Ltd., London, pp. 44-146.

447. Forster, J.R.M. 1976. The development of compound diets for prawns. Proc. of the First Intern. Confer. on Aquaculture Nutrition, Sea Grant Program, Rehoboth Beach, Delaware, Oct. 14-15, 1975 (in press).

448. Kittaka, J. 1976. Food and growth of penaeid shrimp. Proc. of the First Intern. Conf. on Aquaculture Nutrition, Sea Grant Program, Rehoboth Beach, Delaware, Oct. 14-15, 1975 (in press).

449. Heinen, J.M. et al. 1975. A general review of methods of rearing penaeid shrimp larvae. Paper presented at Sixth Ann. Meeting World Mari. Soc., Seattle, Washington, Jan. 27-31.

450. Krantz, G.E. et al. 1975. Culture of pink shrimp, Penaeus duorarum, at Turkey Point Experimental Mariculture Laboratory. Paper presented at Sixth Ann. Meeting of the World Mariculture Soc., Seattle, Wash., Jan. 27-31.

451. Zein-Eldin, Z.P. et al. 1975. Protein quantity and quality in diets of penaeid shrimp. Paper presented at Sixth Ann. Meeting of the World Mariculture Soc., Seattle, Wash., Jan. 27-31.

452. Hysmith, B.T. and L.L. Elam. 1974. Comparison of variables versus stable feeding rates for penaeid shrimp. Shrimp Research, Texas Parks and Wildlife Saltwater Pond Research Study No. 2, Segment 3 Report, pp. 1-62.

453. Hightower, M. and A. Lawrence. 1975. Catabolism of glucose dissolved in seawater, by mysis and post larvae of two species of penaeid shrimp. Proc. of the Sixth Ann. Meeting of the World Mari. Soc., 6:133-146.

454. Kramer, G.L. 1975. Studies on the lethal dissolved oxygen levels for young brown shrimp, Penaeus aztecus Ives. Proc. of the Sixth Ann. Meeting World Mariculture Soc., 6:157-170.

455. Alizade, M.A. et al. 1975. Oxygen consumption of the penaeid shrimp (Penaeus californiensis) as a method to evaluate optimal growth conditions. Paper presented at Sixth Ann. Meeting World Mariculture Soc., Seattle, Wash., Jan. 27-31.

456. Duronslet, M.J., A.I. Yudin, K.S. Wheeler, and W.H. Clark, Jr. 1975. Light and fine structural studies of natural and artificial induced egg growth of penaeid shrimp. Proc. of the Sixth Ann. Meeting World Mariculture Soc., 6:105-122.

457. Lightner, D., C.T. Fontaine, and K. Hanks. 1975. Some forms of gill disease in penaeid shrimp. Proc. of the Sixth Ann. Meeting World Mariculture Soc., 6:347-366.

458. Devers, K. et al. 1975. Evaluation of binder stability in crustacean diets. Paper presented at Sixth Ann. Meeting World Mariculture Soc., Seattle, Wash., Jan. 27-31.

459. Sorgeloos, P. 1975. Research on culturing of the brine shrimp Artemia salina at the State University of Ghent (Belgium). Proc. of the Sixth Ann. Meeting of the World Mariculture Soc., 6:441-444.

460. Griessinger, J. M. et al. (Aquacop), 1975. Maturation and spawning in captivity of penaeid prawns - Penaeus merguiensis, Penaeus japonicus, Penaeus aztecus and Metapenaeus ensis. Proc. of the Sixth Ann. Meeting World Mariculture Soc., 6:123-132.

461. Mock, C.R. et al. 1975. Design and evaluation of waste removal systems for shrimp culture in closed raceways. Paper presented at Sixth Ann. Meeting World Mariculture Soc., Seattle, Wash., Jan. 27-31.

462. Ross, L.A. et al. 1975. Observations on water quality in closed raceway systems used for shrimp culture. Paper presented at Sixth Ann. Meeting World Mariculture Soc., Seattle, Wash., Jan. 27-31.

463. Joseph, J.D. and J.E. Williams. 1975. Shrimp head oil, a potential feed additive for mariculture. Proc. of the Sixth Ann. Meeting World Mariculture Soc., 6:147-155.

464. Milligan, D.J. 1976. A method for obtaining metaphase chromosome spreads from marine shrimp with notes on the karyotypes of two penaeid shrimp, Penaeus aztecus and Penaeus setiferus. Proc. of the Seventh Ann. Meeting World Mariculture Soc., (in press).

465. Conte, F.S., M. Duronslet, W. Clark, and J. Parker. 1977. Maturation of Penaeus stylirostris and P. setiferus in hypersaline waters near Corpus Cristi, Texas. Proc. of the Eighth Ann. Meeting World Mariculture Soc. (in press).

466. Holcomb, H.W., J. Harrell, D. Ramsey, and J. Parker. 1976. Evidence of pH reduction by Rangia cuneata and the resulting effect on growth and survival of Penaeus setiferus and P. vannamei. Proc. of the Seventh Ann. Meeting World Mariculture Soc. (in press).

467. Parker, J.C., H.W. Holcomb, and F.S. Conte. 1976. Growth and production of five species of penaeid shrimp in experimental ponds in Texas 1969-1976. Proc. of the Seventh Ann. Meeting of the World Mariculture Soc. (in press).

468. Morris, J.A. and J.A. Quick, Jr. 1976. Experimental extension culture of penaeid shrimp in a simulated dredged material disposal area. Proc. of the Seventh Ann. Meeting World Mariculture Soc. (in press).

469. Hysmith, B. and B. Colura. 1976. Effect of salinity on growth and survival of penaeid shrimp. Proc. of the Seventh Ann. Meeting World Mariculture Soc., 31 pp., (in press).

470. New, M.B. 1976. A review of shrimp and prawn nutrition. Proc. of the Seventh Ann. Meeting World Mariculture Soc., 8 pp. (in press).

471. Rosa, H., Jr. 1965. A world list of experts concerned with the study of the biology and culture of shrimps and prawns. Food and Agriculture Organization, U.N., Rome, Fisheries Technical Paper 44, 21 pp.

472. Sogandares-Bernal, F. and R.F. Hutton. 1959. The identity of Metaceraria B. reported from the pink shrimp, Penaeus duorarum Burkenroad, by Woodburn et al. in 1957. Journal Parasitol, 45(4): 362-378.

473. Meyers, S.P. and C.W. Brand. 1975. Experimental flake diets for fish and crustacea. The Progressive Fish-Culturist, 37(2):67-72 (April).

474. Vance, M.H., W. Saenz, and D.L. Dubrow. 1959. The use of chlortetracycline in the control of spoilage in ice-stored shrimp. Fla. State Bd. Conserv., Tech. Ser. No. 28, 23 pp.

475. Chamber, C.I., M.H. Vance, and J.E. Alexander. 1957. The use of sodium bisulfite for the control of black spot in shrimp. Fla. State Bd. Conserv., Tech. Ser. No. 20, 19 pp.

476. Delves-Broughton, J. 1974. Preliminary investigations into the suitability of a new chemotheurapeutic, Foranace, for the treatment of infectious prawn diseases. Aquaculture 3(2):175-185.

477. Swingle, W.E. 1970. Survey of the live bait shrimp industry of Alabama. Alabama Dept. of Conservation and Natural Resources, Montgomery, NOAA 72092018, NMFS-2 83 R, 31 pp.

478. Anon. 1976. Development of a brood stock of the jumbo tiger prawn, Penaeus monodon Fabricius. SEAFDEC Aquaculture Dept. Tech. Rept. No. 1, May 1976, Tigbauan, Iloilo, Philippines.

479. Sindermann, C.J. 1970. Principal Diseases of Marine Fish and Shellfish. Academic Press, London and New York, pp. 166-171.

480. Barkate, J.A., C.R. Larimore, Y. Hirnan, and H. Persyn. 1974. Some marine microorganisms related to shrimp diseases. Proc. of the Fifth Ann. Workshop of the World Mariculture Soc., 5:267-282.

481. Johnson, S.K. 1974. Ectocommensals and parasites of shrimp from Texas rearing ponds. Proc. Fifth Ann. Workshop World Mariculture Soc., 5:251-266.

482. Hanson, J.A. (Editor). 1974. Open Sea Mariculture. Dowden, Hutchinson & Ross Publishers, Inc., Stroudsburg, Pa., 410 pp.

483. Hall, D.N.F. 1966. Penaeidae of the east coast of Africa. Mem. Inst. Fondam Afrique Noire, No. 77, pp. 87-97.

484. Sinderman, C.J. (Editor). 1974. Shrimp diseases. Diagnosis and Control of Mariculture Diseases in the United States, Technical Series No. 2, National Marine Fisheries Service, pp. 11-78.

485. Delves-Broughton, J. and C.W. Poupard. 1976. Disease problems of prawns in recirculation systems in the U.K. Aquaculture 7(3):201-218.

486. Baxter, K.N. and R.H. Rigdon. 1970. Pleistophora sp. (Microspolida: Nosematidae): a new parasite of shrimp. Jour. Invert. Path., 16(2): 289-291.

487. Egusa, S. and T. Ueda. 1972. Afusarium sp. associated with black gill disease of the Kuruma prawn, Penaeus japonicus Bate. Bull. Jap. Soc. Sci. Fish., 38(11):1253-1260 (in English).

488. Hanks, K.S. 1976. Toxicity of some chemical therapeutics to the commercial shrimp Penaeus californiensis. Aquaculture 7(3):293-294.

489. Sandifer, P.A., P.B. Zielinski, and W.E. Castro. 1974. A simple airlift-operated tank for closed-system culture of decapod crustacean larvae and other small aquatic animals. Helgolander Wiss. Meeresunters, 26:82-87.

490. Zielinski, P.B., W.E. Castro, and P.A. Sandifer. 1976. Model flow studies of a rectangular shrimp larval tank. Proc. of the Seventh Ann. Meeting of the World Mariculture Soc. (in press).

491. Villaluz, D.K., A. Villaluz, B. Ladrera, M. Sheik, and A. Gonzaga. 1972. Production, larval development, and cultivation of Sugpo (Penaeus monodon Fabricius). Philippine Journal Science, 98(314):205-233 (Sept.-Dec. 1969).

492. Mock, C.R. 1974. Larval culture of penaeid shrimp at the Galveston Biological Laboratory. Proc. of the First U.S.-Japan Meeting on Aquaculture at Tokyo, Japan, Oct. 18-19, 1971, NOAA Tech. Report NMFS Circ.-388, pp. 33-40, edited by W.N. Shaw.

493. Racek, A.A. 1972. Indo-west Pacific penaeid prawns of commercial importance. Coastal Aquaculture in the Indo-Pacific Region, edited by T.V.R. Pillay, Fishing News (Books) Ltd., London, pp. 152-172.

494. George, M.J. 1972. Preliminary experiments in marking penaeid prawns in paddy fields. Coastal Aquaculture in the Indo-Pacific Region, T.V.R. Pillay, editor, pp. 173-177.

495. George, M.J. 1972. On the food and feeding habits of Metapenaeus monoceros (Fabricius). Coastal Aquaculture in the Indo-Pacific Region, T.V.R. Pillay, editor, p. 178 (abstract only).

496. Shigueno, K. 1972. Problems of prawn culture in Japan. Coastal Aquaculture in the Indo-Pacific Region, T.V.R. Pillay, editor, Fish News (Books) Ltd., London, pp. 282-312.

497. Liao, I.C. and T.L. Huang. 1972. Experiments on the propagation and culture of prawns in Taiwan. Coastal Aquaculture in the Indo-Pacific Region, T.V.R. Pillay, editor, Fishing News (Books) Ltd., London, pp. 328-354.

498. Perrot, J. Progress in shrimp culture techniques and the production of juveniles. Proc. FAO Technical Conference on Aquaculture, Kyoto, Japan, May 26 - June 2, 1976, FIR:AQ/76/R.12 (in press).

499. Kurata, H. and K. Shigueno. Recent progress in the farming of penaeid shrimp. Proc. FAO Technical Conference on Aquaculture, Kyoto, Japan, May 26 - June 2, 1976, FIR:AQ/conf/76/r.17, 27 pp. (in press).

500. Hirasawa, Y. and J. Walford. The economics of Kuruma-ebi (Penaeus japonicus) shrimp farming. Proc. FAO Technical Conference on Aquaculture, Kyoto, Japan, May 26 - June 2, 1976, FIR:AQ/conf/76/R.27, 24 pp. (in press).

501. Neal, R.A. and C.R. Mock. A model closed system for aquaculture, incorporating the recycling of wastes. Proc. FAO Technical Conference on Aquaculture, Kyoto, Japan, May 26 - June 2, 1976, FIR:AQ/conf/76/E.22, 6 pp. (in press).

502. Neal, R.A. Penaeid shrimp culture research at the National Marine Fisheries Service Galveston Laboratory. Proc. FAO Technical Conference on Aquaculture, Kyoto, Japan, May 26 - June 2, 1976, FIR:AQ/conf/76/E.23, 8 pp. (in press).

504. Fenucci, J.L. and Z.P. Zein-Eldin. Evaluation of squid mantle meal as a protein source in penaeid nutrition. Proc. FAO Technical Conference on Aquaculture, Kyoto, Japan, May 26 - June 2, 1976, FIR:AQ/conf/76/E.36, 11 pp. (in press).

505. Laubier-Bonichon, A. and L. Laubier. Controlled reproduction of the shrimp Penaeus japonicus. Proc. FAO Technical Conference on Aquaculture, Kyoto, Japan, May 26 - June 2, 1976, FIR:AQ/conf/76/E.38, 8 pp. (in press).

506. Boschi, E.E. and M.A. Scelzo. The culture of commercial prawns (penaeids) in Argentina and the possibilities of producing them on a large scale. Proc. FAO Technical Conference on Aquaculture, Kyoto, Japan, May 26 - June 2, 1976, FIR:AQ/conf/76/E.40, 5 pp. (in press).

507. Cuzon, G., A. Michel, J.M. Griessinger, J.L. Martin, J. Calvas, and J.F. LeBitoux. Results of experiments on Penaeus japonicus: specificity of protein requirement and importance of fatty acids. Proc. FAO Technical Conference on Aquaculture, Kyoto, Japan, May 26 - June 2, 1976, FIR:AQ/conf/76/E.42, 7 pp. (in press.

508. Pérez, P.D. and M.O. Suárez. Experimental rearing of larval stage of the white shrimp Penaeus schmitti and the pink shrimp Penaeus duorarum notialis in the laboratory. Proc. FAO Technical Conference on Aquaculture, Kyoto, Japan, May 26 - June 2, 1976, FIR:AQ/conf/76/E.45, 9 pp. (in press).

509. Caubere, J.L., R. Lafon, F. Réne and Sales. Maturation and spawning of Penaeus japonicus in captivity, controlled tests of this reproduction at Maguelone on the French coast. Proc. FAO Technical Conference on Aquaculture, Kyoto, Japan, May 26 - June 2, 1976, FIR:AQ/conf/76/E.29, 19 pp. (in press).

510. Johnson, S.K. 1975. Handbook of shrimp diseases. TAMU SG-75-603, Center for Marine Resources, Texas A&M University, College Station, Texas 77840, 19 pp.

511. Johnson, S.K. 1976. Chemical control of peritrichous ciliates on young penaeid shrimp, FDDL-S7. Fish Disease Diagnostic Laboratory, Texas Agricultural Extension Service, Texas A&M University (July), 4 pp.

512. Johnson, S.K. 1976. Twenty-four hour toxicity tests of six chemicals to mysis larvae of Penaeus setiferus, FDDL-S8. Fish Disease Diagnostic Laboratory, Texas Agricultural Extension Service, Texas A&M University (July), 2 pp.

513. Newell, S.Y. and J.W. Fell. 1975. Preliminary experimentation in the development of natural food analogues for culture of detritivorous shrimp. Tech. Bulletin No. 30, University of Miami Sea Grant Program, 129 pp.

514. Bland, C.E., D.G. Ruch, B.R. Salser, and D.V. Lightner. 1976. Chemical control of Lagenidium, a fungal pathogen of marine crustacea. University of North Carolina Sea Grant Program Publ. UNC-SG-76-02, 38 pp.

515. Conte, F.S. 1975. penaeid shrimp culture and the utilization of waste heat effluent. Power Plant Waste Heat Utilization in Aquaculture: Workshop I. PSE&G, NSF, RANN, Trenton State College, Rutgers University, pp. 23-46.

516. Conte, F.S. and J.C. Parker. 1974. Texas shrimp farming: a perspective. Proc. Fish Farming Conf., Texas A&M University, pp. 40-44.

517. Brown, A., M.G. Kleve, and W.H. Clark, Jr. 1976. Evidence for the presence of actin in Natantian sperm. Am. Soc. Zool. 16(2):180 (Ab.).

518. Clark, W.H., Jr., A. Brown, Jr., J.W. Lynn, and C. Lu. 1976. Cellular aspects of reproduction in penaeid shrimp. Proc. of the First Int'l. Conf. of Tropical and Subtropical Fisheries, edited by B. Cobb, published by Sea Grant (in press).

519. Brown, A. and D. Patlan. 1974. Color changes in the ovaries of penaeid shrimp as a determinant of their maturity. Marine Fisheries Review 36(7):23-26.

520. Brown, A., Jr. 1970. Food for shrimps and prawns. FAO Aquaculture Bulletin 3(1):4.

522. AQUACOP. 1976. Maturation and spawning of *P. monodon* in captivity. An internal report.

523. Hysmith, B.T. and L.L. Elam. 1974. Penaeid shrimp culture. Texas Parks and Wildlife Pond Research, Study No. 2, Segment 3 Report, pp. 1-62.

Author Index

BOOK II

FRESHWATER PRAWN FARMING (Genus *Macrobrachium*) IN THE WESTERN HEMISPHERE

A State-of-the-Art Review
and
Status Assessment

Compiled and Edited
by

Harold L. Goodwin and Joe A. Hanson

with

William C. Trimble
Paul A. Sandifer

From the contributions and with the cooperation of the participants in the second workshop on the culture of *Macrobrachium* sp. held in Charleston, South Carolina, on June 14 and 15, 1976.

The Oceanic Institute
Waimanalo, Hawaii

January 1977

This work was supported in part by the National Sea Grant Program, NOAA, Department of Commerce, under grant #04-5-158-13 and by The Oceanic Institute, Waimanalo, Hawaii.

We were unable to resist sharing this photo. It is reproduced from a color slide taken by Spencer Malecha at the Tung Kang Marine Laboratory in Taiwan. The species is believed to be rosenbergii. The overall length including chelae is approximately one meter.

PREFACE

In 1974, a Sea Grant[a] to the Oceanic Institute enabled the authors to begin a series of specialized workshops designed to establish the state of the aquaculture art for some key species, and to develop lines of communication among persons and institutions interested in those species.

The first workshop on the culture of the freshwater prawn, Macrobrachium species, was held in St. Petersburg, Florida, in November 1974 with the Marine Research Laboratory of the Florida Department of Natural Resources as host organization. The workshop proceedings were published by the Oceanic Institute with the title, "The Aquaculture of Freshwater Prawns (Macrobrachium species)" by Harold L. Goodwin and Joe A. Hanson.[b]

So much progress was made during 1975 and early 1976 that the second workshop in June 1976 was designed with a different format. Reviewers selected for expertise in a particular field of prawn culture were invited to present papers that updated the state of the art, with discusssion following. It was assumed that all workshop participants were, at minimum, familiar with the prawn culture art at least to the extent described in the first workshop proceedings.

These, the proceedings of the second workshop, follow the workshop format, with each subject introduced by an invited paper, followed by a summary

---

[a]Sea Grant #04-5-158-13

[b]Available only in microfiche ($2.25) or a "blowback" paper copy ($5.00) from the National Technical Information Service, U.S. Department of Commerce, Washington, D.C. 20230, order #PB-246-657/AS.

of the discussion taken from tapes of the proceedings and the authors' notes. The authors used simple criteria for choosing the discussion to be included: First, were the comments based on experience, or were they simply speculative? Second, were the comments first or second hand? To elaborate a bit on this second point as an example: there was much discussion of regulations and attitudes of the Food and Drug Administration, but no representative of that organization was present, nor were copies of regulations at hand. Therefore, only those elements of FDA regulation and attitude expressed by Tom Billy, head of the National Marine Fisheries Service Products Inspection and Safety Division, who works closely with FDA, were included. In other words, no matter how provocative or illuminating a comment, it had to be authoritative for inclusion. The third criterion, of course, was relevance.

With this second prawn workshop, the activities of the Goodwins and Hanson in getting the "prawn people" together ended, and the task was passed to the World Mariculture Society.

The conveners and the attendees owe deep thanks to the Marine Resources Research Institute of the South Carolina Wildlife and Marine Resources Department. Paul Sandifer and Doris Akers carried much of the burden for MRRI, but the Institute staff came close to dropping everything else to be peerless hosts and to make the workshop not only technically successful, but warm with traditional Carolinian hospitality. Special thanks go also to Paul Zielinski of Clemson University who carried most of the burden of arranging a recording setup that resulted in the best workshop tapes of any of the Goodwin-Hanson series. Thanks also to Bill Trimble and Paul Sandifer for allowing us to include their extensive bibliography.

H. Goodwin
E. Goodwin

January 1977

CHAPTER I
OF PRAWNS, PEOPLE, AND SOY SAUCE

The aquaculture of freshwater prawns (*Macrobrachium* species) is a rather special part of the whole vast mass of aquaculture research and development. There are a number of reasons for this author's (Goodwin) affection for the subject and here are some of them:

- Prawn aquaculture is so new that its entire history and the names and faces of its originators are known to us. They are colleagues with whom we interact regularly.
- The prawn culture community of the United States is still so small that we can all crowd into the auditorium of the Marine Resources Research Institute, with enough room for our Latin-American colleagues.
- While prawn culture clearly is successful on a small commercial scale, enough problems and challenges remain so that it will not be routine for many years.
- To feast on the succulent prawns is still infrequent enough, even for the culturists, so that it is a treat to be anticipated.
- The future looks pretty bright, and progress is constant and quite rapid, as aquaculture progress goes.
- Because prawn culture is new, and the species of choice has no history in the United States, it is free from the social, political, and economic burdens faced by those who culture salmon or the American lobster.

By way of introduction, we elaborate briefly on some of those reasons, starting with a bit of history, about two decades ago when Dr. Shao-wen Ling of mainland China, now of Miami, Florida, was sent by FAO to Southeast Asia as a fisheries specialist. Dr. Ling came across a "big, beautiful animal" in the marketplace. It was his first look at the Malaysian prawn, Macrobrachium rosenbergii (de Man). He saw the prawns again in Ceylon, and finally in Malaysia when his FAO assignment took him there, and he began to satisfy his curiosity about their life history. It was not difficult to breed the prawns, or to produce larvae. But after five or six days "all the babies died." They continued to die, for many months. Shao-wen Ling's frustration grew. He began to take out his frustration by drawing prawns in typical, delicate Chinese brushwork style. He also began to experiment with nearly any substance that came to hand, for example, adding a few drops of tea to a container of larvae, or a bit of fish served by his wife for dinner. Still, the larvae failed to survive--until the day when he added a bit of soy sauce to a larvae container. And, "the babies lived!"

The soy sauce was the clue that led to salt as a necessity for larval development, for the prawns are catadromous creatures that spend their lives in fresh water, returning to salty water to hatch their eggs. At least that's the basic fact, but there are emerging variations that make the term "freshwater prawn" little more than a convenient but imprecise identification. As will be noted later, Sandifer et al are growing prawns in brackish water. Further, the ability of Macrobrachium lar, the Tahitian prawn, to traverse the rough oceanic channels of Hawaii so that small plantings on two islands have led to prawns on all principal islands, indicates an affinity for full seawater at some stages of development. There is also observational evidence that not all prawn individuals migrate to strictly fresh water; some apparently stay in the estuaries. More hard data on this aspect of prawn life would be useful, because there are many places where fresh water is scarce, but brackish water relatively plentiful, particularly in third-world countries.

At any rate, the soy-ploy used by Shao-wen Ling resulted in ability to control the life cycle of Macrobrachium rosenbergii, and planned, controlled culture of the prawns began. Actually, prawn culture was not new. Across the tropical belt from Malaysia and Thailand to Ceylon and India, prawns were grown based on recruitment of wild juveniles, but the level of activity was very minor compared with the natural fishery for the many prawn species. Ability to control the full life cycle opened new opportunities for culture, and the opportunity was grasped very quickly by Takuji Fujimura, Chief Aquatic Biologist of the Anuenue Fisheries Research Center of the Hawaii Department of Land and Natural Resources.

Fujimura, who is now Director of the Center, succeeded first in mass culturing larvae and bringing them through to juveniles, then initiated grow-out practices starting with the now-famous Ota's Ponds, which have been in continuous production for more than six years.

Prawn culture developed primarily because the Anuenue Fisheries Research Center shared freely its techniques and even its broodstock and post-larvae. The Hawaii pattern established by Fujimura and the prawn farmers who cooperate with his Center is used, with variations, by most prawn growers.

The aquaculture of the Malaysian prawn is not yet an industry. There is no clear definition of what constitutes an "industry," but no definition is needed to know that prawn culture does not rate so grandiose a term. In his paper on marketing, Chapter XII, Ian Chaston forecasts the marketing problems to be faced when prawn culture does reach that august level. For the present, however, prawn culture is small business.

The primary problem facing the culturist is that of food. Larval culture is dependent on a supply of brine shrimp (*Artemia salina*), and because the nauplius stage is the critical one, this means that culturists really are dependent on the supply of *Artemia* cysts, so-called "eggs." In a few instances, success has been achieved without feeding brine shrimp, but much more remains to be done to find an acceptable, convenient, inexpensive substitute.

In feeding during grow-out to marketable size, we still do not know how much of the prepared feed (marine ration, broiler starter ration, catfish chow, trout chow, et al) is consumed by the prawns and how much goes to fertilize the pond to produce the natural foods which obviously play an important role in prawn health and growth. Experiments are in progress, and we should have data before long. But even when quantitative information is available, proper management of a pond grow-out system probably will not be much different than it is now. It seems likely that significant change will only come when we have sufficient information on prawn nutrition to structure the most effective and cheapest food, with texture, stability, and attractants that ensure its complete consumption. Ponds will still produce natural foods, but cheaper fertilization than prepared foods will be used.

For those of us in the United States, temperature is and will continue to be a problem. The Malaysian prawn, which is the species of choice, is a tropical animal. A tropical environment either must be created or tropical summer temperatures used for a one-crop-per-year prawn farm.

Disease, which plagues growers of other animals, has not been a major problem for the prawn farmer. It can become one. See Chapter IV.

Problems for those who would culture animals for which there is a natural U.S. fishery, and hence a political constituency, do not exist for the prawn farmers. There is no opposition to prawn culture, at least none with the emotional irrationality of the opposition to salmon ranching or lobster farming. Instead, the irrationality is simply that of government, which imposes so many strictures and constraints that it is easier to go elsewhere, as so many American aquaculturists are doing. But, for those with the endurance and persistence to thread the labyrinthine ways of Federal, State, and local regulations, the constraints are no worse than for other species.

Although there is much to learn about processing prawns, as Virginia Sidwell notes in her paper on processing, we know enough so that prawns are being marketed both as a fresh and a frozen product. They are available in an increasing number of restaurants. This alone indicates that prawn farmers are producing, and we know that some have attained the level of a self-sustaining cash flow, and some are even showing a return on investment. There are no really large prawn farms in the U.S. King Caribe Shrimp in Puerto Rico has the most acreage, and Fish Farms Hawaii is a close second. But others are producing on a lesser scale.

Comparison with culture of marine shrimp probably is inevitable. If freshwater prawn culture ever reaches production levels of hundreds of tons, comparison will certainly be warranted. At present, though, it makes little sense to compare. There is no natural fishery for freshwater prawns in the U.S. except for Puerto Rico and the *M. lar* sport fishery in Hawaii, and the prawns imported from the fisheries in India and Bangladesh are as peeled tails only, with poor quality control. In short, not a very good product, and not directly competitive with shrimp. The cultured prawns are sold, for the most part, at six to eight count, heads on, to the institutional market for serving as a specialty item more nearly approximating the lobster tail than the shrimp. No grower, at least to date, has experienced much difficulty in selling the prawns, and they bring prices in the range of lobster per pound.

Whether the current situation, a relatively happy one as aquaculture situations go, will continue, remains to be seen. But the prospects are bright. Few people have any doubt that the freshwater prawn is here to stay, and an increasing number of gourmets are very pleased that this is so.

CHAPTER II
ABOUT *MACROBRACHIUM* SPECIES

The success of Shao-wen Ling in getting hold of the life cycle of the Malaysian prawn, and of Takuji Fujimura and the Anuenue Fisheries Research Center team in developing techniques for both mass postlarvae production and successful grow-out, brought the Malaysian prawn to the fore as the culture species of choice. Spread of Malaysian prawn culture was accelerated by the generosity of Fujimura and the Hawaiian team in sharing both technique and broodstock.

There is no doubt about the choice of *Macrobrachium rosenbergii* as a fine animal for aquaculture, but the possibility always remains that some other species might prove to be even better. The first U.S. workshop on prawn culture revealed that not much was known about most other *Macrobrachium* species, and that available information did not point to a better culture candidate. As of November 1974, the Malaysian prawn was the species of choice, with little potential competition within the genus. In assembling available data for the June 1976 workshop, Miyajima found some promise in other species, but still no major competitor for *M. rosenbergii.*

A. Prawn Species for Culture:
Lester S. Miyajima, Aquaprawns, Inc.

Macrobrachium culture, at present, is synonomous with the culture of *M. rosenbergii*. Hedgpeth (1949) states that over a hundred species of *Macrobrachium* exist in the world; thus, an introduction to alternative prawn species for culture entails the spectrum of world-wide distribution. Of the known species, 26 are found in the Americas (Holthius, 1952). Referring to Mayer (1964), one realizes that diversity in tropical species is extremely common. Therefore, an animal as "plastic" as *Macrobrachium* may vary in morphometric, meristic, and physiological characteristics within a localized population.

Before beginning selection of possible species for study or propagation, certain considerations are necessary. The first prerequisite is the environment. This consideration must answer the question: (1) does one select an animal to fit available environmental conditions, or (2) does one find or modify the conditions to suit the animal?

Since most Macrobrachium spp. are inhabitants of the tropics, one cannot expect this organism to endure the rigors of temperate climates. A few species tolerate cold better than others, but one must consider that cold water (below 15° C) would at best produce marginal growth. Therefore, the opportunities to match species to the U.S. environments are limited, and suitable strategies must be employed including environmental modification (e.g., by use of heated water).

The next set of considerations is concerned with harvestable size. Inherent size is a prime factor, whether expressed as length or weight. Growth is another; not growth per se, but rate of growth under controlled or semi-controlled environments. Growth rate in conjunction with density and survival determines the production capability or culture worth of the species.

Biological information concerning Macrobrachium is widespread in the literature, but is not very deep or thorough. Until recently this information dealt almost exclusively with distribution, taxonomy, and ecology of the genus. Assessment of the literature shows our extreme ignorance of the genus in the wild, beyond a few species, and except for a handful of papers, reference to commercial or experimental culture is sparse.

Analysis of information for those species which have been cultured is often difficult both to interpret and to compare, due to variations in broodstock, rearing conditions, etc. At best, comparisons among studies have to be based on estimated values. Such is the case with this presentation. In order to compare species of Macrobrachium, it was necessary to select specific characteristics which could be followed. For those species with available culture data, the characteristics were placed in categories labeled larval rearing, juvenile and tank rearing, and pond grow-out. For each category, the available data for M. rosenbergii were included as a reference.

Larval Rearing

The two species which have received the greatest amount of research are M. carcinus and M. acanthurus (Table II-1). Of these two species, greater success has been reported for M. acanthurus. As is evident in Table II-1, Hagood and Willis (1976) and Dobkin (1973) have reported nearly identical results with M. acanthurus. Comparisons with M. rosenbergii reveal that M. acanthurus differs by: (1) requiring higher salinity; and (2) having a longer development period.

Table II-1
Comparison of Larval Rearing Among Five Species of Macrobrachium

Species	Salinity (°/oo)	Temp. (°C)	Stocking Density Per Liter	Days to Meta-morphosis	Survival (%)
M. carcinus					
(Choudhury, (1971)	14-17.5	?	25-33	56-65	2.5
(Dobkin, 1973)	?	30	28	?	2.5
(Dugan et al, 1975)	12	27.5	10	-	0
(Harrell Int. p.c.)	16	28	14.5	30-35	40-50
(Lewis & Ward, 1965)	21	?	?	90	?
M. americanum					
(Arana, 1974)	10	28	?	50-72	?
(Monaco, 1975)	15	29.5	?	53	?
M. acanthurus					
(Choudhury, 1970)	15-20	23-27	33	25-40	9
(Dobkin, 1973)	?	30	21	29-50	20-25
(Dugan et al, 1975)	16	30-32	13-16	41-69	1.6-7.6
(Hagood & Willis, 1976	16-18	26-31.5	10.2-21.7	44	25
M. amazonicum					
(Guest, p.c.)	15	26-27	?	18-21	?
M. rosenbergii					
(Takata, 1974)	15-17	26.6-30	?	35	?
(Maddox and Manzi, in press)	12	28	50-67	23-28	50.5

The number of reported successes with M. carcinus are few. Interesting enough, George Kelley of Jacksonville University and Harrell Internation (Vero Beach, Florida) in personal conversations has reported results with M. carcinus equal to that of M. rosenbergii (Table II-2). Perhaps M. carcinus has been underrated in the past.

Juvenile and Tank Rearing

The data acquired for this section appear in Table II-2. As is evident, tank rearing of M. acanthurus received by far the greatest attention. The work of Willis et al (in press) has shown that growth of M. acanthurus at high densities is erratic; however, those studies which produced high growth rates demonstrated that the production of M. acanthurus may equal that of M. rosenbergii (Table II-3).

Table II-2
Comparisons of Tank and Juvenile Rearing Among Three Species of *Macrobrachium*

Species	Growth Rate (mm/day)	Stocking Density (per m^2)	Survival (%)	Duration (days)
M. acanthurus				
(Dugan et al, 1975)	0.225	49.5	88	133
(Willis et al, in press)	0.114	215	58.2	30
	0.221	215	53.3	60
	0.003	860	47.7	30
	0.177	860	35.5	60
M. americanum				
(Arana, 1974)	0.40	?	?	10
	0.37	?	?	19
M. rosenbergii				
(Fujimura and Okamoto, 1970)	0.21	33-333	89.1	62

Table II-3
Growth Comparisons of *M. acanthurus* and *M. rosenbergii* at Same Densities (Willis et al, in press).

		M. acanthurus		*M. rosenbergii*	
Density (m^2)	Age (Months)	Length (mm)	Survival (%)	Length (mm)	Survival (%)
215	0	13.07		13.22	
	1	19.48	58.2	13.54	58.2
	2	26.30	53.3	24.32	52.4
430	0	13.07		13.22	
	1	18.76	6.4	18.14	17.8
	2	34.15	4.8	29.12	16.0
645	0	13.07		13.22	
	1	15.08	54.9	17.26	11.4
	2	26.54	33.6	31.45	6.9
860	0	13.07		13.22	
	1	13.16	47.7	19.98	16.6
	2	23.68	35.5	32.08	5.2

Pond Grow-out

Table II-4 lists three species of Macrobrachium for which pond grow-out data were available. The results for M. acanthurus and its close cousin, M. tenellum, are nearly identical since growth approximated half a millimeter per day. The results for M. carcinus are somewhat less encouraging. Notice, however, that survival for M. acanthurus and M. carcinus was lower than that reported for M. rosenbergii.

Table II-4
Comparisons of Pond Grow-out Among Four Species of Macrobrachium.

Species	Growth Rate (mm/day)	Stocking Density (per m^2)	Survival (%)	Duration (days)	Avg. Size (mm)	Matu- ration (days)
M. carcinus						
(Dobkin, 1973)	0.09	5.6	5.36	166	28	?
(Kelley p.c.)	0.27	1	?	180	80	180
M. acanthurus						
(Dobkin, 1973)	0.35	10.7	12.08	188	85.6	104
	0.42	8.3	15.38	177	94.3	
M. tenellum						
(Maugle, 1974)	0.53	?	?	180	80-110	?
M. rosenbergii						
(Fujimura, 1970)	0.20	6.6	?	420	117	?
(Ling, 1967)	0.62	1.5	?	210	155	?
(Takata, 1974)	0.20	18.0	38	300	83	

Other Species

Table II-5 lists an example of five species of Macrobrachium for which distribution records are available. Such records may be a reference point in analyzing suitability of a species for culture.

In addition, varieties such as the Palau strain of M. rosenbergii as reported by McVey (1975), as well as recent imports from Thailand, need to be considered for use in hybridization of the Fujimura stock from the Anuenue Fisheries Research Center in Hawaii. [a]

---

[a] See Chapter IV.

Table II-5
Comparisons of Biological Data of Six Species of Macrobrachium.

Species	Growth Rate (mm/day)	Size (mm)	Maturation (months)	Fecundity
M. malcolmsonii (Bhimachar, 1965)	0.02	to 230	12	---
M. idae (Bhimachar, 1965)	---	---	---	2,000-20,000
(Katre & Pandian, 1972)	---	---	---	40-160
M. mirabilis (Bhimachar, 1965)	0.01	62	12	550-3,000
M. vollenhovenii (Miller, 1971)	0.29	to 150	9-10	12,000-45,000
M. macrobrachion (Miller, 1971)	---	75	---	---

In closing, this report has attempted to present an overview of the current status of prawn culture other than that of M. rosenbergii. It is evident that present research in prawn culture has been directed toward the Malaysian prawn, but we must not forget that other species of Macrobrachium are available for study and that some of these, such as M. acanthurus and M. tenellum, have the potential for commercial culture.

B. Summary of Discussion

Even though more than a hundred species of Macrobrachium are said to exist worldwide, supporting quite substantial local fisheries in some places, only a handful have been studied to determine their culture potential. Of those studied, none seems to be superior to M. rosenbergii. So, if the Malaysian prawn has a competitor as a cultured species, it remains to be found, perhaps in tropical Africa.

It is also clear that the taxonomy of the genus is less than perfect.

Macrobrachium carcinus

George Kelley of Jacksonville University and Harrell International has been growing M. carcinus for about eight years. He confirmed their bad disposition, characterizing them as "very cranky during larval life, even crankier as PL's, and almost vicious as adults." Growth rate is slower than that of M. rosenbergii, but they are more cold-water tolerant. A cold snap wiped out Kelley's M. rosenbergii, but not the M. carcinus.

Optimal stocking densities are unknown, but, because of the species' territoriality, are less than that of the Malaysian prawns. Kelley commented that he liked M. carcinus, and added that anyone who can grow M. rosenbergii can grow M. carcinus, if they have patience.

An interesting aside about M. carcinus reported by Craig Kensler: they are hunted near Acapulco, Mexico--with machetes.

Jay Huner suggested the use of M. ohione in polyculture, or sequential culture, with crawfish. This small, native species is hardy, slow-growing, and of low fecundity. Nevertheless, crawfish ponds are not productive in summer and could be used for M. ohione. Presumably, the small prawns would be marketable as bait, not as food; a prawn bait industry fishery exists in Louisiana.

C. Notes on Macrobrachium Species in Mexico: Manuel Guzman, Jorge Cabrera, and Craig Kensler

(The following contribution was sent after close of the workshop for inclusion in these proceedings. Guzman and Cabrera are with the Universidad Nacional Autonoma de Mexico; Kensler with a UNESCO program at the same university.)

At present, the taxonomy of the genus Macrobrachium in Mexico is not definitive but is still being worked on. However, to date, 11 indigenous and one exotic species are recognized in Mexico.

Four species are of major importance in the Mexican commercial fishery for Macrobrachium. M. carcinus (Linneaus) and M. acanthurus (Wiegmann) are found along the "slope" (from mountain to coastal plain) of the Gulf of Mexico, in the states of Tamaulipas, Veracruz, Tabasco, and Campeche. M. americanum (Bate) and M. tenellum (Smith) are found along the "slope" of the Pacific coast, mainly from the state of Sinaloa south along the coast to the state of Chiapas on the Guatemala border. Within this range M. americanum is found mainly in the northern region and M. tenellum in the southern.

Five species are of no great commercial significance, although where fished they are an important part of the diet of the local people. The species are: M. olfersii (Wiegmann), M. digueti (Bouvier), M. acanthochirus Villalobos, M. michoacanus Villalobos-Martinez, and M. heterochirus (Wiegmann). Two noncommercial but biologically interesting cave-dwelling species of Macrobrachium have recently been described from Mexico: M. villalobosi Hobbs and M. coconaensis Villalobos-Chavez.

Finally, in 1973 and 1974, the exotic species M. rosenbergii (de Man) was introduced into Mexico (Pacific state of Sinaloa) from California and Hawaii, respectively, for experimental culture work.

Present Macrobrachium Work in Mexico

Two Macrobrachium projects are in progress at the National University (UNAM) in Mexico City. Both projects are mainly biological/ecological in scope and are not concerned with commercial cultivation prospects at present, although such work is planned for the future. Emphasis in both projects is on M. tenellum. Adults, juveniles, and larvae of this species are found mainly in brackishwater lagoons, where seasonal temperatures can range from 25-35° C and seasonal salinities can range from 0-15 ‰. Average maximum total lengths in nature are about 100 mm for females and 120 mm for males, measured from the tip of the rostrum to the end of the telson. Growth rates in nature average 10 mm per month for both sexes and hence both females and males reach maximum size within a year. The species is found in high densities in natural conditions, is not aggressive or cannibalistic, and can tolerate a wide and fluctuating range of temperatures, salinities, and oxygen concentrations. We feel that M. tenellum is a good candidate for experimental field cultivation in Mexican conditions. Lastly, the two UNAM groups are presently preparing a review of Macrobrachium in Mexico, which will include complete distributions of the various species, biological and ecological notes, detailed commercial fishery data and information regarding the various Macrobrachium programs in Mexico.

Several Mexican Government groups and a private group are also working on Macrobrachium projects. Unfortunately, full details are not available at this time concerning their programs, species being studied, localities of work, etc. However, some information about each group follows:

- Direccion de Acuacultura of the Secretaria de Recursos Hidraulicos. Government group undertaking general biological work with Macrobrachium sp. with a view to possibilities of cultivation and fishery development.

- Fideicomiso para el Desarrollo de la Fauna Acuatica (FIDEFA). Semistate group engaged primarily in "extension" work with

Macrobrachium in inland waters. FIDEFA has reported the distribution of several million Macrobrachium sp. larvae for stocking purposes.

- Compania de Ingenieria Fluviomaritima Sociedad Anonima (CIFSA), Private consulting group working on experimental culture of M. rosenbergii near Mazatlan in the State of Sinaloa.

D. Macrobrachium lar

The subject of M. lar, the Tahitian prawn, was not discussed at the workshop because no one present had any information other than a comment that a researcher at the Hawaii Institute of Marine Biology had succeeded in bringing M. lar through to juvenile. Inquiry resulted in identification of Michael S. Muranaka as the successful culturist. Muranaka's research into the biology of the species is not complete, but he has kindly provided some preliminary information, summarized as follows:

M. lar rearing is more difficult than that of M. rosenbergii. From hatching until the fifth to seventh stages, the larvae must remain in 35 o/oo seawater, after which the salinity is lowered. The required salinity range is not yet firmly established, but may be as high as 25 o/oo. PL's must be kept in salinities at least this high. Muranaka believes that wild PL's remain in highly brackish water for "a couple of weeks" before migrating upstream. Most notable characteristic of the PL is its red coloration upon settling, whereas the larvae appear more or less clear except for prominent chromatophores on the ventral side of the abdomen until the settling stage is reached.

The Hawaii Department of Fish and Game released 94 specimens into a stream on Molokai in 1956, and 27 specimens into an Oahu stream in 1957. By June 1965, M. lar was found on the island of Hawaii, and by 1969 the Tahitian prawn was confirmed on all major islands of the state. This spread was only possible for an animal able to live in full seawater.

Previous attempts to raise M. lar were not successful. Muranaka believes that salinity is the key to consistent results. His future research will amplify and clarify his initial successes, with results published through regular scientific channels.

CHAPTER III
DISEASES AND DISEASE CONTROL

Disease has not been a major problem for freshwater prawn culturists. The report of the St. Petersburg Workshop stated, "Freshwater prawn culturists have had fewer problems with disease than culturists of other crustaceans," and this remains true as of the 1976 Charleston Workshop.

A. Diseases and Disease Control in *Macrobrachium* Culture:

Carl J. Sindermann, Director, Middle Atlantic Coastal Fisheries Center, National Marine Fisheries Service

Abstract

Disease and mortality problems in *Macrobrachium* culture have thus far been relatively minor. Disease entities which have been identified include shell disease, filamentous bacterial infestation, egg fungus infestation, ciliate infestation, and muscle necrosis. Several other diseases of uncertain etiology have been mentioned in published reports, including a systemic fungus infection and a protozoan (suctorian) disease. Most of the diseases recognized seem to be consequences of poor water quality or other stresses, and chemical or other control measures have been developed for most of the disease conditions encountered.

Introduction

The short history of *Macrobrachium* culture has been remarkably free of the kinds of serious disease outbreaks with concomitant mass mortalities that have characterized (and still characterize) culture of a number of other aquatic species. Complacency can be dangerous, however, and it is prudent to assemble as much information as we can about diagnosis and control of those infrequent outbreaks which do occur, so that we may be better able to cope with major events, should they appear.

In the year and a half since the first prawn culture workshop, held in November 1974, several publications have referred to disease problems in *Macrobrachium*. These papers usually describe the condition generally, then report attempts at control and the effectiveness of such attempts. I have tried in the present summary to provide a digest of what we know of *Macrobrachium* diseases and their control from the published literature, hoping that such a digest may serve as a base for discussion of new disease events and new information about control.

To provide some structure, I have selected several disease entities which have emerged in *Macrobrachium* culture. They are:

- shell disease (black spot)
- filamentous bacterial infestation (*Leucothrix*)
- egg fungus infestation (*Aphanomyces* and *Achlya*)
- ciliate infestation (*Epistylis*)
- muscle opacity and necrosis

Additionally, I have considered briefly several other poorly defined disease conditions under the heading "Diseases of uncertain etiology", hoping that more information about them will be forthcoming.

Shell Disease (Black Spot)

Shell disease, also referred to as black spot or brown spot disease, continues to be one of the most common and potentially serious problems in *Macrobrachium* culture. Evidence is accumulating that mechanical disruption of the epicuticle is necessary to provide a portal of entry for chitin-destroying bacteria and other secondary microbial invaders. Chitinoclastic bacteria isolated from shell lesions of *M. rosenbergii* in the U.K. were predominantly *Benekea*, but *Pseudomonas* and *Aeromonas* were also represented (Delves-Broughton and Poupard, 1976). These authors concluded that "Shell disease does not appear to lend itself to prophylactic or chemotherapeutic control. Methods to reduce the incidence of shell disease include more careful handling of the animals, controls to reduce aggression among individuals, and a reduction in stocking density."

Dugan and Frakes (1973) and Dugan et al. (1975) found black spot to be the most prevalent disease of adult *Macrobrachium*. The sequence of events in their study included mechanical damage to the exoskeleton, attack by chitinoclastic bacteria, and finally invasion by a lethal freshwater phycomycetous fungus.

These authors found that the fungus (not further described) could be destroyed with a 15-30 minute dip in 20 ‰ seawater. They also noted that under good environmental conditions, the disease never became a problem.

Filamentous Bacterial Infestation (*Leucothrix*)

Sandifer and Smith (1975) described bacterial infections of gills, pleopods, and uropods of juvenile *M. rosenbergii* in a very intensive rearing study in 1975. *Leucothrix*-like organisms predominated. High mortalities occurred; the disease affected most individuals, but the authors were not sure if the infection contributed directly to mortalities. Several chemical treatments were tried but were ineffective, inconsistent, or toxic. Prophylactic measures--reducing population densities and more careful cleaning of tanks--gave best results.

Filamentous bacterial infestations of *Macrobrachium* larvae, tentatively identified as *Leucothrix*, experienced by Aquaprawns, Inc., in Texas, were controlled by mass water exchange when significant numbers were observed (Dugger, personal communication). Positive but inconsistent responses were seen when Nitrofurazone and Erythromycin were used.

Egg Fungus Infestation

Fungi of two genera, *Aphanomyces* and *Achlya*, were isolated from eggs of ovigerous females of *M. rosenbergii* by Dr. C. E. Bland in 1974 from material supplied by Dr. T. I. J. Smith (personal communications from each). Hyphae completely covered the ova, but as Dr. Bland pointed out, both genera of fungi are common in the aquatic environment, and there is some question as to whether they were actually parasitizing the eggs.

Ciliate (*Epistylis*) Infestation

During the fall and winter of 1975, Aquaprawns, Inc., experienced infestations of larvae by the ciliate *Epistylis*, sufficient to prevent feeding (Dugger, personal communication). Ten different chemical treatments were tested. Formalin caused mortalities, molting deformities, and cessation of feeding. Only sulfa-quinine and acetic acid were effective consistently, and because of the high cost of sulfa-quinine, acetic acid was selected as the control method of choice at 2.0 ppt for 1-minute dip. Some *Epistylis* survived the dip, however, and could reinfest the population.

Epistylis has also been recognized as a problem in Hawaiian *Macrobrachium* culture (Nakamura, personal communication). Up to 10% of harvested prawns may be infested, reducing their marketability.

Fujimura (in Goodwin and Hanson, 1975) reported that water hardness, on the order of 300 ppm, resulted in growth of an Epistylis in Hawaiian ponds, with no solution except to move to other ponds with lower mineral content. (Excessively high calcium carbonate content of culture water may also lead to abnormal deposition on the exoskeleton of larvae--a condition which may result in mortality.)

Muscle Opacity and Necrosis

A progressive opacity of body muscles followed by necrosis has been reported in Macrobrachium postlarvae and juveniles by several workers (Sandifer et al., 1975; Delves-Broughton and Poupard, 1976). The disease condition seems associated with environmental stress. Sandifer et al. found that postlarvae placed in high salinity (35 $^{o}/oo$) turned opaque white and died in about a day. Delves-Broughton and Poupard referred to the disease condition as "white syndrome", and reported that salinity, temperature, oxygen, or other stress would cause Macrobrachium and other shrimps to turn white. The color change began in the tail and progressed anteriorly.

Other authors have associated the condition in penaeid shrimps with stress (Venkatoramaiah, 1971; Rigdon and Baxter, 1970), and found that it was reversible if necrosis had not progressed. Delves-Broughton and Poupard (op. cit.) noted that Macrobrachium with the syndrome had a soft exoskeleton, and exhibited slow growth and high mortality.

A condition with some similarity to the muscle opacity and necrosis described above was reported by Fujimura and Okamoto (1972). Exposure of Macrobrachium to direct sunlight in Hawaii resulted in formation of opaque white spots which gradually spread over the entire body until death occurred.

Diseases of Uncertain Etiology

Delves-Broughton and Poupard (1976) described and illustrated a condition in M. rosenbergii in which black nodules formed in the hypodermis, without any break in the exoskeleton. The authors suggested that the origin of the nodules could be bacterial. Furanace at a concentration of 0.09 mg/l seemed to arrest the disease.

Fujimura (1966) described a condition in M. rosenbergii larvae characterized by small opaque white patches beginning at the base of appendages and then spreading throughout the entire body, producing sporadic heavy larval mortalities. Ling (1969) considered this to be a systemic fungus infection but no description of the causative organism is available. There are other vague reports of fungus infections (Sick and Beaty, 1974) but adequate information has not been released.

Smith et al. (1974) repoted quite a different fungus infestation of larvae, apparently associated with accumulation of decomposing food in the tanks. The fungus grew on the anterior appendages and tails of larvae, and produced some mortality. Control was effected by reducing larval density and removing the excess food.

Fujimura (1966) and Ling (1969) mentioned a "protozoan disease" possibly caused by a suctorian, which caused high mortalities of larvae. Control was effected with malachite green, formalin, and copper sulfate. (Fujimura's) dosages were 6 hrs with 0.4 ppm copper concentration; daily half-hour treatment with 0.2 ppm malachite green; and a daily half-hour treatment with 200 ppm formalin.

Conclusions

The remarkably short list of diseases known to be problems in *Macrobrachium* culture may be due in part simply to lack of adequate examination, but it seems true that those disease conditions which have surfaced thus far seem to be a consequence of poor water quality or other forms of stress, such as overcrowding. The disease organisms recognized so far seem to be facultative pathogens rather than obligate or primary pathogens--facultative in the sense that they exert their effects when physiological or environmental conditions become marginal for the host animal or population.

It is becoming apparent that too often in examining aquaculture mortality problems we focus narrowly on a search for pathogens, when the real cause may be environmental, nutritional or physiological, with the superimposition of infection by facultative microorganisms on weakened or damaged hosts. This does not imply, however, that infectious diseases should be ignored--simply that other causal factors should be considered, and water quality conditions monitored consistently.

It seems clear that there is a tendency for too much reliance on chemical crutches. Chemotherapy should be a method of last resort rather than a method of choice in aquaculture disease control. Of course it is still important to have treatments elaborated for those occasions when other preventive and prophylactic measures that are part of good animal husbandry fail.

The most important conclusion, however, is that disease has thus far not been a major limiting factor in *Macrobrachium* culture. Continued attention should be paid during this grace period to the development of control measures for those disease problems which have been recognized.

B. Summary of Discussion

Epistylis

Durwood Dugger reported that *Epistylis* appeared after temperatures got cooler in the Brownsville, Texas, area where Aquaprawns, Inc., was located. After eliminating other possible sources, it was concluded that the infestation came in through the city water. Because *Epistylis* was found on native animals, it was apparent that the organism was plentiful in the area. Aquaprawns' first step was to clean up the prawn environment, difficult to do because feeding the prawns also meant feeding the protozoans. It was necessary to resort to chemicals. Acetic acid produced the best, consistent results. Dosage was two parts per thousand for a one-minute dip. Above 2 ppt, adult prawns showed some "strange reactions", but recovered. No short-term effects were apparent on treated larvae. Dugger concluded that, if you have hard water with calcium carbonate in the system, you also have Epistylis, a conclusion consistent with previous reports from Hawaii.

Edward McSweeney of Syntex noted an infestation of protozoans of two types, one a solitary variety similar to *Stentor*, and the other an *Epistylis*. The infestation started in one tank and spread to others. One tank with a heavy infestation was subjected to a complete water change; when harvested two days later, there was no visible sign of infestation. In a second tank, filtration and circulation were improved and the infestation cleared in about five days with no water change. The experience again reflects the importance of water quality.

Joseph Quick of Dow Chemical commented that he has studied a number of kills in the natural environment, many on animals like the blue crab. All took place in polluted waters with very high levels of bacteria. Since *Epistylis* feeds on bacteria, the relationship is apparent.

Leucothrix

Quick noted that *Leucothrix* is associated with high levels of dissolved organics. Proper filtration that eliminates the organics also eliminates the bacterium. Goodwin commented that this is consistent with the experience of *Homarus* researchers who found that *Leucothrix* infection was directly related to organic loading of the system.

Black Spot Disease

Commenting on black spot disease from the point of view of the commercial culturist, Kenneth Kato of Fish Farms Hawaii said that pressure to market what is harvested means overlooking black spot disease on harvested animals. He did not consider the frequency of black spot disease important in relation to harvestable prawns.

Jay Huner noted that black spot also is found in crawfish. The animal eliminates the disease by molting the infected exoskeleton, unless the infection has gone too deep to be molted, in which case the animal dies. Goodwin recalled that previous reports from Hawaii also spoke of putting infected animals back in the pond to molt the diseased exoskeleton.

Elliot Maynard, reporting on experiences while with Caribe King Shrimp in Puerto Rico, said that shell disease in broodstock tanks was virtually eliminated by adding tree branches, which were changed periodically, and by improving the substrate. A green alga, *Najas guadalupensis*, also was placed in the tanks, and had to be replenished because the prawns ate it in great quantities. Maynard added that far greater densities of prawns were possible after adding algae. Randy Hagood, hatchery manager for Caribe King Shrimp, adds that branches also were used in the Florida Division of Marine Resources Laboratory to add substrate and so permit greater use of the tank volume and reduce stress caused by density, which he characterizes as a major cause of black spot. Hagood keeps his broodstock in slightly brackish water of 2-3 ‰ to help prevent black spot and other problems (Hagood comments through personal communication).

Muscle Necrosis and Opacity

Joseph Quick reported that he had seen muscle opacity show up in response to certain commercial rations, and characterized it as "literally fatty degeneration of the muscle."

Ed McSweeney agreed muscle opacity or necrosis, "white muscle syndrome, is nutritionally related, but there is another type he has seen in which animals, grown in individual tanks on the same recirculating system, turned completely white except for the eyes. Sometimes this happened to one animal, and sometimes to several. The diseased animals often do not make it to the next molt; of those that do, he knows of none that made it to a second molt. McSweeney said this condition is very different from the physical stress syndrome which comes on rapidly when animals are agitated or their oxygen levels are lowered.

Miscellaneous Diseases and Problems Noted

Michel Cuzon, reporting on Aquacop's prawn culture at the Centre Oceanologique du Pacifique in Tahiti, described problems with two types of filamentous bacteria, one which occludes the respiratory system, and the other which invades the exoskeleton causing a black necrosis common on pleopods and appendages, and sometimes extending into the musculature. During the second year of operations in Tahiti, these bacteria caused mass mortalities. By the third year, strict environmental control and use of antibiotics gave good results, essentially solving the problem. Two tanks were treated with tetracyclene at 0.75 grams per meter^3, and two with Streptobi-penicillin at 2.5 grams per cubic meter. Four tanks were maintained without antibiotics as controls. Within the treated tanks, there was no mortality, no necrosis of appendages. When mortality and necrosis appeared in the untreated tanks, they were treated and the disease cleared up within 24 hours. First post-larvae appeared in 28 days. The experiments were conducted in clear water, without phytoplankton. Experiments with green water conducted at the same time failed; all larvae died in three days with the same disease symptoms of filamentous bacterial growth and necrosis.

In response to a question on how to handle hydroids in recirculating filter beds, Stephen Hopkins of MRRI said that hydroids and the medusae can be knocked out, but the cyst stage is almost impossible to treat except by drying out the whole system, or by using very harsh chemicals.

Sophie Jakowska noted that natural populations of _M. carcinus_, _M. acanthurus_, and _M. faustinum_ she and Idelisa Bonnelly de Calventi had investigated in the Dominican Republic carried what appeared to be microsporidian spores in the abdominal muscles, and had ciliates on gills.

Ariel Gaathon reported on a bad experience at Columbia University with artificial seawater during which adults died and larvae died on the fourth day. When a change was made to Instant Ocean, which has a complement of trace elements, the problem ended, indicating that the presence of some trace elements is important to prawn survival.

To the question of whether antibiotics could be added to prawn rations as well as used in water, Tom Billy of NMFS replied that FDA has not issued regulations permitting antibiotics to be used in the prawn diet, although FDA has approved prophylactics for addition to trout and catfish food.

Bernard Cohlan raised the question of a fungus which causes mortality in brine shrimp and apparently affects other small crustacea. The fungus is _Mishnikovia bicuspidata_, which gets into the body cavity of the _Artemia_. Because prawn culturists use _Artemia_ extensively, they should be aware of the possibility of fungus infection. No one at the Charleston workshop had seen any apparent damage from this particular fungus in their prawns.

Joseph Quick commented that the diseases of _Macrobrachium_ appear to be, with one possible exception, opportunistic or facultative types not specific to prawns but occurring in culture of other animals. The one exception, so far, may be an obligate pathogen, a blood ciliate in the same, or an allied, genus of a hemociliate that attacks penaeid shrimp. The organism attacks and kills adults, some dying outright and others during molt. The disease appeared at the Marine Research Laboratory of the Florida Department of Natural Resources in St. Petersburg. No workable treatment was found, but the disease was finally isolated into one tank. Origin of the disease was unknown, but at the time the researchers were collecting animals (M. acanthurus) from the wild. The ciliate shows up quite easily if the hemolymph is sampled. It is slightly larger than a leucocyte, and there is a very sharp drop in leucocyte numbers with occurrence of the disease. Scott Willis commented that there may be two ciliates. When the system was dried out and changed completely at the St. Petersburg laboratory, Willis found some ciliates that closely resembled but were not identical to those which had caused the disease.

Disease Prevention

The fact that Macrobrachium culture has not been plagued by disease to any great extent does not mean that this happy condition will last. Quick pointed out that specific, obligate pathogens are found in virtually all animals for which culture has been tried. Almost certainly, these specific diseases are to be found in wild stock. When people bring in other strains from distant areas, as is now being done, there is a high probability that pathogens will be brought in, too, coming from the natural stocks that are being recruited.

Quick recommended that newly obtained animals be quarantined for a long period, with care taken not to spread the water or food, or even to risk exchange of microorganisms on tools and equipment. Similarly, those shipping animals should take care, holding the stock until sufficient time has elapsed to assure that the stock is disease-free. In brief, all possible steps should be taken to avoid transfer of disease.

Sindermann strongly supported Quick's comments, adding that control of transfers, of the indiscriminate movement of animals, is a subject of international concern and was discussed at the 1976 Aquaculture Conference of the Food and Agriculture Organization at Kyoto, Japan. Sindermann pointed out that it is difficult to know what to expect or to look for in diseases of marine animals, especially those like Macrobrachium spp. which have not been studied sufficiently from a disease viewpoint. We are really only beginning to understand the range of pathogens that may be involved. With better-understood species, such as the trouts, we begin to know what to look for in inspection.

Sindermann concluded that culturists must move toward some kind of standard examination beyond a "quick and dirty" parasitological examination. Something beyond gross inspection, some evidence of histopathology may be the best clue. While this is hard to do, Sindermann pointed out that Sea Grant is supporting people around the country who are willing to offer diagnostic services of which culturists should take advantage. Tissues and samples can be sent to the diagnosticians.

"We have a grace period," Sindermann said. "We should use it to advantage to understand all we can about prawn diseases based on the few insights we have gathered about diseases, so that we can hope to gain enough technology to cope with outbreaks when our grace period ends."

CHAPTER IV
PRODUCTION OF JUVENILES

Since Fujimura closed the life cycle of M. rosenbergii in captivity several years ago, hatcheries concentrating on that species and several others have come into being in some numbers throughout the temperate and tropical Western Hemisphere. Sandifer et al. (1976) reviewed the current status as revealed by recent inquiry, and this was followed by comment from the floor.

A. Status of Macrobrachium Hatcheries, 1976[a]:

Paul A. Sandifer, J. Stephen Hopkins, and
Theodore I. J. Smith, Marine Resources Research Institute

Introduction

Participants at the first Workshop on the Aquaculture of Freshwater Prawns (25-26 November 1974, St. Petersburg, Florida) concluded that, "while growth and survival of larvae to postlarvae may be improved to some degree and the cost of production may be reduced, the present technology for handling larvae is adequate for commercial production" (Goodwin and Hanson, 1974, p. 21). In this context, our purposes in preparing the present review were to (1) determine the current operating and production characteristics of Macrobrachium hatcheries at both research and production scale, (2) highlight major technological advances since the St. Petersburg meeting, and (3) identify major problem areas.

---

[a] This work is a result of research sponsored by the Coastal Plains Regional Commission under contract #10440040; NOAA Office of Sea Grant, Department of Commerce, under Grant #04-6-158-44009; and the State of South Carolina. Reference to trade names in this paper does not imply endorsement by the Coastal Plains Regional Commission, NOAA Office of Sea Grant, or the State of South Carolina.

Methods and Materials

Information summarized here was obtained primarily from a survey of *Macrobrachium* hatchery operators, supplemented with data from the literature and our own research. For the survey, a short questionnaire (see Addendum) was developed and distributed to all individuals known by the authors to be involved in hatchery operations. Although the emphasis in this survey was on hatcheries in the United States, questionnaires were sent to several hatchery operators in other areas of the world as well.

Survey Results

A total of 34 questionnaires was distributed and 27 (79.4%) were returned. Of these, six respondents indicated either that they were not presently involved with *Macrobrachium* or that they had not yet developed hatcheries; the other 21 (10 from private firms, 11 from public institutions) provided the information summarized here.[a] Although some active hatchery operators may have been overlooked in our survey, we feel that the available returns provided sufficient data for basic comparisons and conclusions. Also, the geographic range of the responses (in the United States, from California, Florida, Hawaii, New Jersey, South Carolina, Texas, Puerto Rico, and the Trust Territory of the Pacific Islands; elsewhere from Honduras, Malaysia, Mauritius, Tahiti, and the United Kingdom) was sufficiently broad to provide a realistic overview of the status of *Macrobrachium* hatcheries.

A few of the responses dealt with or included information on species of *Macrobrachium* other than *M. rosenbergii*, but the following comparisons and conclusions are limited to *M. rosenbergii*.

Culture Systems

Static culture systems with periodic exchange of water are widely used, especially in the larger prawn hatcheries. Recirculating systems also are employed widely (8 responses) but are largely confined to research-scale operations. Data from the survey indicate that average production (expressed as number of postlarvae/unit of tank capacity) from static systems is about 16% greater than that from recirculating systems (Table IV-1). However, maximum production levels achieved in the two types of systems are similar (to 62 postlarvae/liter in AQUACOP's production-scale static hatchery system in Tahiti; 41-59 postlarvae/liter in research-scale recirculating hatchery systems at the Marine Resources Research Institute (MRRI) in South Carolina).

---

[a] The number of responses noted for each item of discussion varies since respondents often did not answer all questions.

Table IV-1
Analysis of Average Production Data (average number of postlarvae produced/liter of tank capacity) from Survey of Macrobrachium Hatchery Operators. (Means given are the arithmetic means derived from the average production figures supplied by respondents; n = number of respondents; production scale = capacity to produce > 1, 000, 000 postlarvae/year; research scale = capacity to produce < 1, 000, 000 postlarvae/year; percent difference = $\frac{B-A}{A}$ x 100).

	A Recirculating	B Static/Flow-Through	Percent Difference
Mean	17.3	20.1	16.2
Range	10-32	7-33	
N	8	8	

	Algae	No Algae	Percent Difference
Mean	17.5	19.3	10.3
Range	7-41	10-30	
N	8	8	

	Production Scale	Research Scale	Percent Difference
Mean	15.3	19.9	30.1
Range	8-30	10-33	
N	8	7	

	Private	Public	Percent Difference
Mean	14.0	21.1	50.7
Range	7-25	8-33	
N	7	8	

Salinity and Temperature Conditions

Rearing salinities and temperatures ranging from 8-17 $^o/oo$ and 24-32° C, respectively, were reported by most hatchery operators. However, most preferred salinities of 12-16 $^o/oo$ and temperatures of 28-30° C. These conditions agree closely with those recommended by Ling (1969).

Use of Algae

Use of algae in larval culture was reported by 10 respondents, while nine others indicated that they reared larvae without algae. Most of those who used algae followed the Fujimura technique (Fujimura, 1966) and worked with "green water" cultures developed from wild plankton. These cultures were generally believed to consist predominantly of Chlorella. Reported concentrations of such cultures ranged from 5 x 10^5 to 2 x 10^6 cells/ml. The use of "green water" is restricted for the most part to static larval culture systems where the water is periodically exchanged. Exchange rates ranging from one to two exchanges/day to a complete exchange about every third day were reported for such systems.

A modification of the "green water" approach involved the use of monospecific cultures of selected species of algae. In this regard, species of Chlamydomonas, Chlorella, Dunaliella, Isochrysis, Nannochloris, Phaeodactylum, Pseudoisochrysis, Skeletonema, and Tetraselmis have been tested (Wickins, 1972; Maddox and Manzi, 1976; Manzi and Maddox, 1976). Best results reported with algae (production of an average of 59 postlarvae/liter) were obtained by Manzi and Maddox (1976) in South Carolina. These authors used Phaeodactylum tricornutum at densities of 340, 000 cells/ml in the larval rearing tanks, and demonstrated the beneficial effects of this alga for the culture of M. rosenbergii larvae in their system. Nevertheless, results of our survey (Table IV-1) indicate that, overall, hatcheries in which algae are not employed are producing about 10% more postlarvae/unit of tank capacity than those using algae (primarily "green water"). This difference is probably not significant.

Feeds and Feeding

Most respondents agreed that Artemia is the major food used in their hatcheries, but in one case, Artemia was reported used only at night. Concentrations of Artemia nauplii provided to the larvae vary among operations, but most culturists feed at rates of 5-15 nauplii/ml. In several cases, the concentration is high initially and is then decreased during the latter half of larval development. In other operations, relatively high densities (on the order of 5/ml) of Artemia nauplii are maintained throughout larval development.

Many foods are used to supplement the Artemia diet during the latter half or two-thirds of larval development. The most widely used supplement is minced fish flesh (primarily tuna), followed by frozen adult Artemia, fish eggs, bits of chicken egg, and other materials. Prepared formula foods were listed as supplements by a few respondents, but no details concerning these formulations were given.

Hatchery Size

Macrobrachium hatcheries surveyed ranged from a small research unit (approximately 0.5 kl capacity) to a large production system with a capacity of 182 kl.

Larval Rearing Tanks

Tanks employed for rearing *Macrobrachium* larvae vary markedly from hatchery to hatchery, so much so that it was difficult to make comparisons on the basis of tank shape, size, and construction material. Tanks used in hatcheries include cylindrical fiberglass tanks with conical bottoms; flat-bottom cylindrical tanks of fiberglass, concrete, and plastic; and rectangular tanks of concrete ("cement"), fiberglass, plexiglass and wood (sealed with resin or fiberglass). The cylindrical tank with a conical bottom appears to be the most efficient and preferred design, and it may be worth noting that one of the highest production levels reported (62 postlarvae/liter — AQUACOP) was obtained in a tank of this type. Also, personal communications from researchers (Randy Hagood, Caribe King Shrimp, Inc.; Jean-Michel Griessinger, AQUACOP) as well as studies carried out in our laboratory suggest that better production results are obtained in tanks with dark rather than light interiors. However, definitive data on this point are lacking.

Sources of Seawater

Ten respondents obtained their saltwater from wells, while seven others used natural seawater from adjacent ocean or lagoon areas or unspecified sources. Only four operators use synthetic seawater exclusively, but two others reported using it to supplement supplies of natural seawater. The most widely used synthetic seawater formulation is "Instant Ocean" (Aquarium Systems, Inc., Eastlake, Ohio 44094).

Production Capacity

Average production values reported by respondents ranged from 7-41 postlarvae/liter of tank capacity; maximum levels were 62/liter in Tahiti (AQUACOP) and 64/liter in South Carolina (Manzi and Maddox, 1976). Estimates of yearly production capacity provided by 13 respondents ranged from 0.1-8 million postlarvae, with most operations capable of producing $\geq$ 0.75 million. For purposes of comparison, hatcheries with an estimated production capacity of $\geq$ 1 million postlarvae/year were defined as production-scale while those with a yearly production capacity of $<$1 million postlarvae were termed research-scale. On this basis, research-scale hatcheries appear to be producing about 30% more postlarvae/unit of tank capacity than are production systems (Table IV-1). This difference may be related in part to scale factors but probably also reflects differences in rearing technology. The survey data further indicate that public hatcheries are producing nearly 51% more postlarvae/unit of tank capacity than private operations (Table IV-1). This pronounced difference cannot be attributed entirely or even in large part to scale factors, since some of the largest as well as the smallest operations surveyed are in the public sector. Thus, it appears that an excellent opportunity exists to improve commercial yields through the extension of information from public institutions to the private sector.

Cost of Postlarvae

The question, "What price do you consider reasonable to pay for postlarvae?", was included to obtain an indication of the range of estimated production costs. Unfortunately, this question was inadvertently omitted from several questionnaires, so coverage was not as broad as we would have preferred. Nonetheless, responses ranging from ~\$2-50/thousand prawns were obtained, with operators of research- and pilot-scale hatcheries generally willing to pay higher prices (≥ \$15/thousand) than managers of production units (≤ \$10/thousand). This disparity likely reflects the differences in scales of operations and purposes of the various hatchery units.

Status of Hatchery Research and Development

The commercial production of postlarvae has been sustained at a relatively high level, indicating that, while larval culture technology is still far from being optimized, it is adequate to allow development of commercial ventures. However, there still appears to be much variation in production from operation to operation, and the production efficiency (i.e., number of postlarvae produced/unit of tank capacity) is still rather low. Improvement in both total production and efficiency is needed.

Pilot-scale production of postlarvae in a recirculating synthetic seawater hatchery has been demonstrated. Several hundred thousand juveniles have been produced in the experimental systems at MRRI, thus providing a new technology which may be applicable to commercial operations. This type of hatchery system may be especially useful in temperate climates and where access to natural saltwaters is geographically or economically limiting.

Work by Manzi and Maddox at MRRI has demonstrated that addition of high densities of monospecific algal cultures, especially of *Phaeodactylum tricornutum*, has marked beneficial effects on survival and growth of *M. rosenbergii* larvae (Maddox and Manzi, 1976; Manzi and Maddox, 1976). These investigators have obtained production levels as high as 64 postlarvae/liter of tank capacity, with 83% average survival, in a small research hatchery. However, mechanisms for the beneficial effect of the algae are not clear. Work in our laboratory indicates that, while algal cells are frequently ingested by larvae, there is no evidence to suggest that larvae derive any direct nutritive value from algae. Similarly, in studies with ^{14}C-labeled Tetraselmis and Phaeodactylum, Cohen et al. (1976) found that the algae were not assimilated by M. rosenbergii larvae. They concluded that algae enhance the growth of prawn larvae only indirectly by removing toxic substances, notably ammonia, from the rearing medium and that where other means of waste product removal are utilized, the algae have no effect and may be eliminated. In contrast, work at MRRI has failed to demonstrate any correlation between the presence of algae and the levels of nitrate, nitrite, and ammonia in the rearing water (Maddox and Manzi, 1976; Manzi and Maddox, 1976). The possibility that the algae provided indirect nutrition for the prawn larvae by serving as food for the Artemia nauplii was also investigated by Maddox and Manzi (1976). However, they found that M. rosenbergii larvae appear to receive more nutritive value from freshly hatched Artemia nauplii than from older, algae-fed nauplii. Another possibility is that the algae add a growth factor to the culture water or act as a bactericidal agent. Certain algae are also known to produce substances which can cause a drag reduction in water similar to that produced with synthetic long-chain polymers (Hoyt, 1966). At least one of these synthetic polymers, polyethylene oxide, has been shown to beneficially affect survival of other caridean shrimp larvae (Palaemonetes spp.) in an unknown manner (Sandifer et al., 1975).

Some progress has been made in the development of formula feeds to supplement and ultimately replace Artemia in the culture of larval Macrobrachium. The most significant such work reported to date was conducted at the Skidaway Institute of Oceanography in Georgia (Sick and Beaty, 1974). Unfortunately, this research has now been terminated. At present, experiments to examine different types of natural and prepared foods as supplements and substitutes for Artemia are being undertaken at MRRI and in laboratories of some private firms. Hopefully this work will rapidly lead to the identification or development of other satisfactory larval rations.

Perhaps the most exciting advances in hatchery research have been made by the AQUACOP team working on Tahiti in French Polynesia. These scientists have produced over 60 postlarvae/liter in 800-liter conical-bottom tanks using a culture technique which involves periodically exchanged clear water (i.e., no algae), heavy aeration, and the addition of bactericides.

Major Problem Areas[a]

Feeds

The dependence of Macrobrachium hatcheries on the expensive and occasionally unobtainable cysts of Artemia salina is perhaps the primary

[a] Identified by the authors.

problem facing hatchery operators. In the United States, the supply of *Artemia* cysts appears to be critically low and may be a major bottleneck constricting development of commercial hatcheries. Sorgeloos (1976) pointed out that while brine shrimp occur throughout the world and over 50 strains have been identified, they are exploited in only a few places, primarily in the United States and Canada. He suggested that the critical supply situation could be alleviated by a combination of factors, including the application of new harvesting techniques, the selection and transplantation of strains to increase the exploitable stocks, improvements in the hatching and utilization of *Artemia* in hatcheries, and the development of intensive culture techniques for brine shrimp. For the time being, however, a definite and critical shortage of *Artemia* exists in the United States. Strong research and development efforts are needed immediately so that this limiting factor can be overcome as rapidly as possible. Such work might involve the implementation and demonstration of better *Artemia* utilization techniques, coupled with research to develop other natural and formula foods. In this latter regard, Dr. V. G. Jhingran recently reported[a] that chironomids have been used to replace *Artemia* entirely as a food for *Macrobrachium* larvae in India.

Rearing Cycle

Because a shorter larval cycle should result in reduced operating costs and increased annual production, efforts to reduce the time required to complete a larval rearing cycle should be intensified. Areas for research include environmental factors, nutrition, tank design and flow characteristics, hatchery management techniques, and selective breeding.

Production Efficiency

More emphasis should be placed on increasing the average production of postlarvae per unit of tank capacity. This should result in improved feed utilization and reduced production costs. The overall average efficiency from the survey was only about 18 postlarvae/liter, suggesting considerable room for improvement. Also, research should be undertaken to identify and remove the causes of the sporadic but fairly common failures of individual larval cultures.

Information Exchange

The two-way flow of information between research and commercial hatchery operations must be maintained and improved. There appears to be a pressing need for extension and demonstration efforts to translate research results to commercially usable technology and for operators of commercial prawn farms to identify critical areas that require research.

---

[a]Comment made from the floor during the FAO Technical Conference on Aquaculture, 26 May-2 June 1976, Kyoto, Japan.

Acknowledgements

We are grateful to the "prawn people" whose cooperation made preparation of this review possible. For supplying information, we thank the Agricultural Engineering Institute of the University of Hawaii; the members of AQUACOP (CNEXO, Tahiti, French Polynesia); Mr. Mark Berrigan (Marine Research Laboratory, St. Petersburg, Florida); Dr. Robert Brick (Texas A&M University, College Station, Texas); Mr. Howard Deese (Micronesian Mariculture Demonstration Center, Palau, Trust Territory of the Pacific Islands); Dr. Sheldon Dobkin (Florida Atlantic University, Boca Raton, Florida); Mr. Durwood Dugger (Aquaprawns, Inc., Port Isabel, Texas); ELSHAR Aquaculture Development Services (Coral Gables, Florida); Mr. Takuji Fujimura (Anuenue Fisheries Research Center, Honolulu, Hawaii); Mr. Eng-wah Khoo (Ocean Land Sdn. Bhd., Selangor, Malaysia); Dr. Allen Knight (University of California, Davis, California); the Oceanic Institute (Waimanalo, Hawaii); Mr. John Parker (Caribe King Shrimp, Inc., Aguada, Puerto Rico); Dr. Dieter Pukatzki (Pioneer Aquaculture of Hawaii, Lahaina, Hawaii); Syntex (U.S.A.), Inc. (Palo Alto, California); Trenton State College Shrimp Aquaculture Project (Trenton, New Jersey); Weyerhaeuser Co. (Homestead, Florida), Mr. J. F. Wickins (Ministry of Agriculture, Fisheries and Food, United Kingdom); Mr. Scott Willis (Marine Research Laboratory, St. Petersburg, Florida); and several others. We also thank members of the MRRI staff, especially Dr. John Manzi and Mr. Mark Maddox, for their assistance.

B. Summary of Discussion of Juvenile Production

Michel Cuzon reported on AQUACOP's larval production methods in Tahiti, where larvae are grown in 100-liter tanks with a green-coated interior, with water at 12 ‰ salinity. The brackish water is mixed and stored in fiberglass, 10-cubic-meter tanks for 24 hours before using. The stored water is filtered continually through a sand filter. Because the storage tank is in the hatchery, it is at the same temperature as the larvae tanks, about 27° C.

Larvae are stocked initially at 100 per liter. The water is changed completely every day without handling the larvae; not handling the larvae and uniformity of water temperature prevent stress. Tetracyclene is added to the water as a prophylactic.

Brine shrimp nauplii are fed during the night at a rate of 5 per cc. On the following day, bonito meat and frozen *Artemia* are fed. The strong aeration is increased in the afternoon to keep the tuna and brine shrimp meat in suspension, especially when the larvae reach stages 7 and 8. The operation is in clear water; experiments in larval production in green water were not successful. Success under the present method is production, beginning at 28 days, of 61 postlarvae per liter, a survival rate of 64.4%. Three hundred eighty thousand PL's currently are produced in 800 liters.

Michael Fujimoto stated that no significant changes have been made in the technique of larval culture at the Anuenue Fisheries Research Center in Honolulu. The next change will be to expand from the present three production cycles during the summer to PL production year round through the addition of temperature-control equipment. The Hawaiian center currently operates at ambient temperatures only.

Sandifer's comments on the price of juveniles stimulated discussion. Fujimoto said that the Anuenue Center would sell PL's to participating prawn farmers at $5.00 per thousand, this amount representing essentially operating costs, with about 30% the cost of labor. Spencer Malecha noted that prawn farmers in Taiwan and Thailand pay from $50 to $80 per thousand, these amounts representing hatchery costs. Durwood Dugger calculated Aquaprawn's costs at from $50 to $70 per thousand, all costs figured in.

On the subject of larval nutrition, current dependence on brine shrimp was noted. Shleser pointed out that we're not really talking about brine shrimp, but the nauplius stage when the animal still carries an egg yolk; he referred to work by Persoone and others which demonstrated that the materials that make brine shrimp nauplii so valuable are in the yolk, not in the whole animal. George Piggot stated that he had 10,000 larvae which had survived for six weeks on dried krill flakes; the krill are the small Pacific species obtained in the vicinity of Seattle. Piggot also said that he did not believe juvenile production was a problem, that a real multidisciplinary approach would result in the best artificial food and technique as such an approach has done for other animals.

ADDENDUM

Macrobrachium HATCHERY INFORMATION

1. Name: ______________________________

2. Title or Position: ______________________________

3. Organization: ______________________________

4. Address: ______________________________

____

5. Does your organization operate a hatchery for Macrobrachium prawns?

_____ Yes _____ No

6. Are you working with any species other than *M. rosenbergii*? If so, please name and give the following information for each species.

7. Is your hatchery system a:

_____ continuous flow-through system?

_____ static system with periodic exchanges of water?

_____ recirculating system with major water reuse?

_____ none of the above. Describe briefly: ______________________________

____

____

8. What conditions of salinity and temperature do you maintain?

_____ o/oo S _____ ° C

9. Do your culture practices involve the use of algae? ____yes; ____ no. If yes, please give whatever details you can (species composition if known, "green water", cell densities, renewal frequency, etc. ______________________________

____

____

10. Is Artemia the major feed used in your hatchery? ____yes; ____no. If no, what is used? ______________________________

11. At what densities do you feed Artemia (i.e., 10 nauplii/ml) and is this changed during larval development? ______________________________

12. What feeds, if any, do you use as supplements to the Artemia nauplii?

____

13. Can you quantify the amounts of these feeds used? ______________________________

____

14. How big is your hatchery (number and sizes of tanks, total rearing volume)?

____

15 What kinds of tanks do you use (shape, material, etc.)? ______________________________

____

16. What is your average production in number of postlarvae per liter (or per gallon)? ______________. What is the maximum yield you have achieved in numbers of postlarvae per liter or gallon? ______________________________
What is your estimate of your hatchery's total production capacity per year in numbers of postlarvae? ______________________________

17. What is the source of your salt water supplies? ______________________________

____

18. Do you wish to be identified as a source of information? ___yes - by name; ____yes - by organization name; ____yes - but only as "industrial representative" or "researcher"; ____no.

19. Would your organization produce its own postlarvae if there were another reasonable supply available? ______________________________

20. What price do you consider reasonable to pay for postlarvae? ______________

____

Thank you.

CHAPTER V
ADVANCES IN PRAWN SYSTEMS ENGINEERING

Paul Zielinski first reviewed recent advances made by the work in which he is involved at Clemson University and elsewhere, and then several of the prawn people commented on their own experiences.

A. Advances in Prawn Systems Engineering:
Paul B. Zielinski, Clemson University

Rectangular larval culture tank studies have been completed and some of the results have been reported (Zielinski, Castro, Sandifer, 1976). The best conditions for maintaining all food particles in suspension and permitting controlled small deposits occurred in a 3.7m x 1.2m x 1.2m tank when a single inlet manifold was used. The three, four, or five inlet openings were one-half inch in diameter, equally spaced across the manifold inlet with the holes oriented parallel to the sloped floor. Tank streamlining in the corners showed a definite advantage. Depending upon the weight and size of particles of the food used, it required several tank volumes of flow to retain the majority of particles in suspension and to permit deposits in a small compact area.

Cylindrical tanks have also been under investigation. The flow patterns appear to be more sensitive to the type of inflow and outflow conditions used. By withdrawing flow at the top center of the tank through a 1 1/4-inch pipe and reintroducing it at the outside wall surface tangentially, a better resuspension of the material occurs than if the outlet structure is enlarged in diameter. Figures V-1 and V-2 show the comparative test results. The purpose of utilizing flow circulation is to distribute the food, larvae, and oxygen throughout the tank volume and maintain uniform temperatures, but permit settlement of food particles in a small area for easy removal by siphoning.

Airlift pump studies, conducted at Clemson University, have been conducted in two series: (1) the hydraulic performance data and (2) the quantity

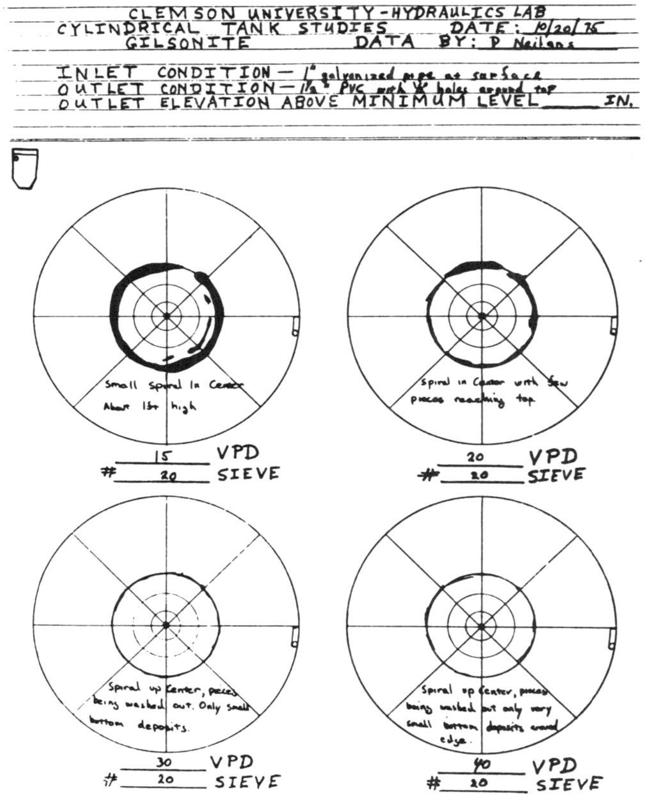

Figure V-1
Clemson University Cylindrical Tank Studies, 10/20/75.

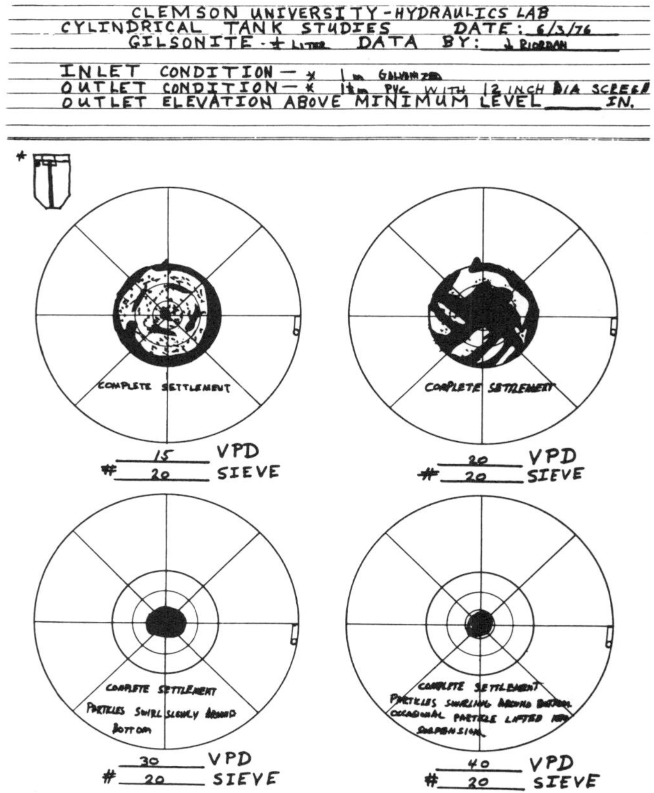

Figure V-2
Clemson University Cylindrical Tank Studies, 6/3/76.

of oxidation induced into flow by airlift pumping. The hydraulic performance was presented partially in a paper (Castro, Zielinski, Sandifer, 1975), while the oxidation studies are in progress and nearing completion. Maximum oxidation of water occurs at optimum hydraulic performance and at approximately 70 percent submergence, while some oxidation is evident for any pump condition. One subtle point that was observed is that the type of airlift pump footpiece apparently does not make a difference on hydraulic performance, but aeration is enhanced by introducing a distribution of fine air bubbles into the footpiece.

Studies of culture water treatment are also in progress at Clemson University. Tests of a five-foot-diameter Swirl Sedimentation chamber have been conducted using Petrothene particles (spec. gravity 1.01 and no. 20 sieve) in flow rates from 3.89 lb. per sec. to 10.5 lb. per sec. These flows converted to a daily basis ranged from 40,000 to 100,000 gallons per day. It was determined that the hydraulic efficiency was excellent, and that the sedimentation efficiency was greater than 90% for the full range of flows.

Additional work by an environmental engineer on the determination of the effectiveness of the biodisc to remove nutrient materials from culture water is also in progress. The nutrient removal capability of the biodiscs has been reported elsewhere in regard to other types of wastes. Design criteria for animal culture wastes is now underway with a small biodisc unit. Devices known as protein skimmers have also been used for nutrient removals in culture wastes.

All devices tested or developed in our program are not considered complete until they have been thoroughly tested in culture systems by biologists. An intensive program of testing all accomplishments to date will be initiated in the near future.

In other areas, the possibility of using waste heat from steam-powered generating stations has long intrigued culturists. In one of these facilities operated by Public Service Electric and Gas Company at Trenton, New Jersey, a trial Macrobrachium growth pond is under investigation. According to Al Eble, potential culturists should consider the following conditions of operation as prerequisite before considering utilizing waste heat in power plant effluents. These conditions are that the temperature differential (Δ T) should be in the range of 10-12° C; the power plant should have at least three units so that at least one is continually generating; cleanup of stream water should be done mechanically rather than chemically, if possible; and finally there should be three sources of water: the river water at ambient temperature, the heated effluent, and a quality source of well water to be used particularly for cooling purposes.

Future efforts by engineers will cover such areas as the problem of low oxidation levels in ponds, planning and building commercial-scale hatcheries and total culture facilities, and refining intensive culture systems for retaining large quantities of broodstocks. Since culture operations require energy for heating, water movement, and oxidation, an effort will be made to review potential inexpensive energy devices such as solar radiation and others.

Materials for use in culture have been a problem; this is an area where more work needs to be done. Most of the problems related to materials have been identified, such as metallic toxicity, and culturists have learned to work around the problem, but the solutions have not been the ultimate obtainable. Improved solutions in the form of nontoxic materials may also improve the cost effectiveness in such functions as heating the larval culture tank water. Optimum sizes of larval tanks, hatcheries, automatic feeders, broodstocks, juvenile retention basins, and workable pond sizes remain as questions to be answered.

B. Summary of Discussion: Advances in Prawn Systems Engineering

Much of the discussion centered on the use of power plant heated effluent for prawn culture. Elaborating on points in the Zielinski paper, Albert Eble noted that the Delta T of 6 to 8 degrees available to culturists at the Mercer power plant near Trenton, New Jersey, effectively produced the same culture environment as that of Charleston, South Carolina. If only ambient river water is used in New Jersey, prawn juveniles can be stocked the first week in June and harvested the second week of September. But with the available power plant heat, prawns are stocked in mid-May and not harvested until the last week of October; in short, the heat means almost a solid five months of grow-out as compared with 3.5 months at ambient temperatures.

Bruce Godfriaux urged southern growers to seek out power plants, pointing out that, of the 800 power stations producing more than 400 megawatts, there must be a number in the right locations for turning southern climates into tropical environments for prawn grow-out.

On the question of expanding aquacultural use of power plants, Goodwin commented that his discussion with power company officials indicated that the companies are interested in aquaculture for public relations purposes, and only then if the costs are modest and there is no interference with the primary mission of selling power. Godfriaux added that a 1,000-megawatt nuclear plant costs over a billion dollars, and to add an aquaculture enterprise is of such relative minor importance with its one- or two-million-dollar costs that companies may not take the potential seriously, unless their boards of directors can be persuaded of aquaculture's importance. He added that there is still a lot of potential in existing power plants where aquaculture enterprises could be added without much interference. As for modifying condenser systems and other elements,

Godfriaux stated that no aquaculture activity could pay back the costs of such modification and still be viable.

John Bennett made a plea to engineers working on above-ground systems to keep a sharp eye on cost-economics. He reviewed the Syntex experience where the goal was to develop a sophisticated, intensive above-ground system to minimize the cost of capital works. Computer analysis of the economics showed that the cost of an above-ground system must be based on a container that holds 20,000 gallons of water, and that for successful economics, the container should not cost more than $1,500. To produce 100,000 pounds of prawns per acre in such a sophisticated system would mean 90 of the 20,000-gallon tanks and a cost of $300,000. Syntex concluded that it is not yet practical to plan for an intensive, high-technology, above-ground system.

CHAPTER VI
POND GROW-OUT SYSTEMS

The State of Hawaii was first in the Western Hemisphere to achieve commercially successful culture of the freshwater prawn. This was due partly to the tropical marine climate of our island state and mostly to the achievements of the State's Anuenue Fisheries Research Center under the leadership of one Takuji Fujimura. There are other supporting factors which might be cited, of course, but the key credits must go to Fuji and the cooperative State/private enterprise plan which came into being after the life cycle of the animal was brought under control and postlarvae became available in commercial quantities.

The cooperative approach employed in Hawaii supports prawn farmers with free postlarvae and expert advice from the Anuenue center during grow-out. In return, the Center retains a degree of control over the grow-out process and is able to gather data which lead to refinements in the grow-out technology and methodology. It is by no means an understatement to say that every person actively engaged in prawn farming in Hawaii is a contributing and benefiting member of the statewide prawn-farming team. This is a model other states might consider following. It does work.

The following paper (titled by its authors, by the way) summarizes the grow-out technology which has evolved through the cooperative program of the State of Hawaii led by Fuji and the Anuenue Fisheries Research Center.

A. An Idiot's Guide to Prawn Ponds:
Michael Fujimoto, Takuji Fujimura, and Ken Kato

Introduction

The basic grow-out system for Macrobrachium rosenbergii in Hawaii has been the same for the last five years. We use unlined earthen ponds with a sluice gate outflow system. Water quality of the ponds is determined by Secchi disc visibility and DO meter, and controlled by altering the flush rate, algicides, and oxidizing agents. The farmer feeds approximately 30 pounds of poultry broiler starter or catfish chow per acre per day.

Physical Requirements of Farm Site

Physical properties considered when selecting a prawn farm site are essentially the same as those considered in selecting a site for a fish farm in any locality. Water quality and quantity, soil type, and topography are among the most important considerations.

In Hawaii, average annual waterflow requirements generally range from a minimum of 15 to a maximum of 20 gpm per surface acre. This flow rate is increased as much as threefold to flush the ponds of excess phytoplankton, and is decreased when phytoplankton densities fall. Ponds are also flushed at maximum flow rate if the behavior of the animals indicates stress (i.e., large numbers of prawns circling the perimeter of the pond, or gathering at the intake) and flushing is continued until the stress behavior subsides.

Chemical parameters of the soil and water required for grow-out ponds are not yet clearly defined. Addendum A lists chemical properties of water supplies and pond water from some of our farms. Of those investigated, only the Kihei farm on Maui experienced difficulties in growth and survival. The problems were attributed to the high mineral content of the pond water, probably the result of leaching from the soil; however, this assumption requires verification through further research.

So far, only the effects of $CaCO_3$ on growth and survival have been investigated. Preliminary results indicate that high CO_3 content can retard growth and may influence molting frequency (see Addendum B).

The basic Hawaiian prawn pond was designed by considering two important factors: (1) economics of the grow-out system and (2) behavior of prawns. Behavioral characteristics determine conditions required for fast growth and good survival. Economics, in terms of capital amortization and operational cost determine whether the "preferred" conditions, based on behavior should or should not be applied. In Hawaii, prawns are harvested by seining the ponds; and since harvesting is the most costly activity in prawn farming, harvestability is given prime consideration in pond design. Preferred conditions are considered as long as they do not raise operating costs and negatively influence the profit margin.

The following briefly outlines some of the important requirements of a Hawaiian prawn pond.

1. The pond bottom is almost perfectly flat and well compacted. This feature facilitates harvesting with a net but contradicts prawn behavioral preference--prawns prefer crevices and irregularities and survival is known to be better when such features are provided.

2. Water depth is three to four feet. This depth is believed to prevent much light from penetrating to the bottom at moderate phytoplankton densities, thus preventing benthic algae or aquatic weed growth, which is known to be beneficial for growth and survival of prawns but interferes with harvesting.

3. The pond berm is 1 to 1.5 feet above the water level and 10 to 12 feet wide. It is high and wide enough to remain dry, and support normal traffic under all weather conditions. Higher freeboard merely adds to labor costs; for example, more distance must be covered to go in and out of the pond, and more grass must be cut.

4. The outflow system (sluice gate) allows for rapid discharge of water, easy water level regulation, and minimal maintenance.

5. Vegetation along the pond's bank minimizes bank erosion, provides shelter and food for the prawns, and habitat for forage organisms. No quantitative or qualitative analysis of the direct or indirect (through the food chain) benefits of bank foliage has been done yet; however there is evidence indicating that it is impossible to raise prawns at the density and production rates presently realized in earthen ponds on a chicken feed diet. For example, in our nonvegetated fiberglass tanks, growth is slow, survival poor, and feed conversion is low. Selection of "proper" vegetation is difficult because, although the presence of vegetation on the berm and at the water's edge benefits productivity and minimizes erosion, it hinders harvesting and requires continuous maintenance. In Hawaii, *Paspalumn* is commonly used as a berm cover, and thus far it appears to fulfill the difficult requirements.

All Hawaiian prawn ponds are stocked with gambusia prior to stocking with PL's. This is done to control the initial bloom of predatory dragonfly nymphs, and later the fish become prey for the prawns and a stable fish population is maintained.

Stocking, Feeding and Water Regulation

The stocking rates utilized at present are between 1.5 and 2 PL's per square foot annually. The method of enumerating PL's stocked basically involves making a "visual standard" by counting out 2,000 PL's and placing them in a bucket with three liters of water. Identical buckets are then filled with three liters of water and placed adjacent to the standard. An "estimator" then adds PL's to the buckets of water until he visually estimates that they contain the same number of PL's as his standard. The estimated buckets are then emptied into the shipping containers and refilled until the required number of estimated buckets are filled.

The prawns are then fed a daily ration of poultry broiler starter or catfish chow. In the beginning, the farmers are instructed to feed 1/2 gallon of chow per acre per day, and to increase feeding if the food disappears within 24 hours, or decrease if there are leftovers. This feeding regime is mainly to build up the productivity of the pond, and the bulk of the feed in the initial stages of rearing is consumed by the gambusia. The feeding rate is stabilized at 25 to 40 pounds of feed per acre daily when in full production. The difference in feeding rates between farms, and even between ponds within a farm, is attributed to differences in natural productivity between ponds, and harvest rates.

Waterflow is adjusted according to phytoplankton density. Flow is increased if the pond has a Secchi disc visibility of 25 cm or less and decreased if it is 40 cm or more. In the case of Pacific Aquaculture Corporation's Laie farm and Ota's ponds, this average annual flow is about 20 gpm/acre. Addendum C shows levels of some water-quality parameters resulting from this waterflow scheme in one of our ponds.

Harvesting rate is used to manipulate prawn biomass for maximum production. In essence, the scheme is to allow the biomass to increase up to a critical level (the point before prawn interaction is noticed); then marketable prawns are harvested. Ideally, total prawn biomass is maintained at maximum levels at all times. Difficulties are encountered though, since the critical biomass varies between ponds, locality, and age of the population; and because harvesting represents the largest fraction of the labor cost, economic rather than biological factors are considered first on a commercial farm. Farmers in Hawaii seine an entire pond at three- or four-week intervals, or one-half of a pond at two-week intervals, thus maintaining prawn biomass below a critical level.

The prawns are harvested by carefully moving a 2" mesh seine through the pond. The seine is specifically designed to harvest prawns and it differs significantly from seines used for harvesting finfishes. The minimum size caught by this seine is 1.5 to 11 cm (eye to tip of telson). Addendum D gives details on the seine.

B. Summary of Discussion: Grow-out

Responding to a series of questions, Kenneth Kato added detail on pond construction and operation. Ponds at Fish Farms Hawaii range from a quarter acre to about three and a half acres. Kato prefers larger and deeper ponds with the ponds deep enough to keep down growth of filamentous algae. Plankton blooms are maintained primarily for shading purposes and to keep down filamentous algae, although plankton also contributes to the food chain to some unquantified degree. The larger, deeper ponds are better able to buffer change; e.g., overfeeding becomes less of a hazard because of greater water volume.

Fish Farms Hawaii stocks only once a year in the summer months. After the initial growing period, the ponds are harvested on a sustained basis, at about three-week intervals per pond, with seines sized to take only marketable (6-8 count) prawns. It takes four men four hours to harvest a 3.5-acre pond, with the harvest ranging from 400 to 600 pounds of prawns.

Pond sides are steeper than for catfish ponds, with about a 2:1 slope and grass planted to reduce erosion. At harvest, the pond level is lowered below grass level. Floating plants were tried by Fish Farms Hawaii, but they become a serious nuisance at harvest time. Waterflow rate is about 15 gallons per minute per acre and is increased to about 60 gpm for flushing purposes.

As for production levels, Kato stated that the once-yearly stocking and sustained harvest method results in dips in production, but over any 12-month period, production averages out to about 3,000 pounds per acre. If production is measured only for that period after stocking when the main population exists and productivity is highest, a 12-month extrapolation would be more than 4,000 pounds. Therefore, given year-round growing conditions and a steady supply of juveniles, two tons of prawns per acre/year would be a minimum expectation.

However, Goodwin noted that no place in the United States has ideal growing conditions, with Puerto Rico having the best year-round temperatures. Nevertheless, substantial geographic areas have a growing season that can be utilized to grow prawns. How to make the best use of the growing season is the object of research at the South Carolina Marine Resources Research Institute at Charleston.

Theodore Smith of MRRI described the intensive hatchery and juvenile-rearing system, using recirculated, artificial seawater, to produce juveniles during the cold months for stocking when the weather warms. Stocking with juveniles as small as .25 grams can result in six months in a harvest of 2,000 pounds per acre.

The South Carolina team has learned that prawns are hardier under poor environmental conditions than at first expected. Prawns stocked under mean conditions of water hardness as low as 5 mg per liter, a DO of 1.7 mg per liter, and a pH of 5.5 survived and grew.

Larvae stocked in South Carolina ponds had a survival rate of from 76 to 81% although harvestable biomass was naturally much less than when larger juveniles were stocked.

Brackishwater stocking also was tried with significant success. Salinity was 7.2 $^{o}/oo$. Some males in the brackish pond grew from less than a gram to over 100 grams in five months.

Sandifer, Smith, et al. have conducted a number of experiments with substrates. Smith reported that the "edge effect" was so pronounced that from 70 to 80% of the prawns were on the substrate edges. Finally, the team used strips, with plants overhead, which resulted in relatively even distribution of the prawns throughout the tank. If no plants were used, the prawns always preferred the lower areas.

The South Carolina ponds were harvested only once, at the end of the growing season. The ponds were drained, and the prawns collected in a sump. The degree of success in moving prawns to the collecting point is in proportion to the smoothness and regularity of the pond bottom. The harvest produces prawns of assorted sizes, posing a marketing problem. Presumably, with enough volume, sorting would result in several size classes that could be sold in accordance with normal consumer preferences.

ADDENDUM A

ANALYSIS OF WATER AND SOIL EXTRACT FROM FIVE *MACROBRACHIUM* PONDS IN HAWAII

The samples were collected by Mr. Michael Cripps, Department of Animal Science, University of Hawaii, and analyzed by the Hawaiian Sugar Planters Association.

Well water samples were collected at the inflow; samples for soil extract were collected from the pond bank in a section thought to be representative of the area.

Table VI-1
Analysis of Water Samples: Macrobrachium Study. Samples received 8/1/74.

Lab No.	pH	NH_4	Na	K	Ca	Mg	Cl	NO_3	SO_4	PO_4	HCO_3	SiO_2
		ppm										
5921 Laie - Fish Farms Hawaii, No. 3 Pond	7.8	.017	31.4	1.0	8.8	11.2	40	.39	8	.25	85	51
5922 Yee - Wadsworth Yee's Pond, Haleiwa. No. 1 Pond	7.3	.023	74.0	3.3	6.3	13.8	97	1.72	16	.57	95	77
5923 Ota - Punaluu No. 1 Pond	8.0	.017	22.9	1.7	12.0	10.0	55	.28	7	.12	49	34
5924 Woods - Ewa Beach No. 1 Pond	7.2	.023	74.0	2.5	26.0	32.0	208	1.11	63	.32	85	64
6294 Kihei, Maui Spld. 8/6/74; rec'd. 8/7/74	8.0	.088	368.0	18.3	58.8	44.4	472	2.67	33	.14	479	44
Analysis of Soil Extracts[a]												
1096 Laie - Fish Farms Hawaii, No. 3 Pond	5.2	.29	9.9	1.6	40.0	1.1	8	.11	25	.03	0	5.8
1097 Wadsworth Yee's Pond - Haleiwa No. 1 Pond	3.2	14.0	7.2	5.3	17.0	25.8	20	.83	1080	.03	0	25.4
1098 Ota - Punaluu No. 1 Pond	8.2	.31	11.5	0.2	15.0	3.4	3	.17	7	.06	67.0	15.1
1099 Woods - Ewa Beach No. Pond	8.3	.23	34.6	15.7	13.0	6.2	44	.28	25	1.56	70.0	20.5
1135 Kihei, Maui	7.9	.35	3675.0	177.5	950.0	481.3	7250	5.45	1908	.28	34.0	17.3

[a]A mixture of part of 2 mm screened air-dry soil and 10 parts of water shaken for 30 minutes, allowed to stand overnight, shaken again for 30 minutes and filtered.

ADDENDUM B

THE EFFECT OF $CaCO_3$ IN THE WATER ON GROWTH OF *MACROBRACHIUM ROSENBERGII*

Michael C. Cripps and R. M. Nakamura

Macrobrachium rosenbergii were raised in water containing varying concentrations of $CaCO_3$ in three trials. The durations of the three trials were 86, 27, and 90 days, respectively, and prawns 5, 13, and 14 months old, respectively, were used.

Statistically significant reductions in average daily gains of weight and length were seen with increasing concentrations of $CaCO_3$ in the water. This inversely linear relationship between average daily gains and concentrations of $CaCO_3$ resulted, at times, in a five-fold reduction in weight gains when comparing the low concentration (65 ppm) to the highest concentration of $CaCO_3$ (500 ppm) used. Average daily weight gains were more sensitive than average daily gains in length in detecting changes in growth rate.

A trend toward increasing molt interval was seen with increasing concentrations of $CaCO_3$ in the water; however, these observations were not statistically significant, probably due to the relatively short durations of the three trials.

In hematological tests of prawns in water at varying concentrations of $CaCO_3$, it was found that:

1. Low $CaCO_3$ groups had lower blood calcium than prawns in high $CaCO_3$ at 5 days after molting, but the difference was not seen at 15 days after molting.

2. Higher levels of inorganic phosphorus were seen in prawns in lower concentrations of $CaCO_3$ but this finding was not consistent.

3. High $CaCO_3$ groups had lower refractive indices.

4. No significant differences were seen for the other blood tests: sodium, chloride, and potassium tests.

These results suggest that higher concentrations of $CaCO_3$ in water inhibits prawn growth, and $CaCO_3$ concentrations of the water should be one of the criteria considered in pond site selection.

Table VI-2
Average Daily Gains in Length and Weight of *M. rosenbergii* at Various Concentrations of Calcium Carbonate in Growing Water.

$CaCO_3$	Average Daily Gain, mg/day			Average Daily Gain, mm/day		
(ppm)	Trial 1[a]	Trial 2	Trial 3	Trial 1	Trial 2	Trial 3
65	2.05	1.64	4.67	0.091	0.038	0.056
100	0.93			0.022		
200	1.16			0.052	0.037	0.035
250		1.10	3.17			
300	0.68			0.018		
500	0.40	0.46	2.22	0.013	0.012	0.025

[a]Trial 1, 2, and 3 using prawns at age 5, 13, and 14 months old, respectively, and duration of trial of 86, 27, and 90 days, respectively.

ADDENDUM C

A SEVEN-MONTH STUDY OF 24-HOUR FLUCTUATIONS OF SELECTED WATER-QUALITY PARAMETERS OF A PRAWN (MACROBRACHIUM ROSENBERGII) REARING POND, OAHU, HAWAII

Thomas Iwai, Jr.

A series of seven monthly 24-hour water-quality measurements of biotic and abiotic parameters of a M. rosenbergii pond (0.73 hectare) were taken from September 1974 to March 1975. Measurements for temperature, dissolved oxygen, pH, total alkalinity, and nutrient analysis (i.e., NO_3-N, NH_4-N, PO_4, Si, etc.) were collected every four hours, over a 24-hour sampling period. Daily fluctuations of chemical oxygen demand (COD) were estimated from chlorophyll levels (Boyd, 1973).

Daily fluctuations of both temperature and dissolved oxygen were observed, with the peak temperature and dissolved oxygen readings generally occurring at 1700. Water temperature ranged from 19.8 to 27.8° C, while dissolved oxygen levels ranged from 5.4 to 11.6 mg/l. Lower oxygen readings were repeatedly observed during the early morning hours.

Peak chlorophyll-a pigment levels were generally observed at 1300, four hours preceding peak dissolved oxygen readings. Chlorophyll-a pigment levels ranged from 19.2 to 170.8 μg/l. Estimated COD values also fluctuated over 24 hours, with peak readings generally observed at 1300. COD values ranged from 24.5 to 73.4 mg/l. The reported correlation of chlorophyll-a pigment levels to the COD (Boyd, 1973; Leary and Iwai, 1974) also suggests the important contribution of phytoplankton to the organic load of a pond system.

Total alkalinity ranged from 34.3 to 76.1 mg $CaCO_3$/l, well within the range of 25 to 100 mg $CaCO_3$/l reported typical of waters of moderate productivity (Shaperclause, quoted in Hickling, 1971).

HCO_3^- form of carbon was dominant throughout the study. Diel patterns of CO_2 utilization were also apparent for all the sample periods. Peak CO_2 utilization (1700) took place approximately four hours after the peak chlorophyll levels were observed (1300). CO_2 readings were also inversely related to oxygen readings.

pH values ranged from 7.21 to 8.84 for all the sampling periods, well within the desired pH levels for pondfish production (Swingle, 1961). pH values were lowest in the early morning (0500), while peak levels were repeatedly highest at 1700, probably due to the peak removal of CO_2 with photosynthesis.

Total dissolved and inorganic nitrogen ranged from 0.1172 to 0.2204 mg N/l and 0.0039 to 0.473 mg N/l, respectively. Decreasing trends of total inorganic nitrogen (particularly NH_4-N) from 0900 to 1300 were also observed. Peak chlorophyll levels at 1300 support the idea of the active uptake by phytoplankton of NH_4-N and CO_2 during these initial hours.

Dissolved organic phosphorus (DOP) and dissolved inorganic phosphorus (DIP) ranged from 0 to 0.0189 mg P/l and 0.0003 to 0.0065 mg P/l, respectively. Decreasing trends of DOP observed between 1300 and 2100 (after the peak of photosynthetic activity) were followed by increasing trends of DOP during the early morning hours. Kerr et al. (1972) attributed the removal of organic phosphorus and nitrogen during darkness to heterotrophic activity by bacteria. Dissolved inorganic phosphorus levels were found to decrease from 1700 to 2100.

Dissolved silica levels, which ranged from 13.1 to 19.8 mg Si/l, fluctuated very little over the 24-hour sampling period. Silica levels were slightly higher than the 1 to 10 mg Si/l level considered typical of natural waters (Lind and Owen, 1934).

ADDENDUM D

PLANS, SPECIFICATIONS, AND COST OF A TYPICAL SEINE FOR HARVESTING PRAWNS

Rope

1. Three-eighth-inch polypropylene for the floater line and 1/2-inch nylon for the sinker line. (Polypropylene is favored because it is light and floats, and is stiff enough to minimize sagging; nylon, because it is soft and conforms well to the contour of the ponds; and a 1/2-inch diameter line because smaller lines (such as 1/4 inch or 3/8 inch usually found on seines this size) tend to sink into the mud.)

2. The ropes are soaked for 12 hours and wet-stretched and dried to prevent twisting.

Netting

1. Monofilament webbing is a must. Considerably less effort is required to pull it through the pond than a multifilament webbing.

2. The netting is double-knotted, 17-pound test, 2-inch stretched mesh. This mesh size will retain an average minimum-size prawn of approximately 10.5 cm (measured from posterior margin of the orbit of the eye to the tip of the telson).

3. Sixty-pound test monofilament was used to secure the netting to the floater and sinker lines by tying a "double clove hitch" at every third eye.

4. Net ends are reinforced with a heavy strand of nylon to prevent tearing.

Sinkers

1. Sinkers are attached to the sinker lines in a way to keep each sinker's cross-sectional area to a minimum. Commercial slide-on sinkers are not recommended because their large cross section will make the net difficult to move longitudinally in the pond mud.

2. We use molded U-shaped leads (3 mm thick, 37 mm long, and 60 grams in weight) that are pressed onto the sinker line to form a sheath with a diameter of 18 mm. The lead's mold was fabricated out of aluminum stock by J. Murphy of the AFRC and the press to squeeze the lead onto the sinker line was fabricated by M. Williamson of the University of Hawaii Agricultural Engineering Department. Sinkers cut to size from a 3 mm thick sheet and hammered onto the line would suffice.

Floaters

Use enough floaters to reduce float line sag. The prawns can crawl over low spots in the net. We use 2 1/2-inch-diameter by 2 1/2-inch-long cylindrical plastic foam floats spaced at 11-inch intervals.

Bag

The bag is detachable from the rest of the net and is similar to the cod end of a beach seine except that the top was left open (to simplify construction) and the spacing between the floats was reduced to prevent prawns from crawling over the bag. The floor dimension of the bag is 15 feet by 9 feet and it tapers down to a four-foot "mouth". The bag will hold 500 pounds of live prawns.

Cost

The materials and price list reflect Honolulu prices, March 1976. The seine has four wing sections, each approximately 130 feet long, and one detachable bag.

Figures VI-1 through VI-5 illustrate the various components of the harvest seine.

A. F. R. C. Prawn Harvest Net and Bag
(Materials, Sources and Costs)

A. Webbing

Nylon: monofilament
Twine: #6 (17 lb.)
Mesh: 2" stretch, double knot
Depth: 42 meshes (7' stretch)
Length: 250' stretch

Source: K. Kaya Fishing Supplies
901 Kekaulike Street Tel: 538-1578

Cost & Quantity: 5 lengths, each at $34.95
TOTAL COST $174.75

B. Floaters

Brand: N. G. R.
Model: 110S; red foam, cylindrical, 2 1/2" x 2 1/2"

Source: Oceanic International Co.
917 Kekaulike Street Tel: 521-3188

Cost & Quantity: 700, each at $0.38
TOTAL COST $266.00

C. Lines

1. Floatline
Type: 3/8" poly/prolene
Cost & Quantity: 600'
TOTAL COST $ 59.50

2. Leadline
Type: 1/2" nylon
Cost & Quantity: 600'
TOTAL COST $150.90

3. Tie down line
Type: Monofilament
Twine: 60 lb.

Source: K. Kida Fishing Supplies Co., Ltd.
212 Kamani Street Tel: 537-1739

Cost & Quantity: 1,000 yds.
TOTAL COST $ 43.50

SUBTOTAL	$703.65
Sales tax	28.75
TOTAL	$731.80

Figure VI-1
Bag of harvest seine. One this size will hold 100 pounds of prawns.

Figure VI-2
Wall section of harvest seine.

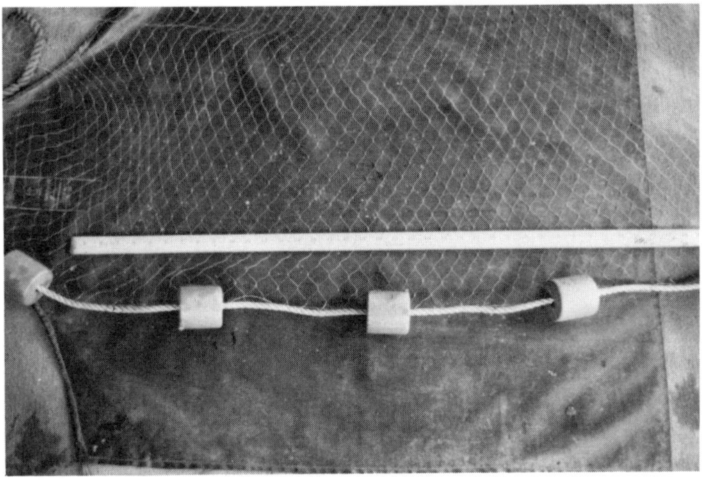

Figure VI-3
Floater line (3/8-inch polypropylene) with foam floats at 11-inch intervals.

Figure VI-4
Sinker line (1/2-inch nylon) with pressed-on lead sinkers.

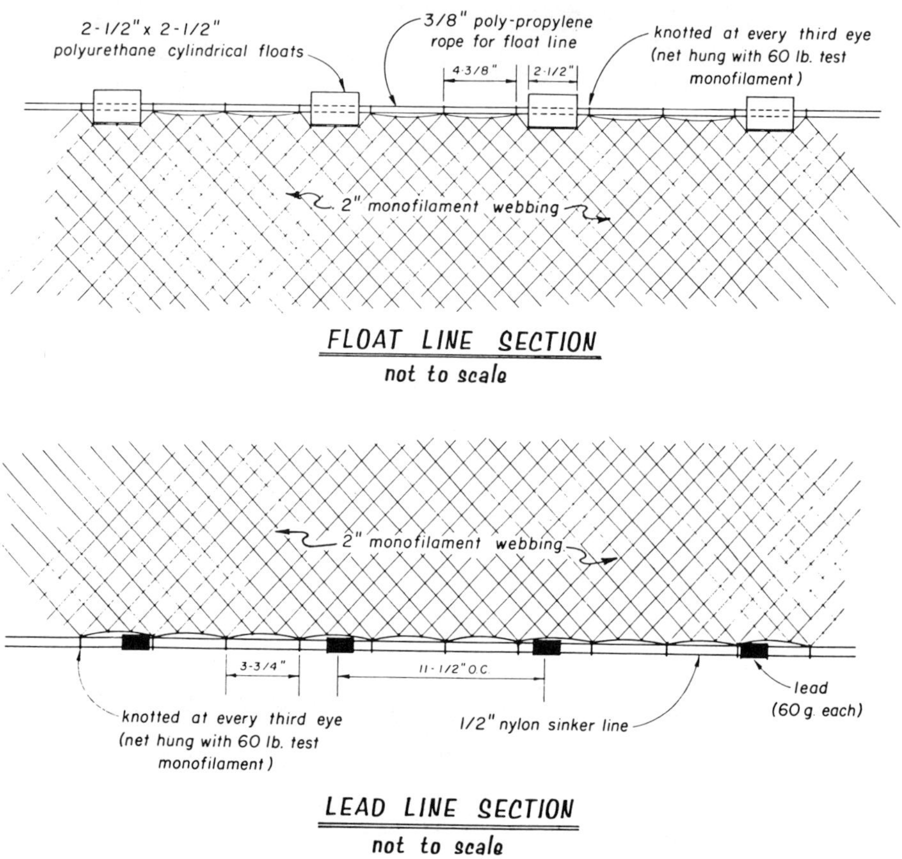

Figure VI-5
Diagrams of float line and lead line sections (not to scale).

CHAPTER VII
INTENSIVE CULTURE SYSTEMS

Poultry farming developed from business into big business through the simultaneous advent of genetic manipulation on the one hand and the "broiler-house" on the other. With visions of massive aquatic protein production before us, almost all aquaculturists dream of the day when aquatic broilerhouses will be achieved and the intensive culture of desirable aquatic animals will be brought almost entirely under human control. Yet, in all cases, aquaculturists are dealing with animals which are genetically fresh from the wild where their adaptations favor low population densities for the most part. In this chapter, E. S. McSweeny first reviews recent developments in efforts to culture fresh-water prawns intensively. This review is then augmented by comments from the floor.

A. Intensive Culture Systems:

Edward S. McSweeny, Aguaculture Sciences
Division, Syntex Agribusiness, Inc.

Introduction

The eventual goal, in the culture of any organism for food, is to maximize production while reducing the unit cost. Intensive culture has become an established fact in the production of some food organisms such as poultry. Although modern technology has been applied to aquaculture only quite recently, a portion of the effort has been directed toward establishing intensive techniques right from the start, without progressing through the historical extensive methods first. This has been particularly true with shrimps and prawns, and relatively ambitious attempts to culture *Macrobrachium rosenbergii* intensively began almost as soon as the life cycle was closed. It is my task to try to summarize the present state of knowledge in this area.

In preparation for the workshop, I distributed a questionnaire among persons engaged in the culture of *Macrobrachium* who might reasonably be expected to be investigating intensive techniques. The response indicated that most researchers are not culturing *Macrobrachium* beyond the early juvenile stage. Commercial ventures are all extensive in nature. So far as I can determine, there are only four projects engaged in growing out *M. rosenbergii* using intensive techniques. These include, in addition to our project at Syntex, the joint project between Trenton State College and Public Service Gas and Electric Company of New Jersey; Solar Aquafarms of Encinitas, California; and the project here at the Marine Resources Research Institute, Charleston. Dr. Albert Eble will discuss the Trenton project; Steve Serfling has provided some information on the work at Solar Aquafarms, and will contribute to the discussion; and Ted Smith and Paul Sandifer have made available results of their investigations here in Charleston. In addition, information provided by Durwood Dugger of Aquaprawns, Inc., Port Isabel, Texas, shows that they have experimented with some intensive techniques in their ponding operation. I will mention some of these below. Also, Ron Wulff, of Ralston Purina, made available results from the limited work they have done with *Macrobrachium* at the Crystal River facility. If I have failed to mention any venture which is practicing intensive culture, I hope the omission will be corrected during the discussion period.

Neal (1973) defined intensive culture as "the rearing of aquatic animals in man-made ponds, raceways, or tanks where environmental control is exercised". For the purpose of this workshop, where we are examining in detail the various facets of the culture of a single species, this definition is too broad; some degree of environmental control is exercised in nearly any fish-farming venture. I would define intensive culture as "propagation and rearing of aquatic species in artificial systems which exhibit three characteristics: a high degree of environmental control; an artificial means to increase production significantly over natural or extensive systems; and minimal water requirements. As regards the last criterion, it is possible to practice intensive culture with a flow-through system, but only at the sacrifice of some degree of control. I have included this as part of the definition because I believe environmental considerations demand maximum water utilization and minimal discharge. My remarks will be directed primarily to closed systems.

Let's operate from the assumption that pond aquaculture is profitable in the United States. The catfish industry has proven this, although of more significance to us are the *Macrobrachium* operations in Hawaii and Puerto Rico. Given this assumption, why should we proceed to intensive systems, with substantially higher capital requirements? There are several compelling reasons, including the following:

- To provide environmental control for optimum year-round growing conditions.

- For elimination of predators.

- For prevention and/or control of disease.

- For protection from external influences, such as contamination or pollution, and from natural catastrophes.

- To make possible environmentally acceptable waste disposal.

The net benefit of these factors is maximum year-round production on a highly predictable basis, which is the key to successful marketing, and therefore commercial success, on a scale which justifies the investment.

There are two basic requirements for culturing any species under intensive conditions. The first is a knowledge of the environmental requirements of the species, somewhat more detailed than is required for extensive culture. The second is a detailed knowledge of the complete nutritional requirements, which is not required in extensive culture. Environmental requirements of *Macrobrachium* are fairly well known, but lack of nutritional knowledge currently constitutes an obstacle to intensive culture.

I am going to break my discussion roughly into two categories: systems, which provide environmental control, including major hardware, accessories, and filtration; and techniques, or the art of manipulating the system.

Systems

A generalized intensive system (Figure VII-1) contains three major components: a culture vessel, filtration, and a motive force for circulating the water through the filter. Inputs to the system include food, heat, and oxygen. The basic elements of this system are discussed briefly below.

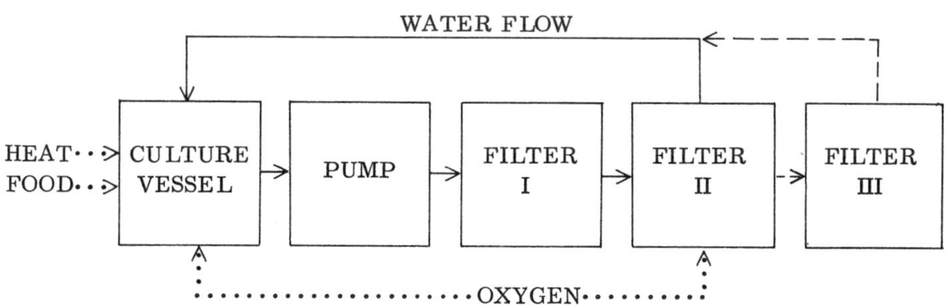

Figure VII-1
Generalized Intensive Culture System.

Culture Container

We are all familiar with culture tanks of various types. It is important to recognize that the tank is merely a vessel to contain the water, organisms, and any accessories. The work of Zielinski, et al. (1974, 1976) has shown that configuration can affect operation, but economics is the strongest factor affecting both material and design. Maximum size will also be limited by the ability to control water quality.

Pump

This represents any means of water movement. The energy source is immaterial, so long as a sufficient flow rate is maintained. Electrical pumps are the choice of the majority, with airlifts running second.

Heating

M. rosenbergii is a tropical organism and, as such, requires relatively warm temperatures for optimum growth. Conventional energy sources and their control systems are familiar to most of us. With increasing energy costs, alternative sources are being investigated for aquaculture. This takes three forms: solar, geothermal, and waste heat from industrial sources. Any form of heat is acceptable, so long as it maintains the required temperatures without affecting water quality. At Syntex, we use solar energy to heat 5,000-gallon tanks for about nine months of the year; I believe the technology and economics exist to make solar heating practical in a commercial venture today. There is quite a lot of interest in geothermal, but no actual projects underway that I know of. The Trenton project utilizes waste heat, so Al Eble is better qualified than I to discuss that application.

Oxygenation

Probably the single most important factor in intensive culture is maintenance of the oxygen content of the culture water. Methods of supplemental oxygenation range from simple bubbling of atmospheric air to sophisticated systems for injecting liquid oxygen. Mitchell, et al. (1976) discussed the types of aeration devices available and their relative performance. Although this work was concerned with pond usage, many of the remarks are applicable to intensive culture. Information from respondents seems to indicate that the aeration device used is determined by the size of the system, with airstones and porous tubing most commonly cited. In current projects, supplemental oxygenation is often applied to biological filter beds, as well as to the culture tank.

A minimum safe continuous oxygen level has not been determined for Macrobrachium, to my knowledge, but low levels that are adequate for maintaining the organism may have deleterious chronic effects. The dissolved oxygen level will vary with fluctuations in filter performance, dissolved organics, and other factors, and at high biomass loadings, the situation can change rapidly.

Feeding Devices

Commercial feeding devices are available which are directly applicable or easily converted to intensive culture of Macrobrachium. Any automated device needs to be critically evaluated from an economic standpoint. Of the respondents, only Solar Aquafarms indicated automated feeding.

Water Treatment

Water treatment can be separated into three stages. The first stage consists of particulate removal. This can be by any one of several methods available. The purpose of this stage is to deliver the water to a biological bed free of suspended material. This is important for long and trouble-free operation of the biological filter. The choice of methods is dependent on compatibility with the rest of the system.

The second stage represents the biological filter. There are two key requirements of this filter: first, that the bed contain a sufficient number of bacteria (i.e., area of bacterial substrate) to oxidize the nitrogenous wastes to which it will be subjected; and second, that the design and operation is such that the bed is always abundantly supplied with oxygen. Siddal (1974) recommends a minimum level of 6.5 mg/liter.

Determining the required volume of a biological filter is something less than a science at the present time. Every type of filter medium will have a different surface to volume ratio. Carrying capacity of a given medium is best determined experimentally, although capacity can be augmented by various techniques, such as suggested by Siddall (1974). This is important in intensive culture to avoid excessively large filters. Production of wastes varies with age and size of prawns and the filter should be sized for the maximum expected loading.

Optimal flow rates through the filter are determined by culture tank volume and rate of accumulation of nitrogenous wastes. Removal rate must be at least equal to accumulation rate, so flow rates are best specified as exchanges of tank volume per unit time. Higher flow rates are important in supplying oxygen to the bacterial bed, but cost more to achieve. Highest turnover rates reported by culturists were nine and ten times daily by MRRI and Purina's Crystal River facility, respectively. At Syntex, we maintain flow rates above a minimum of one exchange per six hours.

The third filtration stage mentioned represents additional treatment methods available, including foam fractionation, UV, ultrafiltration, ozone, etc. This stage is optional, but can be beneficial. Treatment at this stage is generally expensive.

Culture Tank Substrates

Most culturists use some form of artificial substrate to increase the area available to the prawns. There are two basic reasons for this. In the first place, the species exhibits a certain amount of territoriality. By increasing the area available to a given population, the effective density is decreased, thus counteracting aggression due to territoriality. The second reason is to provide an essentially benthic organism with substrates which occupy a large proportion of the water column, eliminating otherwise unused space and concentrating the culture operation in a smaller area.

The simplest form of substrate is a vertical mesh panel. Probably the most widely used form is stacked horizontal layers. These may be porous (as in netting) or solid, but netting is the most commonly used, so as not to impede water circulation. The ultimate in the artificial substrate concept is a three-dimensional, porous, compartment system for housing individual prawns, as described in the patent issued to Serfling (1975).

Relative effectiveness of different types of supplemental substrates is not completely established. All are effective to some extent. Vertical panels are at least equally as effective as the horizontal type in the early stages of growth. There is some indication that they are less effective at later stages, and we are currently retesting this at Syntex. Honeycomb, egg-crate, or tube-type habitats, which provide cavities for the use of individual prawns, are relatively ineffective with *M. rosenbergii*, which is not a burrowing or hole-dwelling animal by nature. It should be remembered that substrate cost can be a significant factor in an intensive system.

When augmenting the substrate area, it is common practice to make portions of the substrate of a solid (or less porous) material, where food can be deposited. In this manner, the total feeding area is distributed throughout the tank, both horizontally and vertically. Food is readily available to prawns at any level in the tank. I am not sure that this provides a real benefit, and would appreciate comments by anyone who has established the necessity or lack of it.

Techniques

After the system is operating, it is the methods used which determine the yield obtained. There are several aspects of husbandry which have a bearing on production. These include stocking densities, density changes and culling, feeding rates and intervals, harvesting methods, etc.

Hygiene

Hygiene and careful husbandry in intensive systems bear an importance out of all comparison to extensive culture. Not only will disease spread rapidly in abnormally crowded conditions, but when it occurs, it is easily spread to adjacent tanks or systems. The loss in this situation includes not only the crop, but the time required to sterilize the system and restart the filter.

Hygienic practices include efficient prefiltration followed by backflushing to remove particulates from the system; rigorous precautions to prevent cross-contamination between systems; care in handling of feedstuffs; and general cleanliness in regard to work areas, tanks, and equipment.

Stocking Densities

M. rosenbergii is a rapidly growing organism which shows a definite growth inhibition at a variable, size-related density, and stocking densities become extremely important in achieving optimal growth. The limiting factor appears to be spatial, so the area available to the prawns must be taken into account in stocking.

Different workers measure density in slightly different ways. At Syntex, we use number of prawns per unit of horizontal area, or combined vertical and horizontal area when using vertical substrates. Solar Aquafarms uses number per unit of surface area, while MRRI uses number per unit area of tank floor; these are presumably the same, and represent the method of most culturists. Aquaprawns measures density on the basis of weight per unit of surface area. These methods have the common fault of defining a stocking rate which is applicable only to the particular configuration. It would be extremely useful to have a formula for stating density which is independent of configuration, allowing direct comparison of results between workers. Factors to be considered in developing such a formula include biomass, available area, system volume, and perhaps turnover rate.

Optimum density is inversely proportional to size, and for best use of tank space it is necessary to make periodic reductions in the population. This can be associated with culling, which will be discussed further below. At Syntex, we use a six-month, three-phase grow-out, reducing density after two and four months of growth. Aquaprawns has experimented with a similar three-phase system, with nearly double the density, but a longer grow-out period.

Culling

Associated with a multistage grow-out where density is reduced from stage to stage, culling, or the removal of the small, weak, or otherwise undesirable prawns, is an obvious way to improve average weight. However, separating the undesirable animals on a large scale is an unresolved problem. We cull by hand at Syntex, and have found that this is faster than the various mechanical devices, such as minnow graders, which we have tried. This method is not practical on a commercial scale. Culling methods which we have tried are damaging to the prawns, and it may be that even the handling necessary for periodic weighings slows growth.

Although culling can substantially increase the average weight of restocked prawns, we do not have any evidence that the final average is increased proportionately.

Harvesting

We are usually dealing with containers of moderate volume in intensive culture. Harvesting by hand is often the simplest solution if the tank can be drained and access be had to the interior. I do not know of a more efficient method of harvesting which is economically justifiable. If the tank is not accessible for any reason, then another solution is required. Drain harvesting is not a simple matter, as is often assumed, due to the rheotaxic behavior and strong swimming ability of *Macrobrachium*. This method is possible, however, if the tank is properly constructed. In a raceway configuration, electrofishing methods might serve to concentrate the prawns. A fish pump is another potential harvesting tool. Although there has been little work done on harvesting methods, I do not foresee this as a major problem.

Feeding

Quantitative food requirements of *M. rosenbergii* change constantly during growth. Determination of precise feeding schedules are important to an intensive operation in terms of both water quality and economics. Feeding schedules are constructed on the basis of biomass remaining at any particular moment, and it is essential that this be predicted with a high degree of accuracy. Biomass is estimated on the basis of a projected growth curve, corrected for expected mortality. Mortality may vary with different systems. At Syntex, we have found that 0.33% per day, or 10% per month, is a reasonable maximum mortality. This can easily be absorbed in scheduled density reductions. Daily feed is calculated as a percentage of biomass, on a sliding scale which decreases with age.

Most respondents to my questionnaire reported a feeding rate of 3% of biomass. However, Scott Willis, who provided information on the program at Florida's Marine Research Laboratory at St. Petersburg, indicated a variable rate during the first month after metamorphosis. They start feeding at 20% of biomass daily, decreasing to 15% the second week, and 10% the third and fourth weeks. This is very close to the rate determined at Syntex. We continue decreasing the rate throughout grow-out, feeding 2.5% after five months. Rates are continually monitored by visual observation and can be adjusted as necessary.

M. rosenbergii is an intermittent feeder, and better utilization of feed can be achieved by delivering the daily ration in several portions. When feeding several times daily, very careful control is required of the amount to prevent overfeeding.

Feed conversions reported by respondents ranged from 1:1 to approximately 4:1. Most were between 2 and 3:1. In actual practice, conversion figures fluctuate considerably between trials. However, our work indicates that overall feed conversions of 2:1 are achievable and may in fact be necessary to the economics of intensive culture.

Growth Rates

Figure VII-2 is a semilog presentation of the growth curve obtained in the Syntex systems. We believe the curve is very nearly representative of the normal growth of *M. rosenbergii*. This curve was constructed by Victor Mancebo of the Syntex staff, who is preparing the supporting data for publication. It compares closely with published growth data taken from large samples. Whether this curve can be substantially improved is uncertain. There is probably some gain to be realized from better nutrition, perhaps as much as 50 percent. Beyond that, we will probably have to rely on genetics.

Summary

In summary, the two areas which need improvement if intensive culture is to become profitable are growth and systems cost. Average prawn weight must be improved to yield at least a ten-count (45g) whole animal, both for total production weight and to put the product into the premium price bracket. Systems cost hinges primarily on containment costs. It is my opinion that above-ground systems are unlikely to be profitable, solely because of tank costs. In-ground systems show promise since containment can be effected relatively cheaply. The real future of intensive techniques lies in a middle ground between fabricated tank systems and extensive ponds.

B. Intensive Culture Systems, Discussion

Albert Eble stated that Public Service Electric and Gas Company is working with Trenton State College, with Rutgers University, and with Long Island Oyster Farms. The PSE&G project is for the use of thermal effluent from the Mercer Power Plant near Trenton, New Jersey. Prawns (*Macrobrachium rosenbergii*) are grown during the warm months, and rainbow trout during the winter.

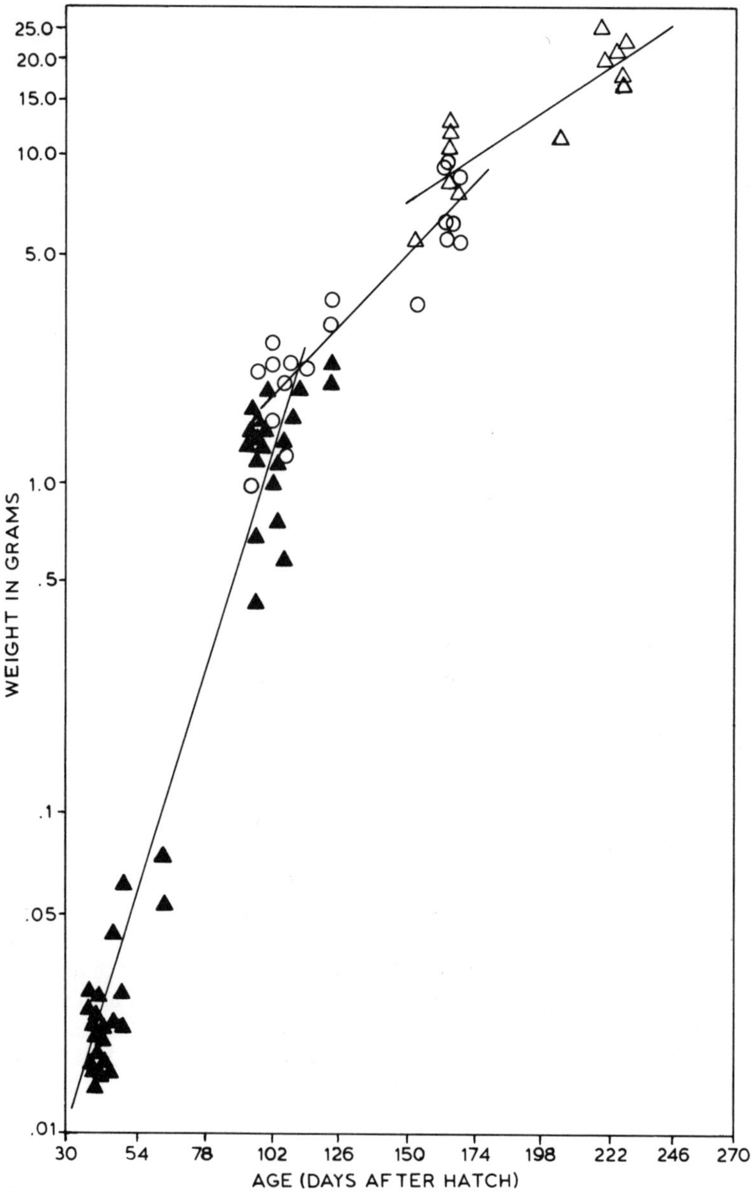

Figure VII-2
Growth of Macrobrachium rosenbergii in intensive systems.

To develop a maximum high-intensity system, the PSE&G team had to use the entire water column. A number of substrates were tried, including horizontal shelving, open and closed compartments (condominiums), and draped netting. None was superior in the amount of surface available to the prawns, and the group chose draped netting for ease of handling.

Nets are draped so they sag, but not quite to the bottom, so prawns can move freely across the bottom. Pond banks are steep and lined with plastic, so the prawns cannot climb out on the banks to molt. Instead, they climb to the netting top--behavior also noted in South Carolina when screen substrates were used.

PSE&G tries to get juveniles in January and grow them to stocking size (2.5 to 3 centimeters) in indoor heated nurseries. Stocking takes place about mid-May in the outdoor ponds, and the prawns are left until nearly the end of October. With this headstart and the approximately five months growing season, a 35-gram prawn should be produced. Stocking density is at the rate of five prawns per foot^2 of bottom, but because of the netting substrate, density is actually about 1.5 to 2 prawns per foot^2 of available surface. Harvested prawns average about one to 1.5 per foot^2 of total surface, or about 3 prawns per foot^2 of bottom surface.

The conclusion after two years of experiments is that the netting assists materially in increasing density.

Because there is a substantial flow of water from the power plant, dissolved oxygen is never a problem. The water is consistently saturated. Water is changed six to eight times a day and could be changed more often if necessary.

Ralston-Purina marine ration #20 was fed, but was not satisfactory, and a change was made to #25 with better results. To see the effect of a high feeding rate, an experiment was conducted with the prawns fed at a rate of 15% of body weight. Conversion results were as poor as 7:1. The fast flow rate permitted use of bluefish heads and other fish residues ground up as a supplement, with prepared food fed in the morning and the fish fed in the afternoon. The prawns "came out of the woodwork" to get at the fish.

Nutrition is expected to be a continuing problem on which the Trenton State College scientists will work with the Rutgers scientists. In all probability, two or three different rations will be required depending on the size of the animal.

Steven Serfling of Solar Aquafarms has developed a closed polyculture system in which prawns are grown in individual compartments. The system is basically a raceway in a greenhouse with a solar heat exchanger to keep the water warm. Gas heat is used as a backup on cloudy days. Water is recirculated and the same water can be reused for three or four months with new water added as the prawns get bigger. Larvae are produced through normal procedures and stocked in the individual cells at 6-8 centimeters long. Feeding is automatic, and wastes drift down to the bottom of the raceways where they are consumed by amphipods, snails, and other organisms which are in turn consumed, at least in part, by the prawns. Wastes also are recycled through floating aquatic plants, e.g., water hyacinth and duckweed.

In response to a question about cannibalism in intensive systems, John Bennett replied that the experience at Syntex was that cannibalism resulted largely from diet deficiency. When Syntex improved the diet, cannibalism dropped dramatically, and survival increased from 40% to 80-90%.

Persons working with intensive systems tended to agree that the size distribution among animals in intensive systems was more uniform than among those in extensive systems; nevertheless, the spread remained very great with the larger animals about 100 times the size of the smallest ones. Suppression of growth through dominance by the large "bull" prawns appeared somewhat less. On the whole, more work remains to be done on growth phenomena taking into account such factors as chemical communication (e.g., molt pheromones).

C. Notes on Mortality, Cannibalism, and Competition in *M. rosenbergii*:

John B. Peebles, Hawaii Institute of Marine Biology

Because prawn behavior, while important in any system, becomes critical in intensive systems, the following notes from one of the very few behaviorists working with prawns was included in this session.

Abstract

Mortalities related to behavioral interactions were observed on 19 occasions over a two-year period. In all cases, the corpses showed evidence of damage to both sensory and locomotor systems. This type of injury is associated with aggressive behavior received during the postmolt period. Data is presented that supports the hypothesis that this molt-related intraspecific aggressive behavior (MRIAB), and not cannibalism, is a major cause of behaviorally related mortalities.

Introduction

The raising of prawns, *M. rosenbergii*, has been divided into two phases, (1) larval rearing and (2) pond grow-out, with mortalities occurring in both (Fujimura, 1974; Dugan et al., 1975). Survival to market size varies from

47 to 61% (Fujimura, 1974) and appears to be partially related to stocking density (Forster and Beard, 1974; Fujimura, 1974). Factors often mentioned as contributing to mortality are competiton, susceptibility to injury of the postmolt (Fujimura, 1974), aggression, and cannibalism (Forster and Beard, 1974; Dugan et al., 1975).

I am presently engaged in a research program designed to determine the behavioral factors influencing the distribution and abundance of M. rosenbergii. During the course of part of my experiments, I have observed a number of mortalities. Some of these were not related to social interactions but were caused by a failure of a prawn to successfully exit its exoskeleton during ecdysis. Others were directly related to both molting and intraspecific aggressive behavior. The behavior observed during this latter type of mortality I call "molt-related intraspecific aggressive behavior" (MRIAB). In this paper I will present data that supports the hypothesis that MRIAB and not cannibalism plays a primary role in the causation of certain types of mortality in confined populations of M. rosenbergii.

Methods and Materials

The data reported in this note originated from both field and laboratory observations. My experiments were not specifically designed to examine molt-related mortalities; however, this type of death did occur. When this happened, I tried to gather as much information on each death as I could. Most of the deaths occurred in holding tanks. Three occurred in experimental situations. Most of the time the sequence of events leading up to death went unnoticed; however, I was able to observe the before, during, and after MRIAB on six occasions.

The holding tanks varied in size from a plastic pool (diameter 106.6 cm by 15 cm deep) to a rectangular tank (68 cm by 143 cm by 20 cm deep). The densities ranged from a maximum of $11/m^2$ to a maximum of $8/m^2$. The experimental tanks also varied in size from a plastic pool (diameter 138 cm by 33 cm deep) to an eliptical tank (longest diameter 138 cm, shortest diameter 75 cm, water depth 34 cm). All animals were fed a pelleted diet (see diet #5, Balazs et al., 1973) approximately four times a week. Water conditions were maintained by an airlift filter.

I observed 19 molt-related aggressive deaths over a two-year period (2/74 to 3/76). On 14 of these, I recorded: day of death (firm or estimate, smell); carapace length (base of eye to dorsal posterior lip of the carapace); sex; and degree of body damage. The degree of body damage was coded as follows: any damaged or missing body part received a value of one; thus, antennules, antenna, thoracic legs, pleopods, eyestalks, abdominal segments, and the rostrum were all inspected for damage. The numbers were then summed to give a body damage index (BDI). The remaining five deaths reported above either occurred in the field (one) or before I began to code for degree of body damage.

Results

No correlation existed between the length of time an animal was in the tank after it died and its BID (Table VII-1, $r_s = 0.28$, n.s.). Strictly interpreted, this means that no cannibalism or subsequent damage occurred after death. I believe that it also indicates that cannibalism was not the primary cause of molt-related aggressive deaths. An examination of the corpses supports this hypothesis. In the cases where I coded for body damage, the data clearly shows that every animal had some damage to both sensory and locomotor appendages (Table VII-1). This implies that a stereotypic pattern is involved in certain types of prawn mortalities.

The sequence of events leading up to death was never fully documented; however, I was able to observe certain incomplete sequences on several independent occasions. The composite results of these observations have lead me to propose a general scheme describing a molt-related aggressive mortality. I suspect that an animal in the premolt condition is actively seeking a habitat different from the one it exploits while in the intermolt state. Two types of observations have lead me to that hypothesis. First, freshly molted *Macrobrachium lar* can be found in the shallows associated with a series of deep pools. Hard *M. lar* are found in the pools (personal observations). Second, I have observed, on one occasion, a number of *M. rosenbergii* aggressively chasing one particular *M. rosenbergii*. This animal was pursued by several animals as it moved around the tank. A high level of aggressive interactions (number of interactions/minute) was observed between most animals in the tank as this animal swam from one side to the other. This animal was then observed to climb out of the water and rest on top of the shelter provided for the prawns. This prawn's entire dorsal surface was out of the water. The animal molted within two days of that observation.

A late premolt prawn that cannot find refuge where it can molt probably has no choice but to molt in the presence of other intermolt animals. I observed this type of situation from molt until death on five occasions. Generally, mortality occurred within 48 hours; on one occasion the animal was dead five hours after it molted. Usually either the chela or the antennules and antenna were the first appendages damaged or lost. It is interesting to note that the chela are among the first appendages to harden after ecdysis. Following this damage the thoracic legs, then the pleopods, and later the uropods received injury. The rostrum and the eyestalks were usually the last body parts to receive damage. An animal in this state could be found on the bottom of the tank, lying on its side, and pumping its gill bailer. Death followed within 24 hours.

Table VII-1
Coded Degree of Body Damage for 14 MRIAB Mortalities.

Sex	Carapace Length (cm)	Locomotor: Abdominal Segments	Locomotor: Pleopods	Locomotor: Thoracic Legs	Rostrum	Sensory: Antenna	Sensory: Antennula	Sensory: Eye	Body Damage Index	Days in Tank After Death
M	4.65	1	7	4	1	2	4	1	21	1
F	3.9	3	10	10	1	2	4	2	32	2
?	4.0	1	10	10	0	2	4	2	29	4
M	4.87	3	10	10	0	2	4	2	31	3
M	4.96	3	7	8	0	2	4	1	25	1
M	5.0	0	2	10	1	0	4	2	19	2
M	5.5	1	0	1	0	2	3	1	8	1
M	4.55	1	4	6	0	2	2	1	19	1
?	?	2	8	8	1	2	4	2	27	1
M	5.2	0	0	6	0	2	4	1	13	3
M	5.3	0	0	1	0	2	1	0	4	1
?	1.32	2	10	10	0	2	4	2	30	1
M	4.91	0	0	3	0	2	1	0	6	1
?	1.7	3	10	10	1	2	4	2	32	1

Discussion

In another paper I offered the hypothesis that M. rosenbergii uses its chela to test its opponent's weight during competition for shelter (Peebles, unpublished manuscript). It was also noted that 48% of the observed aggressive behavior could be accounted for by an initiation-chela push-chela push-termination shunt. The results of the data presented here in this communication indicate that the chela push might have two other functions. These functions would be (1) to determine the opponent's molt state (soft or hard), and (2) to see if the opponent is capable of pushing back. If an opponent was soft (early postmolt) or if chela were lacking, the prawn performing the pushing would have a distinct advantage. An early postmolt opponent is not capable of engaging in offensive behavior. Every early postmolt prawn I have observed would flee the attack of a hard-shell opponent. An animal lacking chela has no means of defending itself. It cannot push back when pushed. These chelaless animals usually have their antenna reduced to a fraction of their original length. In this state other animals can approach undetected and strike at undefended thoracic legs with their chela.

M. rosenbergii has been called a pugnacious species that occasionally engages in cannibalism (Forster and Beard, 1974; Ling, 1969). Gut contents have been shown to contain bits of crustacean appendages (John, 1957; Ling, 1969). Forster and Beard (1974) reported cannibalism accounting for 69% of the mortalities in a dense population (166 prawns/m^2) with "hard-shelled corpses" contributing 5% and "escapes" the remaining percentage. These authors reported cannibalism as "partly eaten remains of soft-shelled (newly molted) prawns"; the "remains" consisted of "the abdomen and part of the cephalothorax"; "hard-shelled corpses" were lacking eyes or thoracic abdominal appendages (Forster and Beard, 1974).

The exact cause of mortalities is an important factor, describing ecological relationships within a population. Cannibalism infers the utilization of a conspecific as a food resource. The behavioral sequences leading to cannibalism can contain intraspecific aggressive components. Apparently such is the case when an adult ghost crab, Ocypode ceratophthalmus, cannibalizes a juvenile. Cannibalism also can be a feeding response. Adult guppies, Poecilia reticulata, treat fry as mosquito larvae. The newly emerged young swim in a jerky fashion. Both male and female adults orient to this movement by assuming a pose (predatory) different from behaviors usually associated with intraspecific aggression (Baerends et al., 1955; Peebles, 1973). Adult yellow perch, Perca flavescens, are also cannibalistic, responding to a seasonal decline in Daphnia occurring simultaneously with a transition in size and abundance of young yellow perch (Tarby, 1974). I suspect that a large number of mortalities observed in confined populations of M. rosenbergii are caused by MRIAB. MRIAB has motor components that are observed in other nonfeeding situations (Peebles, unpublished manuscript); however, it is only during the early postmolt period that aggressive behavior leads to mortality. Thus MRIAB probably serves to eliminate competitors when those competitors will not (or cannot) emigrate.

Cannibalism in adult prawns is probably not a major cause of mortality in well-fed uncrowded populations, and did not occur during the course of the observations presented in this paper. However, it should be noted that these observations were not designed to determine if *M. rosenbergii* will eat one another. Nor do the results answer three specific questions concerning cannibalism: (1) will a hungry prawn kill and then use a conspecific as a food resource; (2) are the body parts that are lost during an intraspecific agonistic encounter used as food by one or both of the participants; and (3) if a prawn happens upon a dead prawn, will it eat it?

Acknowledgments

This study was in part supported by Sea Grant No. 04-6-158-44026 to the University of Hawaii. I would like to thank Mr. Paul Atkins for his critical comments on the manuscript.

CHAPTER VIII
THE NUTRITION OF FRESHWATER PRAWNS

Of Macrobrachium sp. in pond grow-out systems, it remains true that we don't know with certainty what the animals eat. Prepared diets are fed and the prawns grow and prosper. But are they eating the prepared formula feeds or do these go to fertilize the pond biota which the prawns eat in turn? Without doubt, both processes are in operation, but in what relative proportions? Dietary variables are critical in controlled larval-rearing, and prawns fed only formula feeds under laboratory conditions grow less rapidly than pond-reared animals. So endeavors to understand the animals' nutritional requirements and to formulate inexpensive techniques with which to satisfy them go on. In this chapter, G. N. Biddle, Jeanne Joseph, Margo Stahl, and Douglass Conklin explore this so far wide-open issue.

A. The Nutrition of Macrobrachium Species:

Garfield N. Biddle, National Marine Fisheries Service, Southeast Utilization Research Center

Introduction

Adequate nutrition is a prerequisite for the survival, normal development, and proliferation of any animal species. In the natural environment, the giant prawn, Macrobrachium rosenbergii, is able to satisfy its particular nutrient requirements from a variety of sources which include aquatic worms, insects small molluscs, crustaceans, plankton, and organic detritus (Ling, 1969). However, under the restrictive and rigidly controlled conditions of intensive culture, it will become necessary to provide the prawn with appropriate levels of nutritionally balanced feeds.

In recent years, several empirical diets have been evaluated under both laboratory and culture conditions. The majority of these formulations have proved rather unsatisfactory with the exception of a very limited number of proprietary feeds which are commercially available to the prawn culturist. Unfortunately, even these are not totally satisfactory in promoting growth to the same degree as that achieved when prawns are fed live or fresh food. Therefore, many culturists continue to rely on the use of live/natural feeds, particularly for the production of juveniles for stocking purposes.

Feeds and the cost of labor associated with feeding represent one of the largest running expenditures of a *Macrobrachium* culture operation. Hagood and Willis (1976) produced *M. rosenbergii* juveniles using standardized conditions in which the larvae were fed fresh-hatched *Artemia salina* nauplii and supplemented with freshly ground fish. Cost estimates were expressed on the basis of 1,000 juveniles produced per 40-day period. Such estimates did not include amortization of capital outlay, building lease expenses, maintenance and other fixed expenditures which vary considerably with respect to personnel, location, and hatchery design. Total costs were found to equal \$3.56 per 1,000 *M. rosenbergii* juveniles, of which feed costs (dry *Artemia* eggs and ground fish) accounted for \$.61 (17% of the total). The labor costs associated with *Artemia* culture, preparation of the ground fish, and larval feeding were calculated to equal \$1.58 (44% of the total) per 1,000 juveniles. Based on these estimates, approximately 60% of the total production costs were related to feeds and feeding. Thus it is clear that an urgent need exists for the development of feeds which are not only nutritionally complete and highly digestible, but also economically feasible.

Water stability is also a key factor to consider when formulating an aquatic feed, for a feed which is rapidly broken down in the aquatic environment is of little value either to the crustacean nutritionist or the commercial culturist. Thus, the importance of appropriate feed processing methodologies is obvious. Several studies have been reported on the use of various dietary binders (i.e., gums, gels, alginates, etc.), the effects of different processing variables, and the production of diets of different physical forms (i.e., pellets, flakes, extruded or microencapsulated products). The results of such research have been adequately discussed elsewhere (Meyers and Zein-Eldin, 1972; Meyers and Brand, 1975; New, 1976). The use of dietary attractants to improve acceptability and ingestion of formulated feeds has also been examined with *M. rosenbergii* (Sick and Beaty, 1975; Sick, 1976). These studies revealed that freeze-dried *A. salina* and the egg white-derived proteins, ovalbumin and ovoglobulin, exhibited strongest attractant properties. Although the addition levels (15%) of such attractants would be cost prohibitive in commercial feeds, the results thus obtained demonstrate the positive benefits which can be realized by altering the chemosensory properties of crustacean diets.

As suggested by Joseph and Meyers (1975), there is a priority need for practical research on the formulation of crude or semidefined feeds which will promote acceptable prawn growth and permit the industry to achieve some degree of economic stability. Ultimately, however, the optimization of feed quality and cost considerations will be realized only after the specific nutrients required for maximal growth of the prawn (at various stages in its normal physiological development) have been clearly defined.

At this juncture, one might logically ask, "What do we know about the nutrition of Macrobrachium spp.?" It is hoped that the ensuing discussions will provide some insight into this matter.

Protein/Amino Acids

Protein is one of the most expensive singular ingredients in animal feeds and therefore it is not surprising that more time, resources, and effort have been devoted to investigations of protein (amino acid) requirements of prawns than any other dietary component.

Since the early work of Subrahmanyan and Oppenheimer (1969), Kanazawa et al. (1970), and Deshimaru and Shigueno (1972), numerous studies have been reported on growth rates, feed efficiencies, etc., of various crustaceans fed different levels of dietary protein. Many of these studies are discussed in considerable detail by New (1976). Rather than paraphrase such discussion, an abbreviated table has been prepared to illustrate the wide range of dietary protein levels studied. Based upon the information presented in Table VIII-1, it would appear that there is still considerable uncertainty as to the prawn's quantitative requirements for dietary protein. At one extreme there are those studies that suggest increased growth rates and efficiencies with increased protein even at levels in excess of 50% of the diet. At the other extreme are the studies suggesting that optimal dietary protein levels are more in the range of 20 to 30%. In this latter category, several of the reports show no additional growth response at the higher levels and in some instances indicate decreased growth rates/efficiencies when protein exceeds 40% of the diet. Possibly the combination of high dietary protein in the presence of low levels of nonprotein energy may force the crustacean to deaminate significant portions of the protein thus yielding the carbon fragments required for cellular energy metabolism. As a result of this altered metabolism, the animal could be expected to show reduced growth rates and efficiencies. Such an occurrence has been described for fish by Lovell (1975). According to New (1976), a consensus of opinions indicates that the optimal level of dietary protein for different species of shrimp is between 27 and 35%. In the case of the young, juvenile M. rosenbergii, the requirement may be somewhat higher.

When evaluating the reliability of such recommendations, some caution should be exercised since there are numerous factors (which are not strictly of a dietary nature) that can interfere with or even cancel the true response of an organism to dietary protein. For instance, the experimental conditions under

Table VIII-1
Summary of Dietary Protein Levels Evaluated in Feeding Studies with Various Crustacean Species.

Species	Crude Protein Levels Studied (%)	Optimum Levels Suggested (%)	Reference
Penaeus setiferus	14 to 28; 28 to 52	28 to 32	Andrews et al., 1972
Penaeus japonicus	63 to 76	> 60	Deshimaru and Shigueno, 1972
Palaemon serratus	12 to 69	30 to 40	Forster and Beard, 1973
Penaeus aztecus	18 to 35	23 to 31	Shewbart et al., 1973
Homarus americanus	0 to 60	≥ 60	Castell and Budson, 1974
Penaeus japonicus	25 to 50	50	Deshimaru and Kuroki, 1975
Penaeus aztecus	40 to 80	40 ≤	Venkataramiah et al., 1975
Macrobrachium rosenbergii (juveniles)	15 to 35	≥ 35	Balazs and Ross, 1976
Macrobrachium rosenbergii (larvae)	10 to 20	15 to 20	Sick, 1976

which a feeding study is conducted could have a significant bearing on final results, especially if one considers variables such as water temperature, salinity, dissolved oxygen levels, dissolved organics/harmful metabolites, biomass densities, potential space restrictions of the holding systems, length of the study (short versus long term) and others. The growth stage of the prawn and its physiological state are also highly influential on growth rates and efficiencies (i. e., larval versus postlarval; juvenile versus the sexually mature adult), as are the inherent differences in growth potentials between species as well as within a species. Once these factors have been established, there are several nutritionally related issues that must be resolved.

Beyond the actual levels of protein under investigation, we know from studies with domestic animals that requirements are often affected by the quality (amino acid composition) of the offered protein as well as its digestibility by the animal. Considerations should also be given to the possible sparing effects of one nutrient upon another (i. e., protein x energy). The effects of nutrient imbalance on protein utilization must be evaluated to ensure that excesses or deficiencies of other dietary nutrients (lipid, minerals, vitamins) are not interfering with or contributing to the animal's response to the dietary protein. The physical characteristics of experimental diets have a direct bearing on consumption rates and digestibility of ingested food. These characteristics include stability, form, texture, size, density, and palatability. For additional information on several of the variables cited above, the reader is referred to the comprehensive review by New (1976).

In general, the overall role of dietary protein is to provide sufficient quantities of those amino acids which are essential to the normal health and survival of an animal. Traditionally, we think of an essential amino acid as one that either cannot be synthesized at any level within the body, or is synthesized in only limited quantities which are inadequate for meeting total body needs. Under either condition, it becomes necessary for the particular amino acid to be supplied in the diet of a consuming organism. Therefore, in order to establish the quantitative amino acid requirements of the prawn, one must first demonstrate the qualitative (essential or nonessential) need for the specific amino acid in question. Using radioisotopic tracer techniques, Zandee (1966) demonstrated the inability of the crayfish (*Astacus astacus*) to synthesize the ten amino acids which are known to be essential in the diet of domestic animals. Since that study, radiometric techniques have been utilized in several investigations on the qualitative amino acid requirements of crustaceans. Results from a number of these experiments are summarized in Table VIII-2 (for comparison, similar data from two studies with molluscs are also presented). Since no activity was detected for tyrosine in several of the labeling studies reviewed, investigators have concluded that a qualitative requirement exists for this particular amino acid. However, as demonstrated in the crayfish (*A. astacus*) by Zandee (1966), tyrosine is synthesized from phenylalanine via a nonreversible hydroxylation reaction. This biosynthetic pathway involves the enzyme phenylalanine monooxygenase as well as

Table VIII-2
Summary of Qualitative Amino Acid Requirements of Several Crustacean and Mulluscan Species as Determined by Radiometric Techniques.

Amino Acid	1. *Cancer magister* Dungeness crab	2. *Uca pugilator* Fiddler crab	3. *Palaemon serratus* Common prawn	4. *Penaeus aztecus* Brown shrimp	5. *M. ohione*	6. *M. rosenbergii* Giant prawn	7. *Mytilus californianus* California mussel	8. *Haliotis rufescens* Red abalone
Lysine	+	+	+	+	+	[-]	+	+
Histidine	+	+	+	+	+	+	+	+
Arginine	+	+	+	+	+	+	+	+
Threonine	+	+	+	+	+	ND	+	+
Methionine	+	+	+	+	+	+	+	+
Leucine	+	+	+	+	+	+	+	+
Isoleucine	+	+	+	+	+	+	+	+
Valine	+	+	+	+	+	+	+	+
Tryptophan	+	+	+	+	ND	ND	+	+
Phenylalanine	+	+	+	+	+	+	+	+
Tyrosine	-	[+]	[+]	[+]	[+]	[+]	[+]	NC
Proline	-	-	-	-	-	-	[+]	-
Glycine	-	[+]	-	-	-	-	-	-
Hydroxyproline	-	ND	ND	-	ND	ND	ND	ND
Cystine/Cysteine	-	-	-	-	-	-	-	-
Serine	-	NC	-	-	-	-	-	-
Alanine	-	-	-	-	-	-	-	-
Aspartic acid	-	-	-	-	-	-	-	-
Glutamic acid	-	-	-	-	-	-	-	-

Reference

1. Lasser and Allen, 1976
2. Claybrook, 1976
3. Cowey and Forster, 1971
4. Shewbart et al., 1973
5. Miyajima et al., 1975
6. Watanabe, 1975
7. Harrison, 1975
8. Allen and Kilgore, 1975

Table Key	
+	Essential
-	Nonessential
ND	Not determined
NC	No conclusion
[ ]	Contrary to classical amino acid requirements

other factors (Lehninger, 1970). Other somewhat unexpected requirements were suggested for glycine in *Uca pugilator* (Claybrook, 1976) and proline in *Mytilus californianus* (Harrison, 1975). It is of some interest to note that both of these amino acids have been shown to be essential for the chick. The most striking deviation from classical qualitative amino acid requirements is the apparent non-essentiality of lysine in the giant prawn (*M. rosenbergii*) as suggested by Watanabe (1975). This highly unique observation has been further evaluated by Watanabe, Ahearn and Stahl (1976). During a 30-day growth study, juvenile *M. rosenbergii* were fed purified diets which contained constant amounts of all nutrients except lysine which was added at levels of 0, 0.7, 1.4, and 2.8% of the diet. After 30 days, the first three levels of lysine supported significant growth while the fourth level (2.8%) did not. The best growth occurred at the 1.4% lysine level followed closely by 0.7 and 0%. Unless an unidentified source of lysine was available to the prawns receiving the 0% lysine diet, it would appear that *M. rosenbergii* may indeed be capable of lysine biosynthesis. In evaluating such a possibility, one should consider the potential influences of (a) the nutritional status of the juvenile prawn prior to the start of the study, (b) the probable existence of nutrient reserves in the prawn's body, (c) the ability of an animal to preferentially conserve a specific nutrient when that nutrient is deficient in the diet, (d) the possibility of adequate quantities of lysine being made available for absorption from microbial synthesis in the prawn's gut, and (e) the short-term nature of the study (30 days). Under those conditions, it may have been possible for the prawn receiving the 0% lysine diet to mobilize sufficient quantities of lysine (from body protein reserves deposited in the pre-experimental period) to support minimal growth during the 30-day feeding study. Once such reserves were depleted, it is most probable that growth would cease. Thus we have the interesting question of whether growth would have continued in the prawn receiving the 0% lysine diet over an extended (i.e., 90 to 120 days) period of time.

With the exception of the recent work by Watanabe et al. (1976) in which a tentative requirement was established for lysine (1.4% of the diet), quantitative amino acid requirements for prawns remain undefined. In several attempts to quantify such needs, investigators have fed experimental diets supplemented with amino acids either singularly or in mixtures patterned after the amino acid profiles of well-established natural food sources or the tissue amino acid profiles of the organism under study (New, 1976). Addition of low levels of amino acid supplements to diets composed of natural ingredients has resulted in somewhat improved growth in penaeid sp. However at higher levels, reduced growth rates and efficiencies were observed,thus suggesting that the ability of the shrimp to utilize crystalline amino acids is rather limited. In the final assessment of the prawn's quantitative amino acid requirements, factors other than the diet composition per se must be addressed (several of these were mentioned earlier in this section).

Energy/Energy Sources

The need to provide proper amounts of readily digestible, high-energy feed sources in the prawn's diet cannot be overemphasized. Dietary energy has been shown to exert a significant influence on the growth rates and nutrient assimilation efficiencies of both domestic animals and fish. In the absence of adequate levels of digestible energy (i.e., low energy: high protein), an animal will utilize other dietary nutrients to satisfy the energy demands associated with normal cellular metabolism. An extreme of dietary energy (i.e., high energy: low protein) is known to increase the amount of tissue fat deposition and, due to energy satiation, the feed intake is thus reduced before sufficient quantities of protein are ingested, thereby resulting in reduced weight gains.

The energy demands of an animal are essentially two-fold. The first requirement is to satisfy the energy expenditures associated with maintenance of tissue integrity and organ functions which are vital to the survival of the animal. Only after these needs are satisfied can remaining available energy be utilized for production purposes (i.e., growth, maturation, reproduction, voluntary muscular activity, etc.). Of various factors which are believed to affect the overall energy requirement of the cultured crustacean, perhaps none are more important (with the possible exception of environmental factors) than the stage of animal development, the form in which the diet is presented, and the specific energy sources in the diet.

Stage of animal development: In a recent progress report, Stephenson and Simmons (1976) reported preliminary findings from studies on the respiration and growth of larval *M. rosenbergii*. Based on respiration data, caloric requirements of the larvae were found to be approximately seven times higher than those for juvenile prawns. According to Stephenson and Simmons, the higher energy expenditure of the larval stages may reflect their perpetual swimming habit and activity associated with food gathering as well as their extremely rapid development. Experiments conducted by Iwai (1976) have shown an increase in the energy assimilation efficiencies of juvenile *M. rosenbergii* with age. The optimum level of assimilation was found to occur at approximately six months.

Diet form: Caloric comparisons were made to assess larval growth of *M. rosenbergii* (Sick and Beaty, 1974) fed different forms of various experimental diets. Six artificial diets were prepared in three forms (freeze-dried, flakes, gel) and fed to stage 7-8 *M. rosenbergii* larvae under laboratory conditions. Results clearly demonstrated highest energy assimilations when diets in the freeze-dried form were fed. Larvae reared on freeze-dried and gel formula diets developed more rapidly than larvae fed other forms of food. Composition of the diet was also shown to exert an effect on assimilation rates.

Energy sources: When considering the total digestible energy of a diet, we normally think of either the carbohydrate or lipid component. As with domestic animals, the source of dietary energy is known to exert a significant effect on the efficiencies with which such energy is assimilated by crustaceans. In a series of feeding studies, Forster and Gabbott (1971) fed Palaemon serratus various carbohydrate sources at a level of 47.5% in test diets with white fishmeal as the only protein source (also included in the diets at a level of 47.5%). With the exception of cellulose, all carbohydrate sources tested were very well assimilated. Wheat starch, oyster glycogen, and dextrin were 100% digested by P. serratus, while potato starch and boiled potato starch had somewhat lower digestibilities (93.1 and 86.2%, respectively). Andrews, Sick and Baptist (1972) demonstrated that cornstarch was far superior to glucose in supporting sustained growth of juvenile Penaeus setiferus. Significant depressions in growth occurred when glucose was added at 20 or 40% levels to the diet. In a subsequent study with Penaeus duorarum, Sick and Andrews (1973) reconfirmed the superiority of cornstarch over glucose. At an equivalent level (40%) in the diet, growth was approximately 68% of that for control animals when cornstarch was the energy source, as compared to only 12% when glucose was used. Thus it appears that crustaceans are able to utilize complex polysaccharides (starches) more efficiently than simple sugars (glucose).

In comparison, the ability of fish to assimilate complex polysaccharides for energy purposes is quite limited. According to Halver (1976), monosaccharides are best utilized followed in order by disaccharides, simple polysaccharides, dextrins, cooked starches, and raw starches. In the case of dietary fat (which is normally regarded as a ready source of energy), fish are apparently capable of using levels as high as 20 to 30% of their diet (Halver, 1976). This unique ability to efficiently metabolize high levels of dietary fat has a concomitant sparing effect on the protein required for maximum growth. Under most conditions, corn oil has been found inferior to fish oil in the diet of fish. Dupree (1975) stated that additions of up to 15% fish oil in the diets of warm-water fishes have resulted in improved growth rates, while corn oil, when added at levels in excess of 5%, was shown to significantly depress growth.

Unlike fish, there is rather convincing evidence that crustaceans cannot tolerate high levels of dietary fat. Andrews et al. (1972) clearly demonstrated that the inclusion of a lipid supplement (1/3 beef tallow, 1/3 corn oil, 1/3 menhaden oil) at levels of 10% or more in the diet of P. setiferus had an adverse effect on growth and survival. In an experiment with P. serratus, Forster and Beard (1973) found no added benefits in growth when a shrimpmeal-based diet was supplemented with 7.5% or 15% levels of cod-liver oil or corn oil. Significant depressions in growth were observed at the 15% level of supplementation for both lipid sources. At the lower level, growth rates were approximately 15% greater for prawns receiving the cod-liver oil supplement. The inferior nature of corn oil as a lipid supplement in the crustacean's diet was further demonstrated by Sick and Andrews (1973) in a twelve-week growth study with P. duorarum. In this study, 10% levels of beef tallow, corn oil, linseed oil, or menhaden oil were evaluated in casein-based diets. Total mortality occurred in the shrimp receiving the corn oil supplement, while survival exceeded 54% in all other treatment groups. The high levels of omega (ω) 6 fatty acids in the corn oil were suspected as being responsible for mortality among the shrimp receiving the 10% corn oil supplement. In experiments with adult lobsters (Homarus americanus), Castell and Covey (1975) reported superior growth when cod-liver oil was incorporated at 5% of the total diet. Although reduced growth was observed when corn oil replaced the cod-liver oil, poorest results occurred when coconut oil was fed. These findings suggest that the essential fatty acid needs of the lobster are best satisfied by ω3 acids rather than ω6 or saturated fatty acids. In continuing studies on lipid and essential fatty acid requirements of M. rosenbergii, Joseph and Williams (1975) and Sandifer and Joseph (1976) reported better growth in postlarval prawns fed diets (either a commercial shrimp food or a semidefined diet) supplemented with 3% shrimp-head oil, which was shown to be a rich source of ω3 acids. Additional observations indicated that dietary fatty acids of the ω6 family (principally 18:2ω6) inhibited, while acids of the ω3 series improved growth. A tentative requirement (18:3ω3) was suggested by Shewbart and Mies (1973), based on improvements in growth of Penaeus aztecus fed a diet augmented with 1 to 2% linolenic acid. However, no information on total lipid levels of the diet was provided; thus the growth response may have resulted simply from the greater energy content of the diet (as contributed by the fatty acid additions).

Based on the preceding information, it is obvious that extensive research will be required to define the dietary energy requirements of M. rosenbergii and other important crustacean species. Emphasis must be placed on the differential needs of the prawn at various stages of growth and development. Currently, the most acceptable energy sources appear to be starches (wheat and corn starch being better utilized than potato starch). Incorporation of such ingredients at levels between 30 and 40% of the dry diet is most common. In assessing the energy needs of the prawn, careful consideration must be given to the protein component of the diet as well as other essential nutrients. With regard to dietary lipids, additions of neutral fat (free oil) supplements should not exceed 5 to 7% of the diet (dry weight basis). These supplements should be characterized by a relatively high proportion of ω3 polyunsaturated fatty acids (as found in fish and shrimp-head oils), and the total lipid content of the diet (residual lipids plus supplements) should not exceed 10%. A critical need exists for intensified lipid nutrition research, designed to (a) better define the role which dietary lipids have in meeting the energy requirements of the prawn, (b) quantify the optimum level of dietary lipid (as well as type) required for maximum growth of prawns, and (c) further characterize the quantitative requirements of essential fatty acids of the ω3 family and their relationships with other polyunsaturated fatty acids in the diet.

Minerals, Vitamins, and Other Nutrients

Although Halver (1975) indicated that little is known about the mineral requirements of fish, the essentiality of calcium (Ca), phosphorus (P), and iodine (I_2), as well as the importance of the balance between sodium/potassium, have been demonstrated. Furthermore, as pointed out by Ketola (1976), dietary requirements for P and I_2 have been quantified for both salmon and catfish. Thus, it would appear that scientists have taken an important initial step toward gaining a more thorough understanding of the role which dietary minerals have in fish nutrition.

For the cultured crustacean, few studies have been reported on the subject of mineral nutrition. Quite possibly, work of this type has been deferred, based on the assumption that most mineral needs can be satisfied by direct absorption from the environment. A limited number of experiments were reviewed (New, 1976) in which penaeid shrimp received diets supplemented with different levels (2 to 19.5% of the dry diet) of mineral mixtures containing a wide range of Ca:P ratios (0.76:1 to 4:1). Considerable emphasis was focused on the importance of the Ca:P ratios in the mineral supplements used in those studies with little if any attention given to the overall ratios in the final diets. Such an oversight should be avoided, since in the final analysis of the diet, the Ca:P ratio may be substantially different from that originally intended due to the contributions of these elements from the major dietary ingredients. For instance, menhaden fishmeal which (on the average) contains 5.3 and 3.0% Ca and P, respectively, has an approximate Ca:P ratio of 1.76:1 (Miller, 1973). If a mineral supplement with a Ca:P ratio of 3:1 is incorporated at a rate of 5% in a diet containing 35% menhaden fishmeal, then the effect of the latter on the final dietary Ca:P ratio becomes obvious.

In short, we know next to nothing about the dietary mineral requirements of the prawn. Until the appropriate studies yield the information which is needed, we must continue to speculate and guess at answers to the following questions: (a) which minerals are required in the prawn's diet; (b) what levels of each mineral need to be incorporated; (c) which form is the best utilized (most available); and (d) are there significant mineral interactions that need to be defined?

Knowledge of the vitamin requirements of crustaceans in general is very scant (Walker, 1975), particularly if one is referring to those candidate species identified for commercial culture. This situation is in contrast to that of several valuable freshwater fishes such as salmon, trout, catfish, etc., for which rather specific requirements have been established for a number of the water- and fat-soluble vitamins, as reviewed by Ketola (1976).

As with present-day mineral mixtures, many of the vitamin premixes used in prawn diets (both experimental and commercial feeds) are obtained commercially and have been formulated according to specific requirements established for domesticated terrestrial species. Nevertheless, a limited amount of prawn research has been reviewed (New, 1976) in which the importance of β carotene and ascorbic acid (vitamin C) in penaeid nutrition was investigated. As yet no quantitative requirements have been defined. In fact, the role of β carotene has been considered more from its involvement in pigment (carotenoid) formation than as a precursor of vitamin A. Whereas P. japonicus has been shown to utilize dietary β carotene in pigment synthesis (Kitayama et al., 1972), preliminary evidence was presented by Conklin (1975) which suggested that β carotene may not be the precursor of pigments in the lobster (H. americanus). A similar conclusion was arrived at by Joseph (personal communication) in a recently completed trial in which M. rosenbergii were fed a diet supplemented with 40 ppm β carotene. However in our laboratory (studies in progress) we have found that as little as 20 ppm β carotene was capable of producing clearly visible changes in the carapace pigmentation of juvenile blue crabs (Callinectes sapidus) fed artificial/formulated diets. These differences in the ability of various organisms to utilize β carotene in pigment synthesis exemplifies one of the major pitfalls associated with blanket assumptions that what is required by one crustacean species is appropriate for another.

The composition and levels of several vitamin premixes used in studies with penaeid and Macrobrachium species are reviewed by New (1976). A cursory evaluation reveals that some enormously high levels of individual vitamins have been fed. For instance, vitamin D, choline chloride, and vitamin C have been fed at levels varying, respectively, from 5 to 6, 5 to 10, and 18 to 500 times normally accepted levels. Since we know so little about the qualitative/quantitative vitamin requirements of crustaceans, the potential for acute hypervitaminosis/antagonisms is markedly enhanced at such high levels of vitamin inclusions in the diet. In the conduct of research to examine such needs, special care must also be taken to account for the losses of water-soluble vitamins (as well as other soluble nutrients) from the test feeds.

Cholesterol: As early as 1962, evidence was presented by Zandee that the crayfish, A. astacus, was incapable of sterol biosynthesis. Subsequent studies confirmed this lack of sterol synthesis in other crustaceans including the European lobster, Homarus vulgaris, (Zandee, 1964); the crab, Cancer pagurus, (Van den Oord, 1964); and the blue crab, C. sapidus, (Whitney, 1970). As a result of her studies, Whitney (1970) postulated that a dietary supply of animal sterol is required for the proper development of C. sapidus. Several reports by Japanese scientists have further identified (as reviewed by New, 1976) the essentiality of dietary cholesterol for the shrimp, P. japonicus. In those studies, cholesterol additions to formulated diets ranged from 0.05 to 5.0%. Although there was uncertainty as to the optimal level of dietary cholesterol, a noticeable reduction in growth occurred at the highest (5%) addition level. More recently, Castell, Mason and Covey (1975) established that the optimum or required level of cholesterol for H. americanus was 0.5% of the dry diet weight. This requirement is identical to that recommended by Kanazawa et al. (1971) for P. japonicus. Details on the utilization of dietary cholesterol as well as other sterols by crustaceans are discussed by New (1976). Without going into detail, cholesterol has been implicated in the crustacean as an important precursor of various steroids, vitamin D, brain and molting hormones, and has been shown essential for elaboration of new subcellular membranes associated with the rapid growth and molting of crustaceans.

For the giant prawn, M. rosenbergii, neither qualitative nor quantitative cholesterol requirements have been reported. However, it is quite unlikely that this particular crustacean differs from others for which a dietary cholesterol requirement has been demonstrated.

Fiber: There is little known about the nutritional significance of fiber in the diet of crustaceans. Yet, as illustrated in Table VIII-3, it has been a common practice among crustacean nutritionists to include cellulose (primarily as a dietary filler) in experimental rations at levels ranging from 0% to as high as 75% of the total dietary dry matter. Although no studies have been reported which address the potential impact of fiber additions to the prawn's diet, the use of this substance at high levels is questionable for several reasons. First, in comparison to domestic animals, the relative length of the crustacean's gastrointestinal tract is considerably shorter. Thus, the amount of time that ingested food is available for digestion and absorption in the crustacean is extremely limited due to rapid passage rates through the gut. In domestic animals, it has been shown that ingestion of high levels of dietary fiber results in a subsequent increase in the rates of gastrointestinal motility. It is reasonable to suspect that a similar pattern would occur in the crustacean when dietary fiber levels are increased, thereby reducing overall nutrient assimilation. Excess dietary fiber could also exert a negative influence on overall ingestion rates. Conceivably, the inclusion of wide ranges of cellulose in test diets could play as much of a role in the final outcome of a crustacean nutrition study as the specific nutrient(s) under investigation.

Although it has been observed (Ling, 1969) that M. rosenbergii is omnivorous in its eating habits, it is unlikely that plant detritus would account for more than 10 to 20% of the prawn's diet. The ability of crustaceans to utilize cellulose appears to be rather limited as suggested by Forster and Gabbott (1971). In that study, when included at a rate of 47.5% in a defined diet and fed to the prawn, P. serratus, cellulose digestion/assimilation was only 20%, as compared to 90% and greater for other carbohydrate sources tested at similar dietary levels. It is most probable that extracellular microbial cellulase present in the prawn's gut was responsible for the limited digestion which did occur; however, the possibility of an endogenous cellulase system in the prawn cannot be totally ruled out. Whereas cellulose may not be well digested, Venkataramiah et al. (1975) indicated a potential benefit in growth, survival, and food conversion efficiencies of P. aztecus when limited quantities of vegetable fiber were included in their diet. The fiber source in this study was dehydrated, finely ground turnip greens and addition levels ranged from 0 to 5.5% of the diet.

Table VIII-3
Use of Non-nutritive Fiber (Cellulose) in Experimental Diets Fed During Studies on the Nutritional Requirements of Crustaceans.

Species	No. of Diets Evaluated	No. of Cellulose Levels Used	Range of Cellulose Levels Used (% of diet)	Reference
Penaeus setiferus	26	13	0 to 75	Andrews et al., 1972
Penaeus duorarum	18	11	0 to 51	Sick and Andrews, 1973
Palaemon serratus	5	1	5	Forster and Beard, 1973
Macrobrachium rosenbergii	6	1	22	Sick and Beaty, 1974
Penaeus japonicus	14	9	0.3 to 23.3	Deshimaru and Kuroki, 1974
Homarus americanus	4	1	15	Castell and Budson, 1974
Penaeus japonicus	9	3	4 to 29	Deshimaru and Kuroki, 1975
Homarus americanus	11	5	13.8 to 28.8	Castell et al., 1975
Macrobrachium rosenbergii	6 (fed to larvae) 9 (fed to juveniles)	1 6	22 7 to 42	Sick and Beaty, 1975
Macrobrachium rosenbergii	21	14	1 to 69.6	Sick, 1976

The significance of fiber in the diet of crustaceans has been a neglected area of research. However, because of the relatively high levels of cellulose incorporation in many experimental feeds used today, appropriate research should be initiated to characterize the potential impact of this particular ingredient in crustacean nutrition experimentation.

Feeding Rates and Efficiencies

Appropriate feeding methods and schedules must be developed whether one is dealing with the rigid, somewhat artificial conditions of the basic nutrition study or the practical conditions of an actual culture operation in which feeding schedules assume significant economic importance. Key factors which should be considered include: (a) biomass/ feeding rates, (b) frequency with which feed is offered, and (c) length of the daily feeding period and the inherent characteristics of the feed, with particular attention to its ability to resist "nutrient leakage," and its palatability to the animal.

Careful estimates of the total biomass to be fed enables one to adjust the rate of feed offered, thereby minimizing costly feed wastage. As pointed out by New (1976), in many of the feeding studies which have been carried out with prawns, researchers have adjusted feeding rates according to the amounts of un-ingested feed remaining at the end of each feeding period. Under actual culture conditions, such an approach would prove highly uneconomical. In most instances, therefore, feeding rates are maintained between 5 and 10% of estimated biomass with the higher level used most frequently during the first 60 days of postlarval existence (Goodwin and Hanson, 1975), when growth rates and nutrient needs are most critical. In studies with *P. setiferus*, Sick, White and Baptist (1973) found that ingestion rates (expressed as a percentage of body weight) were inversely related to animal size. A similar relationship was observed during preliminary nutritional investigations with *C. sapidus* fed an artificial diet (Biddle, unpublished data). Although this relationship is important, one should remember that under practical conditions, the absolute amount of feed required for maintenance and growth in a larger animal is greater than that for a smaller animal (as shown in the somewhat hypothetical example in Table VIII-4).

Table VIII-4
Feed Requirements as a Function of Size.

	Live Biomass Weight	Estimated Ingestion Rate	Total Feed Required
	gms	%	gms
Group A	15	5	7.5
Group B	35	3	10.5[a]

[a]Group B would require approximately 29% more feed.

The differences suggested in the above example may be further exaggerated when one considers that the efficiency with which absorbed nutrients are utilized by a larger animal is lower. This relationship was observed by Balazs and Ross (1976) in recent studies with M. rosenbergii juveniles fed formulated diets containing variable levels of dietary protein. Thus, to achieve an equivalent amount of weight gain, the larger prawn would have to consume greater amounts of feed. This situation adequately reflects the overall consensus that the most cost-critical period in intensive prawn culture (with respect to feeds) is that of the grow-out phase.

The frequency with which food is offered to the cultured organism should not be overlooked. Because of the relatively short digestive tract which is characteristic of most crustaceans, rates of food passage are quite rapid. Thus, the amount of time that ingested food is available for absorption/utilization is limited. In its natural environment, M. rosenbergii has been observed to eat frequently and rapidly (Ling, 1969). Therefore, in order to maximize ingestion and subsequent nutrient utilization, the prawn culturist should provide small amounts of food several (3 to 4) times each day, rather than the total allotment once a day. This approach would also tend to minimize the adverse changes in palatability of the diet as well as nutrient losses due to dissolution in the aquatic environment.

The duration of the daily feeding period has been shown to affect ingestion rates in P. setiferus (Sick et al., 1973). In that study, a significant decrease in ingestion rates occurred after 6 hours of exposure to a given quantity of food, which, according to the authors, was most probably due to changes in the physical and chemical characteristics of the feed. Prolonged immersion in water increases the porosity of the feed particles, thus accelerating the loss of nutrients to the water. Walker (1975) reported that in an experiment which evaluated the leaching rate of vitamins from a pelleted diet, over 90% of the thiamin had leached from the diet within 18 hours of submergence. Such losses of nutrients undoubtedly affect the chemosensory properties the food may have had originally, to say nothing of the nutritional quality (Sick et al., 1973; Provosoli, 1975). Thus, diet stability assumes major significance in the final assessment of appropriate feeding schedules and practices, which in turn exert a considerable influence on the overall assimilation/utilization efficiencies of the consuming organism.

It has been suggested (Walker, 1975) that the "average edible crustacean" possesses a great potential for rapid growth while maintaining a high efficiency of food conversion (1.5 or less). If such efficiencies are to be realized under intensive culture conditions where formulated diets are used, then additional research on the quantitative nutrient needs of the prawn (at various stages of development) must be conducted. Additional research is also required on the influence of environmental variables (i.e., water salinity, temperature, and quality) on overall feeding activities and assimilation/utilization efficiencies of the prawn.

Conclusion

In the introduction of this paper, the question was posed: what do we know about the nutrition of *Macrobrachium* spp. ? Based on the preceding discussion, it is evident that existing knowledge on this subject is extremely limited. To date, the empirical-practical approach to feed formulation and evaluation has been emphasized. While this approach has a definite role in the field of *Macrobrachium* nutrition, the ultimate development of cost-effective, nutritious feeds will depend on a marked acceleration of carefully controlled experiments designed to quantify the specific nutrient requirements of the prawn.

This type of information was recently identified as one of the highest-priority needs for the continued growth of the *Macrobrachium* industry (Anon., 1976). In order to effectively respond to this need, it will require the coordinated efforts of responsible scientists from university, state, and Federal laboratories as well as from private industry. To ensure meaningful coordination, perhaps we should consider the formation of an informal working committee of scientists from appropriate disciplines (i.e., nutrition, biology, food technology). The committee would serve primarily as a catalyst in developing a set of guidelines and recommendations to assist in the expeditious and systematic definition of the nutritional requirements of *Macrobrachium* prawns. As a further extension of this effort, we should consider greater use of existing newsletters as an appropriate medium for dissemination of current information on *Macrobrachium* R&D. Quarterly newsletters such as those of the World Mariculture Society (contact: P. A. Sandifer, MRRI, Box 12559, Charleston, S.C. 29412) or the Fish Culture Section, American Fisheries Society (contact: J. V. Huner, Box 10809, Southern University, Baton Rouge, Louisiana 70813) provide a useful supplement to normal channels of scientific exchange (i.e., journals, proceedings of meetings/workshops) which routinely have a considerable time lapse between presentation and publication.

In concluding this review, a series of priorities for *Macrobrachium* nutrition research are offered for further consideration.

- One of the greatest needs in nutrition research is for the development of standardized diets against which animal response to experimental diets (with varied nutrient densities) can be compared (New, 1976).

- Improved feeding techniques must be adapted to adequately evaluate the importance of consumption patterns on growth in prawns fed diets of varied nutritional quality.

- The establishment of uniform experimental conditions is of key importance in order to facilitate valid comparisons and interpretation of data obtained by different investigators.

- Considerations must be given to the selection of experimental criteria (other than survival, growth, conversion efficiencies) which will allow a more thorough assessment of animal response to specific nutrient densities.

- In the formulation of experimental diets, researchers must make greater efforts to chemically define their diets rather than depend on the information contained in various feed ingredient tables (New, 1976).

- Research on the nutrition of Macrobrachium must consider the individual requirements of prawns at different stages of physiological development.

- Of the various dietary nutrients required by the prawn, current emphasis should continue to focus on definition of protein/amino acid and lipid/essential fatty acid requirements.

- Additional research is required to better define the effects of physical form, individual ingredients, and processing variables on the stability, palatability, and digestibility of prawn feeds.

- Attention must also be given to the physiology of digestion and absorption of nutrients by the prawn with special emphasis on the significance of competition or antagonisms that may exist between specific nutrients for absorptive sites in the gut of the prawn.

These are but a few of the many aspects of Macrobrachium nutrition that require immediate attention.

With the rapid growth of the industry, we can no longer ignore the critical need for carefully defined research on the nutritional requirements of Macrobrachium spp. As stated previously, this is the only way in which highly efficient, economically feasible feeds will become available to the industry.

Acknowledgement of Contributors

The author is grateful to Dr. D. Miller and Mr. P. E. Bauersfeld, Jr., for their critical review of the paper, and to Mrs. R. Knott, Mrs. M. Larson, and Mrs. R. Sanders for typing the manuscript.

Sincere appreciation is also expressed to each of the individuals listed below. Through their cooperation in providing appropriate information (i.e., reprints, preprints, unpublished reports as well as personal comments), the preparation of this discussion paper was greatly facilitated.

Dr. G. E. Balazs, Hawaii Inst. Mar. Biol., U. of Hawaii
Dr. J. E. Bardach, Hawaii Inst. Mar. Biol., U. of Hawaii
Dr. D. Conklin, Bodega Bay Lab., U. of Cal.
Mr. T. Fujimura, Anuenue Fish. Res. Cent., Hawaii
Mrs. J. D. Joseph, Mar. Resource Res. Cent., South Carolina
Dr. A. W. Knight, Water Sci. & Eng., U. of Cal., Davis
Mr. E. S. McSweeny, Syntex, California
Mr. L. S. Miyajima, Sea Breeze Proc., Texas
Mr. M. B. New, Kelvin Hughes Aquacult. Ser., England
Dr. P. A. Sandifer, Mar. Resource Res. Cent., South Carolina
Dr. L. V. Sick, Coll. Mar. Studies, U. of Del.
Ms. M. S. Stahl, Hawaii Inst. Mar. Biol., U. of Hawaii
Dr. S. Willis, Mar. Res. Lab., Florida

B. Summary of Discussion of Nutrition

Jeanne Joseph of the Marine Resources Research Institute has been investigating the effects of lipids and fatty acids on prawn nutrition, and earlier reported on the pigmentation produced in prawns by the addition of lipids extracted from shrimp heads. When Purina rations were impregnated with shrimp-head lipids, the prawns developed good pigmentation very quickly, with a pink color that was stable through cooking. Later experiments used Beta carotene as a precursor, but first results indicated that carotene fed at 40 ppm for 12 weeks did not result in any higher levels of pigmentation than those found in prawns fed with untreated rations.

In growth experiments, various Purina rations were impregnated with oils or purified fatty acids. A few treated diets produced better growth than control diets. Fatty acid analyses showed that the successful diets had an Omega 6 to Omega 3 ratio of 2:1. By chance, twice as much Omega 6 fatty acids as Omega 3 acids were included in the diets. Studies are now in progress to confirm and refine data on the ratios, and on total lipid levels.

Margo Stahl reported on work at the Hawaii Institute of Marine Biology where investigators are examining both the qualitative and quantitative aspects of amino acids in prawn diets as an initial step in producing a high-quality, inexpensive ration. Through use of glucose labeled with Carbon^{14}, the investigators foun that *M. rosenbergii* (in common with roaches and alligators, but few other creatur apparently can synthesize lysine. Although the conclusion may follow that lysine i unnecessary for growth of prawns on a purified amino acid diet, optimal prawn growth occurred when a lysine concentration of 1.4% was included in the prawn diet. Experiments are in progress to determine whether the prawn actually synthesizes the amino acid or whether synthesis is accomplished by gut microbes.

Since the workshop (personal communication), Stahl notes that argenine and methionone also appear "dispensable" for significant growth, but preliminary data indidate that concentrations of argenine of 1.3% and methionone of over 3% give best growth.

In commenting on similarities in crustacean nutrition problems, Douglas Conklin of Bodega Laboratory, University of California, made a plea for inclusion of prawn nutrition workers in a developing cooperative program of standardization. There already is agreement among lobster (*Homarus*) researchers and some shrimp (penaeid) researchers to establish standard techniques and methods of reporting data and results for ease and accuracy of comparison. The researchers hope to show that a reference diet for cultured crustaceans is feasible. Such a reference diet would permit comparison of results among researchers on single species and genera, as well as comparison over the range of cultured crustaceans.

As a consequence of the workshop, President Harold Webber of the World Mariculture Society appointed Conklin as chairman of the WMS nutrition panel with the express mission of agreeing on standards.

CHAPTER IX
THE STATUS OF PRAWN PROCESSING

Actual commercial production of cultured freshwater prawns today is not very much beyond cottage industry in scale except for a few larger producers. And in all cases, the product is sold live or blanched and iced with heads on to local consumers. So far, the freshwater prawn industry has not faced the problems of long-term preservation and mass distribution over a wide geography. But the intent is to make these animals truly commercially successful. This means that the intent is to create just such problems of processing and preservation. In this chapter, Virginia Sidwell reports the results of her examination of this issue and authorities such as Kenneth Kato, George Kelley, and Robert Shleser offer additional information.

A. Status of Prawn Processing:

Virginia D. Sidwell, Southeast Utilization Research Center, National Marine Fisheries Service

Introduction

Macrobrachium is a gourmet's delight; however, this delicacy is enjoyed by only a few. Today, the production is low and the most current limiting factor, therefore, is availability. When production increases, one key limiting factor may be: how can this prawn best be processed and marketed to maintain prime quality and identity ?

The institutional trade will prefer the freshwater prawns alive, so that they can be prepared for the consumer immediately after they have been sacrificed, when they are at their peak of flavor and texture. Alternatively, the animals may be brought to market heavily iced immediately after harvest. Under this condition, the animals are good for only a relatively few days--four to five. If cooking is delayed, the tail muscle rapidly becomes quite soft and undesirable. Such methods of preservation are adequate when the market is close to the supply and the animal is consumed soon after harvesting. In an expanded market, however, these methods may well be too costly and inefficient qualitywise and therefore not practical.

As the art of growing prawns becomes more refined and production increases, it will be necessary for industry to process the raw material using appropriate methods to assure products of initial high quality and with sufficiently long shelf life for both fresh and frozen products to meet market and consumer requirements. The processes must not be too complicated nor too expensive. The frozen processed product must have a shelf life of at least six to eight months, retaining its identity and consumer appeal. The fresh product, if iced, should have a shelf life of at least seven to ten days.

The objectives of this paper are (1) to review the present status of *Macrobrachium* processing methods, (2) to describe a proposed study whereby the effectiveness of some processing parameters can be evaluated, and (3) to express the need for information on the composition of *Macrobrachium rosenbergii*.

Present Status

From telephone conversations and from responses to my letter to the various attendees of last year's meeting, I learned that, since the meeting at St. Petersburg in 1975, the scientists who are now working in aquaculture of freshwater prawns have done very little work on the processing of the prawn to extend its quality and shelf life, fresh or frozen.

George Kelley of Harrell Internation is successful in chill-killing and blast-freezing the animals, then ice glazing and storing them at -30° F. Under these conditions, it is expected that the shelf life can be extended to at least six months.

At Fish Farms Hawaii at Laie, Oahu, the live animals are chill-killed, blanched for 15 seconds in 150° F water, iced, and then taken to market. I understand that this set of procedures does not always langthen the shelf life of the "fresh" nonfrozen prawn. Chilling does kill the animal, but I question if its internal temperature is raised enough by the blanching treatment (15 seconds at 150° F) to deactivate the proteolytic enzymes present in the hepatopancreas. These enzymes may be one of the causes for the rapid deterioration of the tail muscle. Thus, the rather cold dead prawn is placed in 150° F water for only 15 seconds. Each treatment does not last long enough to change the internal temperature of the animal's body significantly, since the chitanous carapace is a poor conductor of heat. In fact, so is protein. The effectiveness of this step is also associated with the size of the animal and the thickness of the carapace. Therefore, the last step is probably useful only as a surface effect to free the animal from residual microorganisms that may contribute to spoilage.

Because of the factors I have mentioned which interfere with the penetration of cold or heat, the procedure as a whole is undoubtedly not severe enough to deactivate the enzymes in the hepatopancreas. Chilling or packing in ice will only reduce their activity to a lower level; they will still be active to a relative degree and will resume their original activity as the temperature of the body goes up. The proteinaceous enzymes must be denatured before their activity is destroyed.

Lester Miyàjima of Aquaprawns, Inc., is involved in a nine-month storage study comparing tails only with partially gutted (allowing for the removal of the hepatopancreas tissue) and with the whole prawn. Half of each of the experimental groupings were frozen and glazed; the other half stored in vacuum packages and tightly sealed. At the end of sixty days, there was no dramatic difference between the glazed tails and the vacuum packed ones. The same can be said for whole animals held under the same conditions. He did not say at what temperature the carcasses were held.

Bruce Morgan, Amfac Foods, reported that the faster the animal is frozen, either with or without preblanching, the better the final product.

Albert Eble, Trenton State College, sacrifices the *Macrobrachium* by plunging it into ice water. The body is then broken between the cephalothorax and abdomen and the abdomen or tail packaged in freezer bags, sealed, and frozen. He claims that the bags must be heat sealed with no leaks or the food product will deteriorate in the freezer, experiencing so-called freezer burn.

Several others indicated that they quickly freeze the carcasses and store them at low temperatures.

In summary, the present most effective processing method for frozen *Macrobrachium* products is chill-killing and freezing the animals very soon after they have been harvested, packaging in airtight containers or glazing with ice, and then storing at a low temperature. This process will lengthen the shelf life to at least six to eight months.

However, the process cannot be used by warehouses, restaurants, and consumers who lack -30° F freezers. Low temperature does not wholly deactivate the enzyme activity. Thus, when the product is held under ongoing commercial conditions, the enzyme systems will undoubtedly become more active and cause the prawn muscle to deteriorate. Also, this approach is not satisfactory for nonfrozen products, even with a brief hot water dip. Therefore, other processing techniques are necessary to assure a high-quality product for an expanding market.

Proposed Processing Study

The main objective of a series of studies proposed by SURC will be to find other processing parameters that will lengthen the storage life of frozen and nonfrozen products and still maintain the desirable characteristics of the fresh prawn to the extent possible. We intend to focus on the use, initially, of a heat blanch to deactivate the enzyme systems for frozen products. It will be necessary for us to find out: (1) whether the animal should be processed whole or headless; (2) how long the animal should be kept in the cooking water, as well as the temperature of the water; and (3) how much salt should be added to the cooking water.

The prawns for this study will be obtained from the freshwater experimental ponds of the Marine Resources Research Institute, Charleston, South Carolina. The carcasses will be processed either whole or headless. For the first series of tests, they will be dropped into boiling water containing 0%, 3%, or 6% salt and the water will be rapidly brought back to boiling temperature. One batch will have no cooking. That is, the controls for this study will be whole or headless raw prawns. Other groups will be either removed immediately from the boiling water or held in it for a total of three or six minutes. They will be removed from the hot water, cooled with ice water, sealed in plastic bags, and frozen.

The results of this study should give us a better means to evaluate the effectiveness of heat blanching, employing several parameters, for processing the prawn and marketing it as a frozen convenience food. The prawns can be split and heated under a broiler or heated whole in a microwave oven. Under these conditions, the identity of the freshwater prawn has not been lost to the consumer. We also plan to subsequently evaluate this same approach for application to the fresh, nonfrozen product.

Composition

Since the data in literature on the chemical and nutritional composition of the freshwater prawn are quite scarce, I had our analytical group analyze the raw and cooked tails of one lot of small prawns grown in southern Texas. There were 19.2 grams of protein per 100 grams of raw muscle. The cooked prawn contained 26.1 grams. The fat content was about 2%. The raw prawn contained about 100 mg of nonprotein nitrogen and the cooked 73 mg.

During the coming year, we would like to generate some additional composition data. As you know, the composition of seafood varies with age, size, diet, and geographical location; therefore, I would like to have a sample of each of two sizes of prawns--small (18 per pound) and large (5-6 per pound)--from as many aquaculturists of freshwater prawns as would like to participate. Also,

I would like to limit my study to one species, *Macrobrachium rosenbergii*. In order to do a complete analysis, it is necessary to have about two pounds of whole animals corresponding to one growing/harvest variable. Presently, the plan is to determine the protein, fat, ash, moisture, micro- and macroelements, fatty acids, amino acids, and cholesterol content of the prawn tails. If at least ten growers participate, it will give us a reliable value for each of the components. The nutritional information can be used as a guideline for the processor who may want to include nutritional information on the label of his product. Relationships of final product composition to feeding and environmental relationships may also be possible.

In short, the two studies we plan to pursue are just the beginning of the extensive work that will need to be done on the processing of the *Macrobrachium* species. We would appreciate any advice or suggestions as to our proposed studies or information on related studies that others may be proposing that could lead to a more coordinated approach.

B. Discussion of Prawn Processing

Ken Kato reviewed the Hawaii experience in handling prawns for market, emphasizing procedures at Fish Farms Hawaii, largest of the state's prawn growers. The harvested prawns, still alive, are carried to the processing area in recirculating tanks, are chill-killed and immediately dipped for 15 seconds into a blanching tank of 150° F water.

The Hawaiian market demands a fresh, chilled product. Experimentation in handling prawns to meet market requirements was conducted by fish markets. Maynard Steinberg of the National Marine Fisheries Service technology laboratory at Seattle first suggested blanching, but the fishmarket operators proceeded carefully because blanching for too long at too high temperatures would result in the typical cooked crustacean color change, and the operators did not want a cooked product. Blanching temperature available was that of a hot water heater, 150° F. By blanching for 15-20 seconds, then chilling, natural color would be nearly restored. Review of the literature by Steinberg disclosed that blanching at temperatures of 130-135° F could deactivate catheptic enzymes, and Kato believes that the Hawaiian blanching method does deactivate the cathepsins in the musculature, although not the proteolytic enzymes in the hepatopancreas. Unblanched iced prawns go mushy in three or four days, while the blanched, iced prawns are relatively stable for six to seven days, which is suitable for Hawaiian production levels and the Hawaiian market.

Reviewing the experience of Harrell International/Jacksonville University's prawn project in Florida, George Kelley noted that heads-on prawns deteriorated quickly, beginning with musculature in the carapace that supports the chelae. The muscle tissue goes mushy, developing the consistency of "shrimp-flavored mashed potatoes." Prawn tails remain stable under icing for up to six days. As standard procedure, Kelley chill-kills the harvested prawns, which also serves to begin hardening them, then freezes the prawns and ice-glazes them. Under this treatment, the prawns remain stable for six months and longer without freezer burn or deterioration.

Several persons have experienced one mushy prawn among many. While definite reasons for the occasional mushy prawn remain to be determined, two considerations were advanced:

Kato said that Hawaiian ponds produce an occasional male with very long chelae and very short bodies, with apparent spacing that has developed between the cephalothorax and abdomen. The animals are lethargic. When cooked and eaten, these animals turned out to be mushy. They may have reached terminal growth.

Shleser reported seeing cooked animals with complete and intact hepatopancreases that could be lifted out upon cleaning, but with occasional animals in which the hepatopancreas had been disrupted and the muscle tissue had gone soft. Shleser also reported discussion with a prawn grower from Mauritius whose practice was to chill-kill his prawns and then remove the hepatopancreas to prevent softening.

Another possibility, of course, is that the occasional animal may be recently molted, with shell hardened but tissues still soft.

CHAPTER X
INDUSTRIAL-SCALE PRAWN CULTURE

Freshwater prawn culture may have achieved the status of profitable business on scattered small scales, but it is not yet an industry in the economic sense of the term. There are today several "industrial" participants, but they all apply the term "pilot" to their current operations. What are the roadblocks to industrial-scale prawn farming and what should industry do to overcome them? In this chapter, J. P. Bennett examines this question and his information is augmented by the experiences of such persons as Albert Saari, Durwood Dugger, Norman Moe, Kenneth Kato, and Peter Halmos.

A. <u>The Role of Industry in Macrobrachium Culture</u>: John P. Bennett, Syntex Aquaculture Sciences

Private industry did not increase its interest in the aquaculture of *Macrobrachium* to a point of involvement until the early 1970's. The development of a *Macrobrachium* aquaculture industry is still in its infancy.

The first attempts to grow *Macrobrachium* for sale in the United States were probably made by Fish Farms Hawaii on Maui and shortly after by APARI in Homestead, Florida. Fish Farms Hawaii later moved from Maui after harvest failures to become Pacific Aquaculture and now has a small number of producing ponds on Oahu. APARI did not fully develop an industrial aquaculture farm and was recently purchased by Weyerhaeuser with the intention of expanding pond farming of *Macrobrachium* in Florida after completion of feasibility studies.

Since then several other companies have entered this budding crustacean farming industry both in pond aquaculture and with intensive systems (Table X-1).

Table X-1

Companies Involved in Aquaculture of Macrobrachium

Extensive:	
Pacific Aquaculture	Hawaii
Caribe King	Puerto Rico
General Mills	Honduras
Sun Oil	Texas, Costa Rica
Harrell International	Florida
Weyerhaeuser	Florida
Florida Power and Light	Florida
Intensive:	
Syntex Aquaculture Sciences	California
Solar Aquafarms	California

However, the number of companies actively participating remains small. In addition, there are a number of entrepreneurs in Florida and California who are active with a few acres of ponds. In pond aquaculture, most of the ventures are still at small-scale industry level (as in Puerto Rico and Hawaii) or are in the pilot phase as in Latin America. There is at present no "aquaculture industry" but prospects for the future appear promising if certain difficulties can be overcome (Table X-2).

Table X-2

Difficulties Toward Successful Industrial-Scale Macrobrachium Aquaculture

- Shortage of *Artemia* and no substitute for larval feeds
- Lack of low-cost diets for adult prawns
- New gourmet food product--high advertising costs
- Heads-on market needs to be developed in U.S.
- Small sizes compete with penaeids
- Price volatility of prawns--high-risk commodity
- Intensive systems--high capital cost
- Pond systems in U.S.--few sites, E.P.A. controls
- Pond systems abroad--government stability, communications, management, weather
- Inadequate harvesting techniques

Constraints to Industrial-Scale Macrobrachium Aquaculture

The biggest constraint to the developing industry is the shortage of *Artemia* for larval rearing and the lack of a suitable live replacement or a reliable low-cost larval dry diet.

Quantities of about 500 pounds of Artemia are required to supply postlarvae for a 30-acre farm, assuming 50% survival of larvae. Continual supplies on a regular and rapidly increasing scale of demand probably cannot be delivered by the major supplier, San Francisco Bay Brand of Metaframe, to support farms of 100, 500, and 1,000 acres as are projected for farming locations in Latin America, Puerto Rico, and Hawaii. The position is so acute that periodic shortages of Artemia might for a while even curtail research in prawn aquaculture.

A further constraint to Macrobrachium culture is the lack of a proven market in the United States for this crustacean as a heads-on animal. Presently the Macrobrachium tails, of low quality, are imported from Asia and sell at a discount from penaeid tails. A large crustacean importer in San Diego has stated that aquaculture-grown Macrobrachium prawns would retail at a higher rate than imports, probably at prices equivalent to penaeids in the Green Sheet.

In intensive systems, it is difficult to produce a successful return on investment if one has to throw out half the harvest weight by selling headed prawns and in addition survive the impact of a 48% corporation tax on profits. It may be equally difficult to survive in pond aquaculture of Macrobrachium if selling headed prawns because biomass yields are smaller than in intensive systems. The requirement is for large areas of ponds (hundreds of acres) with high risk and problems in control and management.

There are small amounts of pond-grown Macrobrachium being sold heads-on to local markets, e.g., by Caribe King in Puerto Rico and Fish Farms Hawaii, but this cannot be considered an industrial-scale operation, which usually means the multimillion-dollar range. Development of a large-scale market of such size will require the skills of advertising, food distributors, and probably retailers with the cooperation of restaurateurs and hoteliers. The United States public has conservative tastes in foods and some unique skills will be required to introduce not only a new crustacean, freshwater product, but to market it heads-on. In Europe, Latin America, and Asia, consumers are used to buying crustaceans heads-on but not in the United States where even lobsters are imported as tails.

Prawn purchases are income-elastic. Both the housewife and institutions resent buying luxury foods at certain price levels. Restaurants will only pay 30-40% (sometimes 50% for the largest crustaceans) for the food items charged in the cost of a served meal. The portion size required to satisfy a customer is 4 to 6 ounces. At the current price of $3.00/lb. for 20 count prawn tails, the cost to the restaurant per six-ounce serving is about $1.13. This figure becomes the approximately maximum amount the restaurant will pay to provide one serving of prawns. Thus the $3-4 cost for prawns may be a limit for prawns delivered to institutions.

Prawns are substitutional food items in many restaurants. If the price of the prawns rises too high, they will use less shrimp per portion or move to other combinations (e.g., prawns and steak as in the Sizzler restaurant chain). They may drop prawns altogether if the price is too high for them to meet the 30-40% margin. Thus, as far as corporate interest is concerned, the market may not be sufficiently stable.

In the past, marked price volatility has occurred in the price of shrimp and prawns (e.g., 40 to 50 count penaeids now \$2/lb. were \$1 two years ago). Such swings can have a decidedly negative impact on the cash flow of an aquaculture venture, perhaps resulting in its rapid collapse.

In seeking to price the product for the market, care must be taken to avoid competition with the penaeid tail prices. A separate price structure must be established for a head-on freshwater prawn of consistent high harvest weight and quality.

Constraints in available technology are also present. We do not possess suitable foods, effective water quality-control systems, efficient harvesting techniques, etc., for consistently profitable ventures. If this is so, who will invest in aquaculture? Large corporations are expected to meet and maintain R.O.I. levels better than can be obtained from investment in commercial paper investment and at low risk. The continued failures in small aquaculture ventures provide an unsuitable climate to presently interest most large corporations in investment in aquaculture

Harvesting technology remains primitive and needs to be improved for large-scale commercial operations. Present systems are either too cumbersome, time-consuming, or manpower intensive, or a combination of these.

There is insufficient knowledge of processing methods for *Macrobrachium*. Plants that have processed these prawns complain that the shell is hard and will not be peeled by existing penaeid peelers. No mechanical heading machines exist that do not also remove the peel. The best way to market will be to split the shell longitudinally by machine and market them whole.

These unknowns in harvesting and processing are further doubts to add to the problems of growing a commodity product at high risk and the lack of a proven market for the product. It is no wonder that many corporations are hesitant to enter the aquaculture business at this time.

The small numbers of corporations involved vary in their activities from pilot-scale pond ventures in the U.S. (e.g., Harrell International) or abroad (e.g., General Mills, Honduras) to highly sophisticated, intensive, closed systems where yields 40 times greater than those in ponds are being attempted (e.g., Syntex).

Some companies see their interests directly in farming and selling prawns, either to their captive corporate-owned restaurants or to the institutional trade. Those companies which retail the harvested prawns through their own restaurants are at a decided advantage in profitability, compared to those who must market through the wholesaler.

Other companies see their returns from marketing the support technology (e.g., superior hybrids, feeds for adults and larvae, filters, water-quality hardware, closed systems for larval and adult grow-out, etc.) to a developing aquaculture industry of which Macrobrachium is but one product.

Whatever the case, there needs to be sufficient profit cushion in operations to take care of swings in the farming market, as for example are seen in the grain or cattle-farming industries. The farming of known high-volume food products has a history of high risk with only big operators managing to survive volatile commodity prices, effects of unpredictable weather, and the vagaries of the public demand. In aquaculture we recognize these risks too but have the added burden of moving into unproven farming methods and unproven markets. It is for this reason that most companies choose to watch rather than commit funds to a high-risk new endeavor.

John H. Ryther at Woods Hole Oceanographic Center has summarized the position of industry in aquaculture very well indeed (Barrons, January 19, 1976): "The embryonic (aquaculture) industry in the U.S. has a dismal track record that threatens the field with extinction before it fairly has gotten underway. Most of the new industry difficulties must be attributed to the ambition, impatience, and naivete of the aquaculture entrepreneur." He also acknowledges that since the disappearance of the small fry, pure aquaculture enterprises have become a scare commodity for investors. Indeed he suggests that as far as publicly held companies are concerned, participation is largely limited to the giants (e.g., Con Agra, Dow, General Mills, Ralston-Purina, etc.) who see fish farming as a sideline or a hedge against the future. For the most part, the large corporations play a waiting game, being content to send representatives to technical meetings but being ready to move into new areas of the business once an adequate research base is available to make the risks acceptable.

What is required to stimulate further corporate involvement in aquaculture research is some acceptance of the commercial risk by government in developing an aquaculture industry and their support, without strings, of defined areas in well-managed ventures, e.g., accepting joint risk of investment in a pilot plant once basic research shows some viability; or giving tax incentives to aquaculture farming. Government action should be to seed a new developing industry without the usual requirement to control patent rights or strangle research with bureaucratic controls after funding.

As the technology builds in Macrobrachium aquaculture, corporations will establish farm units. However, because of growing inhibitory E.P.A. restrictions and shortage of suitable land, inadequate locales with suitable environment and temperature in the United States and high labor costs, these farms will probably be built offshore with prawn products being marketed into U.S. markets. This offshore aquaculture will not ease the problem that two-thirds of prawn products are imported into the United States producing a burden on balance of payments. Only later when breakthroughs develop in intensive systems will a fully developed Macrobrachium aquaculture industry be possible in the United States.

B. Discussion of Industry Participation

On completion of his prepared text, John Bennett called on industry representatives present to comment further, and specifically to state their levels of operation.

Albert Saari of General Mills described that firm's prawn operation in Honduras as an "advance pilot operation" and said the company was nine months to a year away from a decision on the future. About 25-30 acres were in actual production with about 10,000 pounds of prawns being produced monthly. Because ponds were phased in and out, as part of the research and development, precise numbers could not be given. Since then, Red Lobster Inns, a General Mills subsidiary, has taken over the operation and is now testing freshwater prawns in a number of their restaurants.

Durwood Dugger reported that operations of Aquaprawns, Inc., the Sun Oil Company subsidiary, were phasing down in Brownsville and the point of decision was about to be reached. If the company decides to go ahead, a site will be found in a more productive climate, probably in Central America.

Norman Moe stated that the Weyerhaeuser prawn operation at Homestead, Florida, which was initiated by APARI under Paul Mulvihill, was suffering from a lack of grow-out facilities and planned to build about a 25-acre pilot plant. The hatchery is under good control. Moe pointed out that, even if successful in growing shrimp, problems of scale are involved in a major corporation's decision on whether to go into aquaculture. A company used to substantial return on investment from large operations might simply find even a profitable operation too small. Weyerhaeuser probably will not reach the point of decision on its prawn-farming operation for two or three years.

Kenneth Kato described Fish Farms Hawaii as a small company with the current objective--which has been reached--of developing a self-sustaining (in terms of cash flow) pilot operation. Kato believes his technological problems pose no great difficulty and is concentrating on the biologic problems.

Peter Halmos said that Farm Fresh Shrimp is engaged mainly in research at the Turkey Point power plant at Miami, Florida.

King Caribe Shrimp was not represented at the meeting because the firm was in the midst of expanding its prawn acreage. With about 60 acres, and another 50 in prospect, King Caribe remains the largest of the prawn growers.

Goodwin calculated the number of U.S. firms and their approximate production areas and concluded that, as of June 1976, the embryonic freshwater prawn industry was represented by about 125 acres of productive waters, with a probable average of about 2,000 pounds per acre per year.

CHAPTER XI
MARKETING

Ian Chaston of Booth Fisheries, invited reviewer for prawn marketing problems, prepared a paper on the assumption that aquaculture of the freshwater prawn had advanced far beyond its actual state to production of millions of pounds per year rather than thousands, the actual case as of June 1976.

Consequently, Chaston's presentation should be read as a forecast of the problems the "prawn industry" will face when high production levels are reached. He did not intend his paper to imply that small prawn operations, with small production, could not be economically viable, only that problems of scale put a new face on things.

A. Macrobrachium--The Problems of New Product Marketing Where There is No Real Definition of the Market:

Ian Chaston, Booth Fisheries, Division of
Consolidated Foods

Background

A fundamental element in marketing management is the development of a business strategy to provide the basis for the direction of corporate operation. A typical approach in new product marketing is to examine the concept of the product, the nature of the market into which the product is being launched, and the point of entry into the standard channels of distribution. After examining these elements, it is then possible to develop a strategy in which one defines the benefit of the product, the attributes of the product which justify the benefit, and from this basic strategic definition, one can begin to construct a marketing plan for the successful introduction of the new product.

The Product Concept

In the case of *Macrobrachium*, the first problem that one encounters when examining the marketing proposition for the product lies in the area of attempting to define the nature of the product. The previous Sea Grant *Macrobrachium* workshop (1974) correctly identified that a major problem is the definition of "the popular product name" for the animal. Individuals at this workshop centered their discussion upon the problem of developing a name for the product which will be acceptable to the Federal Trade Commission.

A more fundamental point, however, than the question of the name of the product relates to the actual product positioning in the marketplace (i.e., does one market the product as a total new entity or does one consider the item as a substitute for existing products of a similar form?).

In examining the concept of *Macrobrachium* as a total new product entity, it is necessary to consider the cost of launching a product positioned in this way.

If one examines the history of business within the U.S., there are numerous examples of new product introductions in the retail business to provide evidence that significant investment is required to develop the product concept and communicate this concept to the consumer. This is especially true in situations where the product has no association with other existing products which can provide an association of ideas for the consumer to more easily comprehend the nature of the new product.

In fact, when one examines the new product retail marketing success of recent years, one finds that the majority of new products were developed as an extension of an existing concept. Examples of this are instant coffee, disposable diapers, and the very up-to-date situation of frozen donuts.

If one examines the successful new product propositions of recent years, there is evidence that there are major costs involved in launching such products. Although it is never possible to accurately assess the cost of entry of individual products, there are a number of factors of expenditure which one can reasonably estimate accurately. For example, a minimum weight television campaign in the total U.S. would require probably somewhere in the region of $5 million. If one then adds the support cost of promotional allowances, the cost of supporting a sales force and the general administration of the marketing operation within the first two years of the product's existence, one is conservatively expecting to spend somewhere in the region of $15 to $20 million in order to successfully establish the product.

This is obviously a "top end of the market" situation and it is possible to effectively launch a new product at a lower cost although naturally the final total sales of the product will be significantly less. What is demonstrated by this analysis, however, is that if one were to consider Macrobrachium as a distinct new product entity, then the cost of introducing the product would be extremely high. Furthermore, it should be realized that the successful launches of new products in the U.S. have been achieved on items that offer a very high gross margin (e.g., 60% to 70%). Such gross margins cannot be presumed to exist in the frozen fish and seafood business and hence one would probably never be able to generate sufficient gross profits from Macrobrachium to ever fund the expenditure of the type defined above to establish the product as a new product entity with its own distinct market segment.

Having therefore discarded the proposition of considering Macrobrachium as a new product entity, one then concludes that it would be marketed as a substitute for an existing product. In this situation, one then has the question of the nature of the substitution and the two most likely candidates are lobster tails and marine shrimp.

The total U.S. lobster tail market in 1972 had sales of $140 million and by 1974, this had increased to $193 million (i.e., increment of +37.8% in dollar sales over three years). However, when one examines the total number of pounds associated with the sales in 1972, it was 34 million pounds and by 1974 had reached 36 million pounds (an increment of +5.9%). Thus, the conclusion one can draw on the lobster tail market in the U.S. is that growth has been generated on merely an increase in the unit price, reflecting the impact of inflation, and that there has been virtually no genuine growth in the size of the market over the period 1972 through 1974. Although no numbers are available for 1975 or 1976 at this point in time, indications are that this zero growth situation has remained in the lobster tail market over the last two years.

Given that the total volume of sales is quite small, this lobster tail market can always be defined as a premium price category wherein premium price has been generated through the low availability of the product. Thus, if one considers that it is possible to grow 1,000 pounds per acre of Macrobrachium per year, it would only require some 34,000 acres of Macrobrachium ponds to produce a tonnage equivalent to the size of the lobster tail tonnage sold in the total U.S.

In this situation, where one doubles the volume availability of Macrobrachium marketed as a substitute lobster tail, it must be expected that there will be a significant decrease in the unit price of lobster tail products. In this instance, in view of the relatively small differential between cost and final sales price for culture systems at this time, this significant price decline generated by improved product availability could possibly create a situation where production costs for Macrobrachium would exceed the final selling price. Consequently, one must consider that positioning Macrobrachium as a substitute lobster tail is probably an incorrect product proposition for the item.

The shrimp market within the U.S. provides a contrast to the lobster tail market. The total dollar sales of shrimp products in the U.S. (this includes green headless, P&D and breaded shrimp) for 1974 is equal to $1.05 billion and an increment of +32.7% over 1972. The total pounds of sales in 1974 were 4.122 billion pounds (an increment of over 10% on the pounds in 1972). It is therefore apparent that the shrimp market offers a much larger market than the lobster tail market and, furthermore, the growth in pounds for shrimp versus lobster tails indicates that there has been some genuine increase in the market size for shrimp products over the last two or three years.

The advantage of the shrimp market proposition is that it is an extremely large market and therefore the impact of Macrobrachium culture, in terms of total availability of raw material, would probably be less significant in effect than in the case of positioning the products of lobster tails.

For example, it would take over a million acres of Macrobrachium ponds to get anywhere near producing the current production of marine shrimp that are caught from natural fisheries and marketed in the U.S. Consequently, one could expect significantly better price stability for Macrobrachium positioned as a substitute marine shrimp item. Thus, in my opinion, the positioning of Macrobrachiu should be directed toward this area in terms of developing the product's marketing concept.

The exact nature of the positioning does require some further consideration because, theoretically, it is possible to develop a "fanciful name" which differentiates the product from marine shrimp. The advantage of the fanciful name is that it provides differentiation, which the Federal Trade Commission may consider a suitable situation when treating the product as a substitute marine shrimp, but it also provides an even greater advantage in the sense that successful differentiation from marine shrimp may permit Macrobrachium to be marketed at a higher unit price than may be achieved from a pure "marine shrimp substitute" proposition. I make this point because from the limited information currently available on Macrobrachium sold in the U.S. when it is treated as a straight substitute for marine shrimp, the price tends to be somewhere in the region of 30% less than regular shrimp. Thus, it is possible that given an effective fanciful name, one could generate sufficient differentiation from the marine product to achieve a higher unit price for the product than merely treating it as an inferior substitute for marine shrimp.

In proposing the proposition of a fanciful name, one must recognize that there is a certain incremental cost involved in developing recognition of the fanciful name among end users. Further, it should be realized that for a fanciful name to be effectively established and recognized among consumers, there would be a need for the whole Macrobrachium industry to accept a uniform name to be applie to all production. Thus, it would be necessary for all producers to agree on the fanciful name and, therefore, provide a unified front in preparing a recommendation to the Federal Trade Commission requesting acceptance of the proposed name.

The above analysis assumes that Macrobrachium can effectively be utilized as a substitute for marine shrimp. The marine shrimp market can be divided into three product forms; namely, green headless shrimp (which may or may not be served in its final form at the end outlet as a peeled or unpeeled product), peeled and deveined shrimp, and breaded shrimp. The question that arises is whether the Macrobrachium tail can be peeled and therefore utilized as a raw material for either P&D or breaded shrimp. My small exposure to this animal indicates that there may be a significant problem in peeling this product, using known production techniques. In fact, if one is unable to peel this product using normal peeling methods, then it is quite obvious that one could not utilize it as an effective substitute for marine shrimp. In this instance, one therefore may be forced to accept the positioning of the product as an alternative to lobster and this in itself would limit the future potential size of the market for the product over the long term.

The above discussion has made no reference to the international market for this product which exists in a number of countries. There is no reason to believe that one cannot, in fact, market the product as a shrimp substitute in the U.S. and also exploit the product in the form in which it is usually marketed using its "popular name" in other markets such as Hawaii and the Far East.

The Product Form

Assuming that the product is marketed as a substitute for marine shrimp, the question does arise on the size of the tails that one would grow for such a market. The shrimp industry uses green headless shrimp counts ranging from 15 count through to over 300 count tails.

Most of the research that I have seen tends to emphasize the 31/35 count tail because this is the accepted product form that is sold in markets such as Hawaii. The total profit in any sales situation can be defined in terms of Profits = Sales Dollar per Inventory Turn X Gross Product Margin X Inventory Turn. To illustrate this point, let us take an extreme example where one grows large count tails at 1,000 pounds per acre, sold at a price of $3.50 per pound at a 20% gross margin, one crop per year. This would generate some $700 gross profit. Contrast this with the situation where one grows a 100 count tail, same productivity, but because one is harvesting the crop at the more efficient point on the growth curve, is able to produce four crops of product per year. This would result in a gross profit of $800 per year as opposed to the lower gross profit achieved by producing much larger tails. Furthermore, when one adds the concept of risk defined in terms of the probability of crop failure, the production of large versus small tails further widens the difference between potential gross profits.

One final point in this area of production relates to the fact that the growth area of the shrimp business at this point in time is in the smaller count tails which are being utilized as the base raw material for extruded shrimp products. Consequently, there may be some logic to reexamining some of the financial models that people have developed for Macrobrachium in terms of evaluating the possibilities of producing more crops of high-count tails than emphasizing the single crop of 31/35 count tails on the basis of the high unit price per tail.

Market Choice

The frozen fish and seafood business in the U.S. can be conveniently segregated into institutional and retail. The structure of these two markets is entirely different and the common illusion that one's marketing methods are just as applicable in both markets is in fact invalid. This is due to both the nature of the market and also the distribution channels that are utilized in the markets.

In terms of market size, it is difficult to gain accurate data on total U.S. market sizes because (a) there are products sold to apparently retail outlets that are finally sold to institutional end-user outlets and (b) there is a significant volume of institutional-type products that are sold through the retail stores. Accepting these slight definition problems, we estimate that the retail frozen fish and seafood business is worth some $600 million, whereas the institutional market in the U.S. is probably some $1.8 to 2.0 billion in total size. Therefore, as a market, the institutional market is significantly larger than the retail business. Further, in examining growth trends, we expect to see a faster growth rate in the institutional side of the fish and seafood business than we predict for the retail. This situation reflects the greater increase in the consumer use of fast-food outlets and the involvement of outside catering in such areas as schools, universities, hospitals, etc.

Thus, using market sizes as a parameter, one would probably conclude that the institutional market is a more suitable source of Macrobrachium sales. The other factor which makes the institutional area more suitable relates to the fact that one is marketing a premium price species and certainly if one examines the retail business, premium price species definitely have a lesser importance as a percentage of the total sales than they do in the institutional area. This reflects the fact that consumers, when eating out, are prepared to "treat themselves" and therefore are prepared to pay a higher price per unit serving than they are when feeding themselves at home. Thus, items such as lobster and shrimp tend to have a greater share of the total sales volume institutionally than they do in the retail market.

The one possible advantage of the retail business is that, should one decide on a distinct fanciful name for the product, it is easier to communicate the product concept to the consumer than it is through the institutional business. The reason for this is that in the retail business one can effectively influence the consumer through the medium of advertising, whereas such activity in the institutional area is more difficult because of the wide range of differing types of end-user outlets.

The one possible way of overcoming this problem of preinfluencing consumers prior to entering an outlet in the institutional business is to open a restaurant franchise based on Macrobrachium where, of course, the outlet itself is identified with the product. An example of a generic approach to institutional marketing is provided by the General Mills Red Lobster Inns where the emphasis is on lobster products and therefore the consumer is predecided to purchase that type of product prior to entering the outlet.

However, to launch a national fast-food franchise based on Macrobrachium, I feel, would be extremely expensive in its initial years in terms of generating consumer trial of the product through its outlets. Thus, the one advantage that one can perceive in terms of entering the retail business of Macrobrachium is that one can more effectively generate trial of the product if it was marketed under a fanciful name.

The one complaint one does hear on the institutional business (and one is never sure how genuine this complaint really is) is that it is difficult to persuade institutional buyers to accept new products, whereas in the retail area, stores are fully aware of the need to introduce new products to satisfy changes in consumer needs. Thus, if one wishes to define a time span, it would probably take some three or four times as long to introduce a new product institutionally than it does in introducing the product within the retail market.

In considering all of the above, my own viewpoint is that the institutional market (because of both the total size and total importance of premium-priced items relative to total dollar sales) is probably the preferable market choice to initially introduce Macrobrachium. I would, however, not totally discard the retail market because there is possibly a new product proposition available for Macrobrachium in terms of marketing the product as a "lower cost alternative" to marine shrimp. However, in suggesting the retail market, one should be aware that the shrimp business in the retail area is beginning to suffer from consumer reactions to the high prices of recent years. The net result of this reaction has been to reduce the total dollar sales of all types of shrimp products to a larger degree than has occurred in the institutional business. The apparent solution to this problem in the retail area is the introduction of so-called "extruded shrimp products" which use low-cost high-count shrimp blended with some form of "filler" which is reextruded into a shrimp-shape and marketed at a much lower unit price. Personally, I feel this is the growth area of the retail shrimp business and by 1980, I am quite confident that over one-third of total breaded shrimp volume in the U.S. would be in the medium of extruded shrimp products. (Possibly, however, if Macrobrachium can economically be produced in a high-count form [200-300 tails/lb.], then it can provide raw material for this growth area of the retail business.)

Entry Point Decision

In the above discussion, I have assumed that the producer will actually be the final marketer of the product. In actuality, this is not a necessity because it is quite possible for the producer to enter the distribution channel at a much lower entry point.

The standard channel of distribution in the fish and seafood industry is a four-step process from producer to manufacturer + marketer to distributor and then to final sales outlets (retail or institutional).

The simplest proposition for the producer is to either sell to a broker or to a manufacturer of fish and seafoods and, therefore, minimize the marketing costs. I recognize, however, that a number of people feel that vertical integration of an aquaculture company with ownership of the distribution channel is a more profitable proposition than merely acting as a supplier of raw material. Therefore, I feel that there is some benefit in outlining some of the costs involved in executing such a proposition.

In the retail situation, let us assume the simplest case; namely, that we are selling a peeled and deveined product. If one were to launch the product across the entire U.S., this would require a sales force consisting of at least six regional managers plus two head office sales people at a total annual cost of $400,000. One would then have to add the variable costs of brokerage at 3% (it's assumed that one would use a broker in this situation as opposed to one's own direct sales force), a cooperative advertising allowance paid to retail stores--3% and one's own promotional and advertising costs assumed to be set at 5% of sales. As one will probably not achieve more than 20% gross margin on the P&D product, and deducting the 11% previously mentioned variable sales cost, this leaves a net profit to the operation of 9% of sales. Thus, the break-even to fund the cost of sales force is in the region of $4.444 million and, therefore, to make any real profit, one would probably aim to sell $6 million of the product. This equates to a 10% brand share of the total U.S. retail P&D shrimp market and although this sounds like a reasonably small share target figure, its achievement will be significantly more difficult than most people assume. In fact, one of the key problems one will encounter with the concept is that to profitably sell a "single product proposition" of such low total dollar sales is virtually impossible. In fact, if one wishes to entertain the proposition of the producer marketing the Macrobrachium, success will probably only be possible by integrating the item into a fuller line of fish and seafood products.

In the food service situation, one has a much more diffuse sales environment because there are 500, 000 potential accounts (although approximately 140,000 of these being bars, small independent restaurants, and prisons can be ignored). This leaves 360, 000 accounts, although it is possible to reduce the size of this target market by recognizing that 120, 000 of these accounts actually generate 75% of the total sales. Furthermore, by calling at the head offices of 100 major chains and 3, 000 minor chains, one can effectively reach the majority of the 120, 000 key accounts in the U.S. Nevertheless, in order to cover the key institutional outlets, one requires a sales force similar to that described for a retail situation at an annual cost of $400, 000. The variable expense is lower in the food service business; namely, 3% brokerage and one's own adversing/promotion cost of some 2% of sales. Unfortunately, however, gross margin is also lower (probably in the region of 10% of sales), leaving a net 5% gross margin after a variable expense and the break-even point of $8 million.

The comments made above on the management of a retail single product proposition are again applicable to the food service business; namely, if one wishes to consider marketing Macrobrachium as opposed to selling to a manufacturer, one would really need to integrate into a fuller fish and seafood line in order to create a viable marketing proposition.

The one other possibility is to achieve "complete vertical integration" in the sense of having a captive end-user account by establishing a restaurant franchise around the product. This approach has been used with success in the catfish-farming industry and if one wishes to market a single product proposition, I would recommend the concept of a Macrobrachium franchise would be the most effective approach to utilize.

Conclusion

One can make the following conclusions. First, the most critical element in the whole marketing concept of Macrobrachium lies in the area of product positioning and this is an industry level as opposed to single producer question. I recommend that one consider the product as a marine shrimp substitute, although obviously prior to a final conclusion on this point, there is further work to be done to evaluate how effectively the product can technically compete in the arena.

Given an answer to this question, the industry has then to reach a general consensus of the name that they wish for the product. Assuming that one is considering a marine shrimp substitute, I see no reason why the product could not be called "Freshwater Shrimp". This is a title for which I associate no negatives and given that the product has physical attributes capable of supporting effective competition with marine shrimp, then the product should be successful.

The remaining industry question relates to which market one should enter first with the product. From the above, I recommend the institutional market as probably the most suitable during the early life of the product.

The question facing the individual producer relates to the distribution channel entry point. My viewpoint is that the average aquaculture operation is not suited to entering a point well into the distribution channel purely because one is marketing a "single product proposition" which is a difficult concept to effectively execute in the U.S. frozen fish and seafood business. Thus, for most producers, I would recommend the concept of selling the product through either a fish broker or by developing a direct purchase relationship with a fish and seafood manufacturer. It should be realized that in developing such a relationship, there are obvious advantages to seafood manufacturers in purchasing aquaculture products; the most significant being that the product is available on a virtually continual basis as opposed to the extremely seasonal availability pattern one encounters in traditional fisheries. Thus, assuming that the product is an effective shrimp substitute, then the shrimp manufacturers in the U.S. obviously would welcome the opportunity to buy a consistently stable raw material supply. The nature of the purchasing contract will probably differ from that which exists for natural fisheries where at best one can merely indicate an interest in purchasing the product, because of the unpredictability of catch size. This is not the case as far as aquaculture is concerned where one can more accurately predict total crop size over a 12-month period as the basis for developing a well-defined contractual relationship between suppliers and manufacturers from aquaculture sources.

For those operations which feel that they wish to go further into the distribution channels, then I would personally recommend the concept of developing one's own restaurant franchise centered around _Macrobrachium_. If the interest merely lies in entering the distribution channels but not in developing one's own final outlets for the product, then I would strongly recommend the concept of integrating one's farm with a previously existing fish and seafood operation either through acquisition or the development of a joint venture. This would then permit one to include _Macrobrachium_ with other species in a fuller product line, thereby circumventing the "single product proposition" problem.

B. Summary of Discussion of Marketing

A participant not identified on the proceedings tape said that his organization had been following prices on the "green sheet" and that freshwater prawn (shrimp) tails from India and Bangladesh pretty much followed the marine shrimp prices but at a dollar less per pound.

Jay Huner added information on the M. ohione fishery in Louisiana. The fishery once supported a catch of two million pounds annually, and is now down to about 100,000 pounds. Prawns of 100-count (to the pound), heads on, were selling as of June 1976 for $1.20 per pound, with prawns used for bait selling for $24.00 per gallon--about 5 pounds per gallon.

On the question of a name for Macrobrachium rosenbergii, Tom Billy reported on a meeting with Food and Drug Administration officials. At present, FDA will accept a label that carries, as a brand name, BLANK PRAWNS, with the words "large" or "cultured" freshwater shrimp in letters at least half the size of the brand name. (For BLANK, read "Malaysian, Hawaiian, Puerto Rican, Caribe King, Aquaprawns, Jones, or whatever market research or a strong ego may suggest.)

C. Macrobrachium Fishery and Market in Mexico

Jorge Cabrera, Manuel Guzman, and Craig Kensler discussed Mexican prawn fishing and marketing and later supplied the following for inclusion in this publication:

The Macrobrachium fishery in Mexico is based on wild stocks only. The bulk of the annual commercial landing is made up of four species: M. carcinus and M. acanthurus from the Gulf of Mexico region, and M. americanum and M. tenellum from the Pacific Coast region. The commercial fishery is seasonal and, although not considered an important Mexican fishery, amounted to almost 700 metric tons in 1974, valued at $8,360,000 (U.S. dollars).

During early June 1976, the Mexico City market price of Macrobrachium (whole animals) was $14.00 per kilo, or $6.36 per pound. Marketing of Macrobrachium in Mexican markets is always of whole uncooked animals commonly sold by the name "langostinos" and it is never included with penaeid shrimp sales. In the past five years, Macrobrachium spp. (whole animals) have consistently sold at a higher price per kilo than penaeid shrimp tails. At present, all Macrobrachium landed in Mexico are for the national market, i.e., there is no export of the product.

Notes on commercial landings of Macrobrachium during the period 1970-1976 are given in Table XI-1.

Table XI-1
Commercial Landings of *Macrobrachium* in Mexico

	1970	1971	1972	1973	1974	1975	1976
Total landings in metric tons[a]	200	300	400	700	697	---[b]	---[b]
Total value in millions of dollars	1.60	2.88	4.16	7.84	8.36	---[b]	---[b]
Market price/Kg. in U.S. dollars[c]	8.00	9.60	10.40	11.20	12.00	12.80	14.00
Market price/lb. in U.S. dollars[c]	3.64	4.36	4.73	5.09	5.46	5.82	6.36

[a]Source: FAO Yearbook of Fishery Statistics, Vol. 38, 1974.

[b]Data not available.

[c]Mexico City market price of whole animal, any size.

CHAPTER XII
FOOD REGULATIONS AFFECTING THE CULTURED PRAWN PRODUCER, PROCESSOR, AND MARKETER

This chapter contains only the paper presented by Thomas J. Billy and its brief citation list. In editing the proceedings of the freshwater prawn workshop, it was a policy to include comments from the floor only when their authoritativeness was in little serious doubt. At this stage of the development of the freshwater prawn industry, the weight of emphasis remains on productivity; therefore, few of us rank as authorities in the realm of relevant regulations as yet. Thus, the editors considered discretion the better part of valor in this instance, and many thought-provoking comments which emanated from the floor during the lively discussion following Mr. Billy's talk must remain lost to posterity on the cutting room floor. However, we think the reader will agree that Mr. Billy's talk covers the subject very nicely indeed.

A. Food Regulations Affecting the Cultured Prawn Producer, Processor, and Marketer. Thomas J. Billy, Seafood Quality and Inspection Division, National Marine Fisheries Service

Introduction

So far in the workshop we have heard much new information, discussion, and concerns expressed about the breeding, diseases, grow-out systems, some nutrition, feed processing, and marketing of freshwater prawns. Now I would like to briefly discuss another dimension of freshwater prawn culture; one that impacts or can impact on all of the above areas. This dimension is food regulations as they affect the prawn producer, processor, and marketer.

The primary mandatory requirements applicable to fish and fishery products produced in the United States or imported are those contained in the Federal Food, Drug and Cosmetic Act, as amended. All cultured fish and shellfish intended for food use are subject to compliance with the requirements of this Act, in the same manner as are the naturally produced fish and fishery products. Under the Act, there are regulations pertaining to a variety of areas of concern to those involved in the culture of freshwater prawns, including:

1. pesticide residues and
2. food additives;
3. good manufacturing practices (GMP's) and defect action levels; and
4. labeling requirements.

The significant aspects of these regulations as they relate to products of cultured fish and shellfish are the same as those applied to all other food products except meat and poultry, which are covered by somewhat similar regulations set forth by the U.S. Department of Agriculture.

I will briefly outline some of the general regulatory requirements to be considered in the culture, handling, processing, and marketing of freshwater prawns. Then, some of the key concerns such as pesticide residues, food additives, GMP's and defect action levels, and labeling will be discussed. Examples of how these regulations impact on the activities of those attending this workshop will also be provided.

Pesticide Residues

Under the Food, Drug and Cosmetic Act a "raw agricultural product" means any food in its raw or natural state, including fish and shellfish and products thereof. Raw agricultural products such as cultured prawns containing pesticide residues are in violation of the Act unless:

1. The pesticide chemical has been exempted from the requirement of a tolerance; or
2. A tolerance has been established for the particular pesticide and the residue does not exceed the established tolerance.

In the regulation of pesticide residues, the amount or level permitted is covered by a tolerance, guideline, or action level, depending upon the amount of information available and the circumstances surrounding the particular situation.

For fish and fishery products, including cultured prawns, the current situation relative to pesticide residues follows:

Contaminant or Pesticide	Product	Level (ppm[a])	Status
Mercury	Fish and shellfish	0.5	Guideline
PCB's	Raw fish	5.0	Action Level
DDT	Raw fish-edible portion	5.0	Action Level
Dieldrin	Oil of fish	0.3	Action Level
Mirex	Raw fish	0.1	Action Level
Kepone	Finfish	0.1	Action Level
Kepone	Shellfish	0.3	Action Level
Kepone	Crabs	0.4	Action Level

[a]Parts per million

In the culture of freshwater prawns, potential pesticide contamination problems can result from exposure to contaminated water, soil, or feeds, as well as from aerial spraying or runoff.

Food Additives

"Food additives" are substances which may be added to foods directly or they may be unavoidable components. Such additives may be purposely introduced to enhance the flavor, textures, stability, or to retain the moisture of foods. The term includes any substance used in processing, packaging, holding, or transporting of foods. Under the Food, Drug and Cosmetic Act, any source of radiation and radioactivity are also considered food additives. The law excludes from the definition of a "food additive":

1. Substances generally recognized as safe; or

2. Substances having "prior sanction" under the Federal Food, Drug and Cosmetic Act, or other laws.

It is under "food additives" where the so-called "Delaney Clause" comes into play. The Act states that <u>no</u> carcinogenic or mutagenic substances (i.e., cancer-causing agents) may be allowed in foods. This includes the minute amount of radioactive material that may occur in cooling water used in nuclear power plants. Such heated water has been considered for use in grow-out systems and could result in radioactive residues in the fish or shellfish grown in it.

Thus, researchers as well as prawn producers and processors should consider whether the chemicals or other ingredients used in their feeds or other parts of their operations are subject to the safety clearance requirements of the food additives regulations. They are encouraged to seek an opinion regarding the status of these materials prior to use.

Good Manufacture Practices (GMP's)

Processed foods produced in the United States are required to be manufactured under conditions, including physical plants and controls, that will permit compliance with minimum good manufacturing practices described in the regulations.

Procedures to control particulate and bacterial contamination over a range of activities including processing, storage, handling, and distribution as well as other operating practices are covered. The regulations also provide for the rejection of food which contains "filth." For cultured freshwater prawns, care must be taken to ensure the procedures used in handling, processing, packaging, and holding the prawns comply with the applicable requirements.

Defect action levels have been set for natural or unavoidable defects in food for human use if no hazard to health is involved. The action levels are set because it is not now possible, and never has been possible, to harvest and process foods, including fishery products, that are totally free of natural defects. The levels represent the limit at or above which the FDA will take legal action against the product to remove it from the market.

Two key points regarding defect action levels, however, merit emphasis. First, compliance with defect levels will not prevent the FDA from acting against a manufacturer who does not observe applicable GMP's. Unsanitary plant conditions, for example, are a violation of GMP's and render the food unlawful. This applies even though the defect levels may be below the FDA's action level.

Second, the mixing of a food containing some type of defect at or above the current defect level with another lot of the same or another food is not permitted and renders the final food unlawful regardless of the defect level of the finished food.

With respect to freshwater prawns, the defect action level for decomposition in fresh and frozen shrimp products applies to those in the aquaculture business as well.

Labeling Requirements

Details concerning type sizes, location, etc., for required label information are contained in regulations which cover the requirements of both the Federal Food, Drug and Cosmetic Act and the Fair Packaging and Labeling Act. Required label information must not only be conspicuously displayed but it must also be in terms that the ordinary consumer is likely to read and understand under ordinary conditions of purchase and use.

Name of Producer, Processor, Distributor

The name, street address, city, state, and zip code of either the manufacturer, packer, or distributor, and an accurate statement of the net amount of food in the package must be listed on the label. The street address may be omitted by a firm listed in a current city or telephone directory.

Net Weight Statement

The net weight on packages containing one pound (avoirdupois) or more but less than four pounds must be declared first in total avoirdupois ounces, followed by a second statement in parentheses in terms of pounds and ounces, or pounds and common or decimal fractions of the pound.

Common or Usual Name

The common or usual name of a nonstandardized food such as freshwater prawns must appear on the principal display panel in bold type and in lines generally parallel to the base of the package as it is displayed. The form of presentation of the product also must be included unless depicted by vignette or unless the product is visible through the container. To prevent substitution of one kind of seafood for another, it is essential that labels bear names which accurately identify the products designated. Labels of seafoods must bear the common or usual name of the food, if there is one. Based upon recent discussions with representatives of the Food and Drug Administration, an acceptable label designation for cultured freshwater prawns is:

"(Descriptive term)" PRAWN
"Cultured Freshwater Shrimp"

The second line would appear on the label in type half the size of the common name. This designation is based upon the historical practice of labeling "Malaysian Prawns" in Hawaii.

Ingredients

The ingredients in a food must be listed by their common names in order of their predominance by weight. The word "ingredients" does not refer to the chemical composition but means the individual food components. Spices and flavors may be listed as such, without naming the specific material. If the presence of a specific expensive ingredient is promoted, additional information such as the percentage of the expensive ingredient may be required. If permitted and used, artificial colors or chemical preservatives must be conspicuously declared in the labeling. Artificial coloring is not permitted if it is used to conceal damage or inferiority or if it makes the product appear better or of greater value than it is.

Enforcement of Food Regulations

The aforementioned is a brief description of some of the key regulations that should be considered by those involved in the culture of freshwater prawns. Another consideration is how these regulations are enforced.

Enforcement efforts are carried out by the Food and Drug Administration (FDA) on a selective basis. Products most likely to represent a safety or health hazard to the public are given first priority. Second priority is given to products which are decomposed, filthy, or produced in an unsanitary environment, but which do not constitute a hazard to the health or safety of the public. The remaining effort is devoted to deceptive practices or the so-called economic class of violations, including labeling.

Factory inspections and laboratory analysis of samples are the major means employed by the FDA in conducting its enforcement activities. The requirements are the same both for products manufactured in the United States and for those imported. Within the United States, enforcement is maintained through inspections of food-producing plants, through the collection and analysis of samples collected at various stages of the production process, and by legal actions brought against foods found to be out of compliance.

Summary

Much of the discussion during this workshop has been devoted to breeding, diseases, nutrition, and grow-out systems in the culture of freshwater prawns. Participants have been encouraged to "think in terms of the animal" in their research and development efforts. Through the presentation of some of the key food regulations operative in the U.S., however, I hope I have encouraged all of you to be sure you consider another point of view; namely, that the prawns you produce are food that will grace someone's table. Therefore, your practices should ensure that the prawns produced are safe, wholesome, and of the desired level of quality.

CHAPTER XIII
THE ECONOMICS OF PRAWN CULTURE

This chapter, as an artifact of the functioning of the workshop, contains only the comprehensive review paper presented by K. J. Roberts. There was no recorded discussion.

The Economics of Prawn Culture:

Kenneth J. Roberts, Clemson University and
South Carolina Marine Advisory Services

The attendees at the workshop represent a good mixture of aspiring commercial aquaculturists and researchers. A valuable aspect of the discussions will then be the exchange of viewpoints and data from commercial and research culture efforts. This diversity of participants and interests perhaps dictates the broad title for this session. Prawn culture economics should include discussion of international comparative advantage, institutional factors, business organization, production, processing, and marketing. However, the paucity of economic data on these components of prawn culture has never inhibited discussion or decisionmaking about *Macrobrachium* culture. In my presentation, viewpoints on the development of *Macrobrachium* culture will be the main offering with data revelations of lesser importance.

The presentation will focus on delineating the important implications of public policy and business organization as related to *Macrobrachium* development. As a stimulus for discussion, each of the components of these topics will receive some attention. The components of comparative advantage, institutions and business organization suggest that more than accounting and bookkeeping techniques are involved in economic analysis. Economics involves the allocation of resources among competing uses in a manner that meets specified objectives. The allocation of resources is affected by capital availability, institution technology, and management. Thus, to generalize about the economic aspects of aquaculture solely from a base of technological information generated by the biological sciences often elicits dissatisfaction with aquaculture development efforts.

Involvement of economists with aquaculture research teams is an anomaly. Contact with economists currently active in aquaculture research revealed that economists are too often called in ex post facto to analyze a system and that the profession is viewed in a service role. Neither of these roles is satisfactory. Ex post facto economic analysis of experimental data has generated descriptive studies of production systems. Many times the studies are offensively referred to as feasibility studies. As to the second issue, the relegation of economics to a service role in aquaculture development results in important research management and public policy issues being ignored. While economists can continue to generate ex post facto analyses on a request basis, analysis of the aquaculture development process and public policy issues must receive higher priority. In this vein a few thoughts are worth sharing.

Questions of international and regional comparative advantage in the culture of Macrobrachium must be answered. This area of research is critical to Macrobrachium development because of the geographic diversity of culture efforts. Will better water quality, longer growing seasons, fewer permit problems, and other factors encourage the export of Macrobrachium technology? If demonstrated to be important, will these factors offset isolation from markets, fluctuating exchange rates, etc.? Under what circumstances will the United States have comparative advantage in the culture of Macrobrachium?

To operate in a void of such fundamental information will prevent pervasive statements of the public's role in Macrobrachium culture from being developed. We cannot expect to involve the public further without information delineating the alternatives of less public support. This support certainly includes further publicly funded research efforts, participation by the public in hatchery construction and maintenance of superior broodstock (as in agricultural production) and perhaps preferential treatment for culture enterprises in obtaining construction permits in coastal areas. Agricultural food production has been facilitated by incentives from a host of public agencies such as local zoning boards (via lower assessments for agricultural land), state agricultural universities (through their crop and herd improvement units), and the Soil Conservation Service (by soil improvement and pond construction). Macrobrachium culture in the United States will need the assistance of these institutions or new ones such as the developing coastal zone management agencies. Are coastal zone management agencies in your state making provisions for aquaculture development?

A fundamental difference between the economists' and biologists' views of aquaculture is the target of research or analysis. The biologist researching

Macrobrachium focuses on components of a culture system such as the hatchery, postlarvae production, feeding, disease control, etc. The focal point of the economist's attention is not the system but the firm. The system is the productive process which the firm manages in order to meet specified objectives. Analysis of the firm, its management, and its development of a product to meet consumer needs may be the missing links in current aquaculture development work. While it is a certainty that no firm can operate profitably with a technologically unsound production system, promising laboratory results will not compensate for poor business organization and management. The cost information available on experimental Macrobrachium culture indicate that a large percentage of total costs of a commercial operation are fixed costs. These costs are incurred whether production is pursued or not. The fixed or sunk costs represent assets with low alternative values because of their utility for specific culture purposes.

In South Carolina, the first cost and return budget of the grow-out system revealed that fixed costs were nearly equal to variable costs. Thus, research efforts restricted to the production process may appear dramatic in terms of previous results but unimpressive in terms of impact on total cost. Experiments may indicate improvement in the production process because the input-output coefficients are favorable, but total higher fixed costs may indicate a negative result.

The organization of the business should be of interest to the public sector and the economist. Given that a firm can expect to have a significant portion of its Macrobrachium culture costs committed to fixed items, perhaps the business should be organized in a way to encourage flexible use of assets. This may include designing the grow-out facilities in a less than optimum manner for Macrobrachium but with the capability of utilization by an alternative specie(s). Other alternatives include land rental or share-culturing as means of ensuring flexibility if business failure occurs. The point to be made here is that economists would like to see the biologists' quest for an optimum system include optimum flexibility. Ian Chaston indicated that industry anticipates a significant growth in demand for products utilizing 100-200 count tails. Natural supplies appeared to be limited in this respect. Whether this void is real or imagined, it points out that aquaculture systems are not being designed to allow for capitalizing on such opportunities. The species, feeds, etc., planned to meet the 45-gram-per-animal market may be too expensive to meet real market needs in some instances.

As I have indicated, the focus of an economist's thoughts is often the firm. The majority of aquaculture research appears to be suitable for utilization by large vertically integrated firms. This is not necessarily undesirable, but does point out that biological experiments often prohibit commercial use of the information by certain types of firms. In essence, aquaculture development labors somewhat under a corporate image. Macrobrachium development research teams are less guilty of this than research teams working with other species. The cost and return aspects of Macrobrachium culture should include explicit mention of the type of firm and source of data. Without the first, important policy questions may be ignored and misuse of data is likely. Cautioning the economists and those applying the numbers about the experimental nature of the data base is desirable. Experimental data has more usefulness in research management than in decisions concerning feasibility, which is an iterative process.

Elaborating on the firm as the center of the economists' attention is important when considering Macrobrachium culture. The sole proprietorship or small closely held corporation will likely be the prevalent form of business organization utilizing research results in marginally suitable growing regions such as South Carolina. Perhaps this form of business will be encouraged by the public sector for employment or other purposes. It is my view that public participation in areas with marginal growing conditions or high land costs will be necessary if large vertically integrated firms with sufficient capital and legal resources to seek ideal growing conditions abandon such marginal areas. Therefore, the economics of Macrobrachium culture must provide information for incorporation in public decisions. Public support may be necessary to provide postlarvae and superior broodstock for the small firm. The small firms in marginal areas are more likely to operate batch growing systems which will be accompanied by significant processing and marketing problems. Two particular problems appear most important to the firm in this situation:

1. Harvesting to meet market variations in quantity and size of animal will be very difficult, and

2. A small firm with a batch process has less chance to differentiate its product from that of other firms or a similar animal.

The larger vertically integrated firm organized to provide hatchery, broodstock, grow-out, perhaps an intensive closed system, continuous production, processing, and product identity services may exist initially only in ideal geographical locations. The chosen location will exhibit the most advantageous biological conditions and favorable government attitude toward land utilization for culture purposes. The two types of businesses mentioned entail different capital costs, labor skills, production schedules, levels of product differentiation, controls over product marketing, and control over the variables in the production process. The main similarity is that management is an important determinant of feasibility. Management is seldom mentioned as a key element of the productive process by university types. It is in this vein that we must remember a dollar of fixed costs yields different products due to management. We must be cognizant of this fact and begin with the firm when performing feasibility analyses as a prerequisite to successful development of Macrobrachium culture.

The Macrobrachium culture effort in South Carolina was described in Paul Sandifer's paper. An internal report on the economic implications for a small firm of the experimental results was prepared in June 1976.[a] A ten-pond system, each of one acre of water surface, was chosen to provide comparison with Shang's economics work at the University of Hawaii. The highlights of the report are noteworthy. Land with access to marine waters, if used as the location, sold for approximately $3,000 per acre in early 1976. Engineering information indicated that 21 acres of land was necessary to provide 10 acres of water surface. The capital required for investment and operating costs amount to $125,000. Suitable land approximately ten miles inland is available at $700 per acre. The capital needs for such a location would diminish to near $76,000. Thus, the locational aspect of the grow-out operation could reduce capital costs by $4,900 per acre. The impact on production and water costs is unknown at this point. Interest on debt, depreciation, and other fixed costs account for approximately 50 percent of initial total costs. Other components of total costs include feed (16%), electricity (14%), and hired labor, repairs, liming, and other variable costs (20%). The assumptions and conversion coefficients for the end results of the report were made explicit. The result was a zero return price for the owners' labor and management of $1.86 per pound, heads on. This assumes a harvest of 893 pounds per acre and land at $3,000 per acre. Since the system is one of batch harvest, the zero return price is an average for animals of many sizes. Sandifer's previously published papers contain the grow-out results. The zero return price would fall to $1.49 per pound if Sandifer's projected production of 1,173 pounds per acre is attained.

The preliminary information needs to be refined as the Macrobrachium work continues in South Carolina. The figures and those of all aquaculture economics reports need to be embellished by the incorporation of costs reflecting legal services, environmental assessment reports, and additional operating capital due to time delays and crop failure.

---

[a] Roberts, Kenneth J. "Monetary Aspects of Experimental Macrobrachium Culture in South Carolina" — An Internal Report to the South Carolina Sea Grant Program, Agricultural Economics and Rural Sociology Department, Clemson University, June 1976.

CHAPTER XIV

GENETICS AND SELECTIVE BREEDING

Spencer R. Malecha of the Department of Genetics of the John A. Burns School of Medicine of the University of Hawaii and also of the Anuenue Fisheries Research Center of the State of Hawaii's Department of Fish and Game was the featured speaker in this session. Dr. Malecha and his coworkers are virtually the only group working intensively in the genetics of Macrobrachium at this time. Thus, Dr. Malecha's review of his work constitutes, in essence, a review of the entire field of Macrobrachium genetics and selective breeding. For this reason, it seems appropriate that the review be as extensive as it is and that the ensuing discussion be as spirited as it was. Therefore, only a little of the spirit was removed from the discussion as it was recorded for presentation here.

A. Genetics and Selective Breeding of M. rosenbergii

Spencer R. Malecha

Introduction

Most research work on M. rosenbergii that can be evaluated through publicly available information emphasizes culture and husbandry. There is almost no published information bearing directly on genetics and selective breeding efforts in this species. Some information is available that indirectly relates to various aspects of such efforts and I will allude to these throughout the review. Our group in Hawaii seems to be the only one engaged in formal genetic work in M. rosenbergii, so I will give an account of some of our results; but, since this presentation has a "workshop" format, details will be left for formal publications. Since aquaculture genetics itself is in its infancy, I will also mention genetics work on other species where appropriate and will give a brief discussion on the theoretical background to certain research efforts. This review is not meant to be exhaustive and is done only to highlight work that is, or should be, proposed or is ongoing in M. rosenbergii.

Background

General

Aquatic species, despite their demand in world food markets, have not been subjected to intense life cycle control and subsequent genetic improvement because most aquatic protein is (or was) readily available from the wild. Curiously, Japan which leads other countries in producing the greatest percentage of its per capita aquatic protein through aquaculture has not employed genetic improvement programs to increase production. Nonetheless, the yields of almost all of the domesticated organisms important to man can be attributed to some type of genetic selection via the domestication process itself and/or the direct application of the principles of quantitative, chromosomal, and "Mendelian" genetics following the rediscovery of Mendel's work in 1900.

Some genetic research and manipulation has been applied to a few aquatic species, and oddly enough, the goldfish is one of the most genetically "manipulated" aquatic organisms for reasons of aesthetics and basic research. Selective breeding to directly improve commercial productivity in aquatic species thus far has focused on oysters and clams (Longwell, 1975), salmonoids (Donaldson, 1970), and carp (Cherfas, 1969; Moav & Wohlfarth, 1973).

M. rosenbergii and the "Anuenue" stock

Takuji Fujimura and his colleagues at the Anuenue Fisheries Research Center (AFRC) in Hawaii achieved success in culturing *M. rosenbergii* larvae on a large scale in the mid-1960's. Commercial ponds in Hawaii have become the source of the broodstock (hereafter called the "Anuenue stock") for AFRC's yearly larvae rearing program which has been supplying an increasing number of commercial and research enterprises in Hawaii and elsewhere. Indeed as far as I can document, *most* of the commercially aquacultured or researched *M. rosenbergii* in the Western Hemisphere can be traced to Anuenue stock and the latter can be traced to less than twelve founders shipped from Malaysia to Hawaii in the early 1960's.

Our genetic work and other published work has emphasized the Anuenue stock also.

Chromosomes: Sex Determination and Intraspecific Cytotaxonomy

Sexual size dimorphism in preharvest adults limits muscle biomass production in maturing females because they use their nutrient intake for vitallogenesis. Also, the presence of the sexes in pond populations seems to contribute to the amplification of the population size variance due to the heterogeneous individual growth rates that contribute to the "bull-runt" phenomena (Fujimura and Okamoto, 1970).

In crustacea, a variety of sex-determining mechanisms are known (Charniaux-Cotton, 1960), and these are usually associated with a specific chromosonal arrangement with "maleness" due most directly to a hormone produced by the androgenic gland, a separate organ from the testes. However, there are a number of marine invertebrates in which sex is rather labile so that genetic factors play either no role or only a subsidiary one (White, 1973). This may be so for *M. rosenbergii* since there is one published account of sex-reversal in this species (Rao, 1967). Consequently, there is a need to determine whether a specific chromosomal pattern is associated with sex determination in *M. rosenbergii*.

Some work has been done on chromosome analysis in crustacea and the early work of Niiyama (1959) is the most extensive. By and large, all this work utilized simple squashes of genital tissue which served to only crudely visualize *meiotic* divisions. Milligan (1976), however, used osmotically shocked hepatopancreas cells that were arrested in *mitotic* metaphase[a] to obtain clear chromosome visualization and counts in three species of *Penaeus*.

We modified the techniques of Milligan and produced the results in the following table. This data suggests that the 2n (diploid) number in *M. rosenbergii* from the Anuenue, Sri Lanka, Palau, and Australian stock is either 114 or 115.

The chromosome counts in the table were made on hepatopancreas cell metaphase spreads.

Table XIV-1
Chromosome Counts of *Macrobrachium rosenbergii*

Strain	No. of Animals Analyzed	No. of Cells Analyzed	No. of Chromosomes Observed <108	108	109	110	111	112	113	114	115	116	117	118	119	120	>120
Anuenue	9 ♀ & 1 ♂	26				1		1	3	2	12[a]	5	1				1
Australian	1 ♀	9		1				1	1	2	2	1				1	
Malaysian	1 ♀	2								1		1					
Sri Lanka	1 ♀	7						1	1	1	3		1				

[a]Modal diploid number

Although much more work is needed, there is no evidence of sex chromosomes, intraspecific chromosomal polymorphisms, or polyploidy. More samples must be taken to confirm this preliminary result.

[a]By colcemid injection.

Assessment of Intraspecific Variation

Obviously, the gene pool in the prawns that gave rise to the Anuenue stock, presently cultured in Hawaii and elsewhere, has proven to be sufficient for profitable production. However, this is not the only *M. rosenbergii* gene pool available to the aquaculturist. This species, distributed over vast areas of Southeast Asia, India, Australia, and areas around the South China Sea (see Figure XIV-1), is almost certainly composed of intraspecific physiological races (also called "ecotypes") similar to those found in other crustacea (Strong, 1972, 1973), oysters (Hillman, 1964; Loosanoff & Nomenjko, 1951), catfish (Smitherman et al., 1974), and plants (Clausen & Hiesey, 1958; Gregor & Watson, 1961). Geographic isolation, local environments, and community structures cause ecotypes to evolve differences in various aquaculturally important life history parameters, such as fecundity, growth rate, larval development rate, maximum attainable size, and various behaviors (Lewontin, 1965). Ecotypic variation could be used, for example, to develop genetic strains that: (1) were less aggressive, (2) had females who became reproductively mature at or beyond the harvest stage, (3) had a shorter larval development time, or (4) had greater tolerance to lower growout temperatures.

It is imperative for us to search for ecotypes containing useful adaptations that would make prawn aquaculture more profitable. We have obtained two shipments of prawns from locations in Sri Lanka and Malaysia, one shipment each from two Australian locations, and one shipment from Palau, Micronesia. We have collected but not as yet (March 1977) received specimens from near Calcutta. These locations are shown in relationship to a rough estimation of the natural distribution of *M. rosenbergii* in Figure XIV-1. This estimate, shown by the dotted line in the figure, was derived from personal accounts and mapping out the collection locations as described by Holthius (1950), Johnson (1960a, 1960b) and Tiwari (no date). In addition, I spoke to a member of the Zoological Survey of India (Tiwari, pers. comm.) on a trip in June 1976 and he indicated that *M. rosenbergii* is restricted to areas close to or in estuarine waters. In fact, *M. rosenbergii* is not considered a freshwater organism in India. Also, I observed a preserved specimen closely resembling *M. rosenbergii* in the Tungkang Marine Laboratory in southern Taiwan; so I've included a question mark near this country to indicate the possibility of it being the northernmost point in the species range. Likewise, personal accounts communicated to me by biologists working in southwestern Papua indicate *M. rosenbergii* is present there also, so it may also be in the western Solomon Islands. Consequently, another question mark is placed in that location.

The Australian stocks represent two distinct ecotypes which are identical in appearance as juveniles and adults but show definite differences during larval development. These two varieties occur approximately 600 miles apart near the

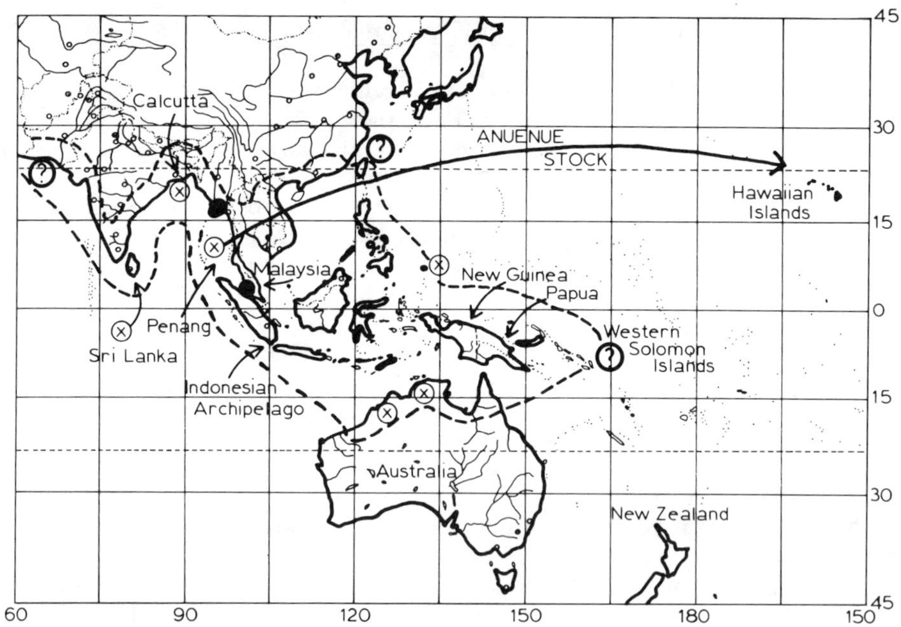

Figure XIV-1
Estimated native distribution of M. rosenbergii (dotted line) and locations (circled x's) from which broodstock have been obtained for evaluation in Hawaii. Closed circles indicate locations for stock obtained by other workers. Data from personal communication, Holthius (1950), Johnson (1960a, 1960b) and Tiwari (no date).

northern Australian cities of Derby and Darwin. Females found in Derby have large green eggs (Aus. A) while Darwin females (Aus. B) have small orange eggs.

In June 1976 one Aus. A and one Aus. B female were mated 20 days apart (we suspect that both females were mated to the same male). One week later an Anuenue female released larvae which were reared under the same conditions as Aus. A and Aus. B. Finally, nine weeks later in September 1976, we received berried females from Palau. These were incubated and the larvae reared in an outdoor culture system.

Table XIV-2 shows each stage of larval development (after Uno and Kwon, 1969) the day on which this stage was first noticed, the day of the first postlarvae, and the time when 90% of the population had metamorphosed.

Table XIV-2
Stage of larval development, posthatch day on which each stage was first observed (3), day of the first "scout PL" (1), and day when only 10% of original larval batch remained (2).

Stock	Stage 1	2	3	4	5	6	7	8	9	10	11	(1) PL	(2) 10%
Australian A.		$5^{(3)}$	7	8		9	10	14	17	18	19[a]	19	23
Australian B.							7	10	11	12	14	16	19
Palauan		3	5	7	10	11	13	15	20	22	25	26	34
Anuenue		3[b]	5	7	8	10	11	14	15	17	20	25	31

[a]11th stage characterized by one large spine on rostrun.
[b]Stalked eyes but no development of rudimentary exopod on telsom.

The Anuenue and Palau varieties followed very closely in development stages with the Australian types a few days behind. The Aus. A grew faster during the later stages.

The Aus. B variety developed the fastest of all and settled in the shortest time (16 days). Both Australian varieties metamorphosed in fewer days than either the Palauan or Anuenue types. Aus. A settled in 19 days from hatch with 10% larvae remaining after 23 days. The Aus. B larvae settled in 16 days with 10% remaining after 19 days. The Anuenue and Palauan settled in 25 and 26 days with 10% remaining after 31 and 34 days, respectively.

Figure XIV-2 shows the growth rate per day of larvae from each ecotype. This estimate was done by taking five of the largest larvae each day and measuring their postorbital lengths. Day 1 is the day of initial release of larvae from the brood.

Initially there was a difference in size among the newly hatched larvae. Both Anuenue and Aus. B type larvae were 1.85 mm, while the Palau and Aus. A types were 2.30 and 2.50 mm, respectively. This difference became less noticeable at the 7th day (3rd to 4th stage). This is probably not significant since from this point on, variation within the population will begin to appear. Convergence on the 10th day did not apply to the Aus. B which was 35% larger than the others on that day. Excluding the Aus. B variety, there was no significant difference between growth rates of the other types.

Australian PL's have distinct yellow and white spotted bands running from side to side across the abdominal segments. These bands begin to disappear at 2-3 cm and the animal becomes mottled with black spots. At 5-7 cm, these yellow and white bands again appear until adult size when the animal loses these bands except for a black spotted band between the thoracic and first abdominal segment.

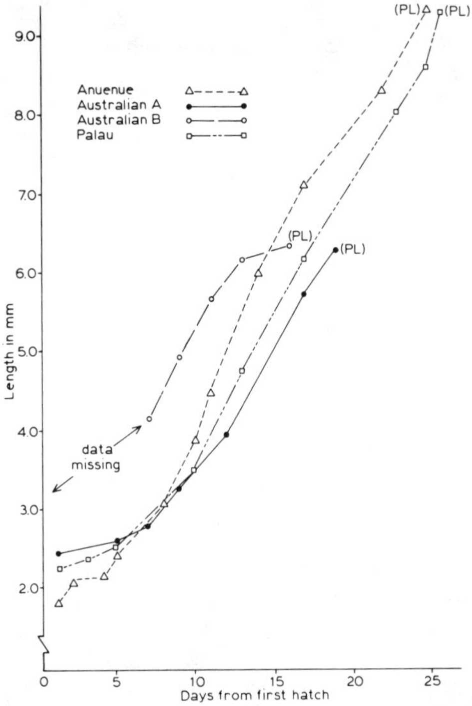

Figure XIV-2
Growth rate of fastest developing larvae (see text) of four "ecotypes".

Larvae of the Palau variety do not differ in appearance from the Anuenue stock. However, from approximately 1 cm, the animal becomes mottled with black spots.

Unlike the Australian type, the Palau animal retains this black spotted marking on the abdomen into the adult stage--which makes identification of this animal very easy. The Anuenue stock has black horizontal lines on the thoracic segment as juveniles but, as adults, are a bluish gray color with no distinct markings.

Genetic Selection

Genes control visible traits or "phenotypes" in many ways but some act in a simple "additive" fashion and may be grouped into those that contribute small incremental additive effects to the phenotype value and, because genes come in pairs, into a set of "noncontributory" genes. The number of contributory and noncontributory additive gene pairs determine the number of phenotype values that a random breeding population will display.

When a number of gene pairs control a trait in this fashion, their genes are collectively called "polygenes" and the traits they control are referred to as "polygenic traits" (Thompson and Thoday, 1974).

Simple mass selection breeding systems take advantage of the fact that economically important phenotypes like size or growth rate are "quantitative" traits most likely controlled by many additive polygenes. If the gene frequencies are the same for contributory and noncontributory genes, then the phenotypes in a random breeding population will be distributed along a continuum ranging from those who do not display one extreme phenotype (e.g., small size) to those individuals who display the other extreme phenotype (e.g., large size). Individuals at phenotypic extremes are more likely to possess a greater number of either contributory or noncontributory genes in their genetic makeup or "genotype".

Genetic selection, as referred to in this review, involves establishing "high" and "control" lines by selecting and crossing parents with large and average performance values and assessing whether a genetic "gain" results between the two lines after a number of generations. If there is no environmental effect, the mean phenotypic value of a subsequent generation will eventually equal mean phenotypic value of the selected parents. This would be the largest possible "selection" response because the progeny mean value has "moved over" from the values of the unselected population to equal the mean value of the selected parents.

When environmental factors also affect the phenotype, then continually selecting parents from the desired end of the continuum can produce a steady, generation-after-generation "selective response" in the desired direction. When the phenotype value is controlled by only a few polygenes and is not affected by the environment, then only a few generations are required to have the mean phenotype value of the progeny equal that of their parents.

When an unselected population contains individuals that have maximum combinations of contributory and noncontributory genes, the population is said to possess enough "genetic variance" so that a selection response could be expected. It is hoped that this is the case with cultured *M. rosenbergii* stock since it is well known that similarly aged individuals in tank and pond populations have phenotypic variances of sufficient magnitudes to permit the possibility of genetic selection. This is shown in Table XIV-3 which has been expanded from one given in Gjedrem (1975) to include other species and h^2 values (see next section). Table XIV-3 demonstrates that *M. rosenbergii* phenotypic variation in orbit length from tanks and ponds is within the range of variation of other economically important characters from other species where the genetic variance estimates (h^2) have been completed (see next section). It is interesting to note that Palauan, Australian,

Table XIV-3
Phenotype standard deviations and coefficients of variation and heritabilities for different traits in different species.

Species	Character	Standard Deviation	Coefficient of Variation (%)	h^2
Cattle[a]	Milk yield	767	15	.30[f]
Sheep[a]	Fleece yield	0.61	13	.40[f]
Swine[a]	Growth rate	70.7	12	.29[g]
	Backfat thickness	2.16	10	.55[g]
Rainbow trout[a]	Body weight[b]	4.37	33	Best "guestimate" for salmonoids h^2 = .10 - .20[a]
	Body weight[b]	7.49	70	
Sea trout[a]	Body weight	6.18	51	
Salmon[a]	Body weight	5.08	102	
Carp	Weight	--	--	.10 - .20[h]
Catfish	Length (5 mo.)	--	--	.12 ± .30[i]
	Length (15 mo.)	--	--	.67 ± .57[i]
	Length (5 mo.)	--	--	.61 ± .35[i]
	Length (15 mo.)	--	--	.75 ± .53[i]
Lobster	Size	--	--	.30[j]
M. rosenbergii (Anuenue stock)	Orbital length[c]	2.35	27	--
	Orbital length[d]	1.16	15	--
	Orbital length[e]	.542	22	--

[a]Gjedrem, 1975.

[b]Different studies as given in Gjedrem (1975).

[c]Preharvest sample taken on 5/23/75 from 1/3 acre previously unstocked commercial pond stocked in 10/12/74 at approximately 1.5 PL/ft^2.

[d]Preharvest sample taken on 5/9/74 from 3.5 acres previously unstocked commercial pond stocked on November 2, 5, & 7, 1973 at approximately .62 PL/ft^2.

[e]Mean values from 11 hatchery tanks containing Anuenue stock PL's from cohorts used to stock pond given in d above. Larvae hatched on or about 8/22/73, sample taken on 11/9/73.

[f]Falconer, 1960.

[g]Lasley, 1972.

[h]Kirpirchnekov, 1969.

[i]Reagan et al., 1976.

[j]Hedgecock et al., 1976.

and Sri Lankan stocks display coefficients of variation values similar to those of the Anuenue stock. Although standard deviation and coefficient of variations are not available, the graphs of the length-frequency (phenotypic) distributions of M. rosenbergii populations grown at different densities as given in Sandifer and Smith (1975) are similar to those from which the M. rosenbergii values in the table are derived. All in all, there appears to be sufficient phenotypic variation in M. rosenbergii to permit the possibility of a response to genetic selection.

As we delve deeper into the genetics of crustacea, there may emerge a lesson drawn from prior experiences with carp and trout. Selection within carp strains produced no gain and one or two generations of inbreeding resulted in severe "inbreeding" effects which precluded any genetic improvement via this means (Moav and Wohlfarth, 1973). Consequently, carp breeders have concentrated on crossbreeding existing varieties for hybrid vigor. By contrast, the fork length of rainbow trout has been increased nearly 50% between 1944 and 1968 by mass selection (Donaldson, 1970). Why did mass selection produce such a dramatic response in trout and nothing in carp and how does this relate to M. rosenbergii? No one can furnish a complete answer since the genetic control of quantitative traits such as size is poorly understood. However, one thing seems clear--the original rainbow trout stock, first subjected to selection in 1944, was undomesticated and, therefore, possessed genetic variation amenable to mass selection. On the other hand, carp have been undergoing intense selection for centuries and have become divided into a number of genetically dissimilar races or strains which now possess a kind of genetic variation not responsive to mass selection.

In a genetically untested species like M. rosenbergii, one cannot say for certain how much of a response selection will produce. However, inasmuch as M. rosenbergii is undomesticated, has a high fecundity, and the mastery of its life cycle permits genetic selection, breeding methods can easily be applied to it with a good chance that a selection response will be observed.

Dobkin (1974) conducted a selection experiment for size in Florida and observed a mean size difference between the third and first generations. However, it is impossible to tell if this was a bona fide genetic difference since no unselected control group was evaluated.

We have been conducting single parent genetic selection in Anuenue M. rosenbergii. The first generation progeny group, derived from selected and unselected parental groups, showed differences in juvenile size, indicating a genetic gain is possible in Anuenue stock. However, we have not demonstrated this as discussed below.

Our genetic stock material was hatched on June 22, 1973 ($\pm$ 5 days) from an unknown number of berried females. The larvae were subsequently raised in 11 large outdoor tanks by AFRC. After three months of grow-out, PL's from these 11 tanks were pooled into size classes, according to the standard deviation of their orbital lengths. After approximately seven months of growth in tanks, the prawns were again classed according to their relative orbital lengths and breeder prawns having extremely large and small phenotype values were then randomly selected from two of the classes. It was assumed that relative size is indicative of relative growth rate, and that the two classes of breeder prawns therefore represented "fast" and "slow" growers.

Three specific matings were obtained from these breeder prawns. The matings, as well as the progeny from these matings, will hereafter be designated as types A, B, and C as follows:

Type A = fast growing ♀ X fast growing ♂ ("high" line)

Type B = slow growing ♀ X slow growing ♂ ("low" line)

Type C = slow growing ♀ X fast growing ♂ (control group: "unselected")

The matings were completed within a six-day period to allow for the concurrent development of the progeny groups which were also treated equally by standardized rearing management conditions.

After all larvae in the groups had metamorphosed into PL's, four random samples of PL's (two of 180 and two of 10) were taken from each group. This resulted in a total of 12 sample populations. The 180 PL's were placed in 5,000-gallon circular tanks and the 10 PL's in 100-gallon rectangular tanks. Uniform conditions in these grow-out tanks were maintained by standardizing feedings and water management. Two of the 180 PL populations, one from group A and one from group B, died mysteriously soon after stocking.

Orbital length measurements for each group taken between 90 and 170 days after hatching revealed differences in relative growth rates (Figure XIV-3). The remaining four populations grown in the 5,000-gallon tanks (A, B, C_1, and C_2) were analyzed using the least squares method for unequal subclass numbers to test two hypotheses: (1) whether the differences between progeny groups were maintained regardless of age (no age X group interaction) and (2) whether the differences between the group means were significantly different from an overall mean of all groups combined. The differences between the means of the progeny groups were found to be unaffected by age and so, holding age constant, a determination was made of the significance of the deviations of the four progeny group means from an overall mean. The means of progeny groups A, B, and C_2 were significantly different from an overall mean while C_1 was not. This makes it impossible to say that the observed differences are due to environmental or genetic factors since the difference observed between the full-sib control groups

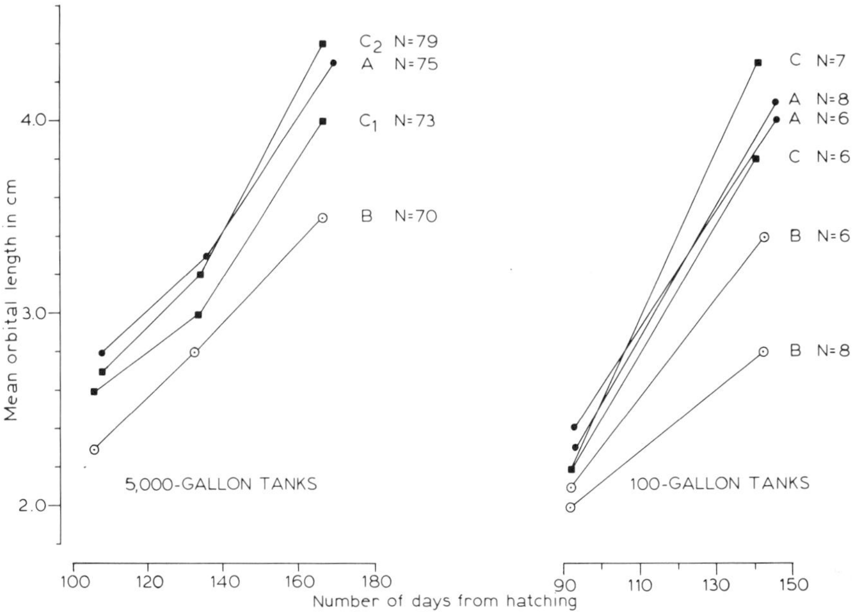

Figure XIV-3
Mean orbital length of first-generation progeny groups from selected matings reared in 5, 000- and 100-gallon tanks. A = "high" line, B = "low" line, C = control group, "high" X "low" line parents. N = sample size (see text).

(C_1 and C_2) suggests that observed differences between progeny groups are highly dependent upon nonheritable environmental differences.

Though it cannot be concluded that the observed differences among the groups are in part due to a heritable response to selection, a comparison of the length frequency distribution of progeny groups A and B (Figure XIV-4) suggests artificial selection has been effective since the two distributions are slightly skewed in opposite directions.

Further studies are currently being conducted to remove certain ambiguities so that a greater understanding of the causal components contributing to M. rosenbergii growth rate variation can be reached. Our immediate aim is to improve the existing pilot study design to facilitate an analysis of variance of the results by incorporating better techniques of environmental control.

Several points of caution must be made here. It is now common knowledge among M. rosenbergii aquaculturists that in mixed-sex populations, similarly

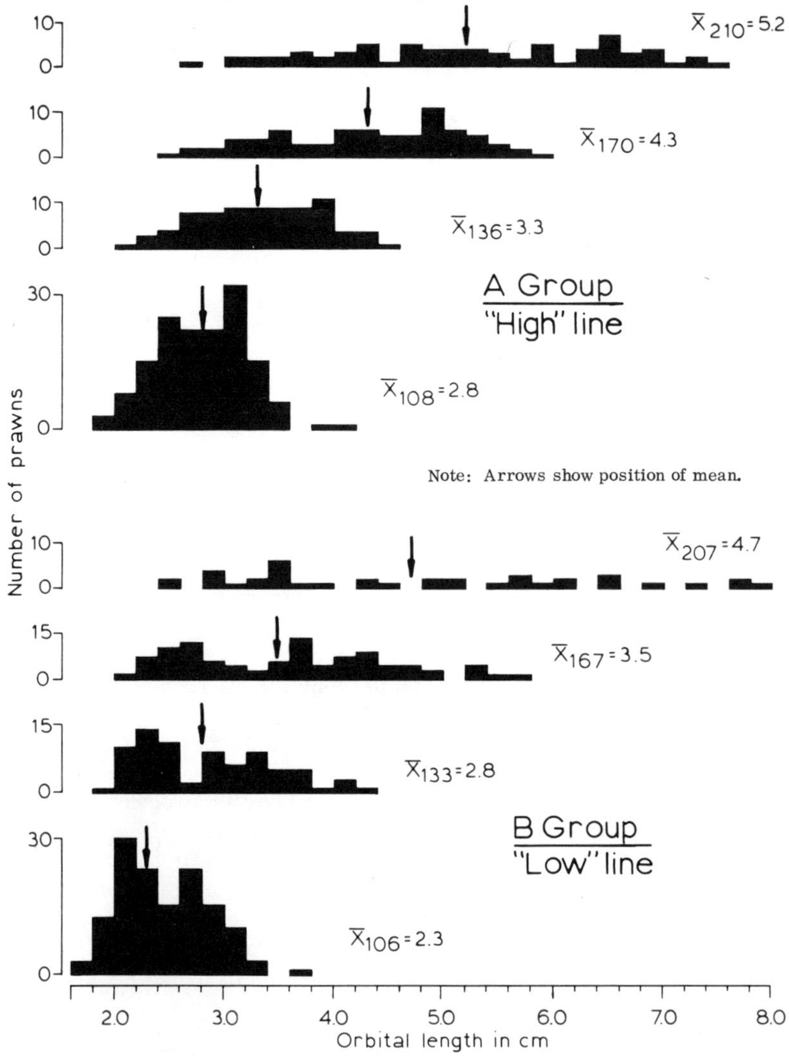

Figure XIV-4
Orbital length frequency distributions of first-generation progeny from selected parents in the A group ("high" line) and B group ("low" line) grown in 5,000-gallon tanks (see Figure XIV-3 and text). $\overline{X}$ indicates population mean at different days (subscript) from hatching; e.g., $\overline{X}_{133} = 2.8$ refers to a mean orbital length of 2.8 cm at 133 days posthatch. The figure demonstrates population variance increase through time and shows that the A group ("high" line) is more skewed to higher orbital lengths than the B group ("low" line).

aged individuals will have heterogeneous growth rates leading to positive skewness in the size distribution in which the skewed tail of the distribution is made up of "bull" animals who are disproportionately larger than their fellows. This "variance amplification" increases the coefficient of variation and can be seen in studying the graphs of growing M. rosenbergii populations (Fujimura and Okamoto, 1970; Sandifer and Smith, 1975, and Figures XIV-3 and -4 in this review). No one knows what effect this bull-runt mediated variance amplification will have on the outcome of a selection program, but there may be only a weak correlation between the size of a bull animal and its genotype. Consequently, selecting these as parents will produce only a negligible response in the progeny.

A second caution has to do with the amount of genetic variance in the Anuenue stock, the one that is cultured most intensely. The Anuenue stock was derived from a very small number of wild-caught animals imported from Malaysia by AFRC, so its derivative stocks represent an extremely limited sample of the gene pool of M. rosenbergii, and it is possible that the total amount of genetic variance available for a possible selection response is limited (Al-Murrani, 1974). Consequently, small selection responses may be observed even though the genetic variance of size in M. rosenbergii as a species is sufficient. Any genetic selection program must be designed to take this into account and additional genetic variance may have to be "pumped into" the stock under selection by means of outcrossing individuals from such stock to wild-caught, unrelated individuals.

A third caution has to do with the phenomena of "genotype-environment interaction" whereby genetic groups give different phenotype values in different environments. It must be realized that a genetic stock produced by a breeder gives it optimum performance within the limits of a particular management system (Moav et al., 1975; Moav et al., 1976b). In this regard, it is very important that stock evaluations are conducted in commercial ponds because the production performance of genetic group will be the end result of the interaction of its genetic makeup with the "environment" in which it lives. For pond-cultured prawns, the "environment" would be the entire management system including water quality, feeding rates, and temperature regimens acting singularly or in concert with the genotype.

Genetic Variance Estimation

In a selection program, if parents are selected on the basis of phenotype values only partially determined by the genotype, then the mean phenotype values of their progeny will not equal the mean phenotype value of the parents and the amount of "gain" observed is proportional to the amount of genetic variation in the population. The amount of genetic variance that is present in a breeding population can be estimated as the "heritability" (h^2)[a] of the trait. If V_g = genetic

[a]Not a power function.

variance and V_t = total phenotypic variance, then $h^2 = V_g/V_t$. A restrictive definition of h^2 is $h^2 = V_{ga}/V_t$, where V_{ga} = the "additive" genetic variance.

Heritability estimates are, therefore, valuable to the animal breeder because they allow a continual assessment of the amount of "genetic variance" available in a broodstock and can be used to predict genetic "gains" and cost/benefits that could be expected under a scheme of genetic selection.

Since heritability values are variance ratios, they can be estimated with ANOVA (analysis of variance) methods.

Table XIV-2, referred to in the previous section, lists h^2 values for several nonaquatic and aquatic species. In general, h^2 values for the latter are lower than in the former. The catfish h^2 values are noteworthy since they show that length and weight decreased dramatically through time. Similarly, Lannan (1972) reported the following h^2 values (too numerous to list in Table XIV-2) estimated by analysis of variance of full siblings for characters in the Japanese oyster, *C. gigas*:

Attribute	Heritability
L (12 mo)	.81 ± .07
L (18 mo)	0
W (12 mo)	1.17 ± .05
W (18 mo)	.10 ± .12
H (12 mo)	.81 ± .27
H (18 mo)	.19 ± .003
Size (12 mo)	.93 ± .28
Size (18 mo)	.15 ± .14
Shape (12 mo)	.31 ± .18
Shape (18 mo)	.13 ± .12

Note: L = Length; W = Weight; H = Height

I list these values because they illustrate the dramatic decreases in heritability that were realized in at least one species. This phenomenon could be relevant to h^2 estimates in *M. rosenbergii* because populations of the latter display significant increases in phenotypic variance during grow-out (see Figure XIV-3). See Sandifer and Smith (1975) and Fujimura and Okamoto (1970).

If significant h^2 decreases through time are observed in *M. rosenbergii* genetic groups, then selection programs (designed to take advantage of genetic variation but based only on the visible phenotype performance values) will have to be conducted accordingly. In this regard, it will be valuable to study the intrapopulational behavior factors that contribute to this variance increase and to further assess whether these behaviors are genetically mediated.

Hedgecock et al. (1976) estimated the variance of growth rate within and between full sibling families in the lobster and demonstrated significant family (i.e., genetic) contribution. These workers were unable to furnish a direct estimate of V_{ga} within their experimental design but used the correlation between full sibs to give them a mean upper estimate of $h^2 = .30$ indicating that a selection program based on growth rate could save more than a month of growing time to market.

No genetic variance estimation has been done in M. rosenbergii to date for many reasons, not the least of which are those related to biological barriers to achieving the numerous single-pair matings needed for genetic variance estimation. We have been working along these lines as described below.

We established a broodstock of over 100 individually housed Anuenue stock females which we surveyed each morning for six months looking for premating molts. We attempted to achieve as many single-pair and repeat matings as possible of different mothers to the same father in order to perform an analysis of variance of larval development time and larval size. We succeeded in establishing the following seven half-sib groups with each group having a common father (sire) and from one to seven mothers (dams). Some dams were mated twice to a sire.

Half Sib Groups	Number of Dams	Number of Full Sib Families
A	7	9
B	6	8
C	6	7
D	5	6
E	3	3
F	2	3
G	1	1

Larvae from each of these full sib families were individually reared until the first "scout PL's" were observed. At this time we recorded the time in days from first hatch to scout PL, took a sample of the larval population and measured the size of individual larvae, and recorded their stage according to Uno and Kwon's (1969) criteria. We then performed an analysis of variance for unequal subclass numbers for scout PL time and larval size. We wished to determine if there was a statistically significant variance component for these variables which could be attributed to the fathers (or sires). We did not find this and we concluded that there is a neglible paternal (i.e., genetic) variance component to larvae development time and size in the Anuenue stock.

However, we observed a large range of scout PL time from approximately 24 to 45 days which we attribute to the mean temperature in our larval rearing tanks: lower temperatures in the tanks during the colder months (in which some larval families were reared) extended their scout PL times appreciably.

During this experiment, which we consider preliminary, we learned several things. First it is important to perform experiments of genetic variance estimations under very controlled environments and to achieve near simultaneous matings of dams and sires so that full- and half-sibling families can be reared at the same time in order to reduce the time-mediated residual error term. This successful sequential mating of one sire to many dams is difficult in *M. rosenbergii*. Furthermore, a large broodstock of reproductively mature females is needed to have the desired number of receptive females on hand for each round of mating. In this regard, mature *M. rosenbergii* females mate only when they molt and they do so roughly every 4-8 weeks. We calculated that molting occurs at the rate of 2-3 females/100/day.

An ANOVA design that meets minimum statistical requirements (Reagan et al., 1976) calls for 20 sires, each mated to three dams. If the female premating molt rate is 2-3 animals/100/day, and assuming all premating molting females mate successfully, then a female broodstock containing between 2,000 and 3,000 individuals is needed to generate the 60 full- and half-sibling families needed for the proper analysis. Obviously a broodstock of this size composed of individually housed prawns is very expensive to maintain so a way must be found to predict premating molts in *M. rosenbergii* females. Perhaps a better alternative would be to perfect *in vitro* fertilization and development methods so that groups of egg masses from rather few females could be fertilized by sperm from a number of males. This would serve to increase the precision of h^2 estimates as well as eliminating much of the expense of maintaining large female broodstocks.

Additional technological constraints must be considered. Following metamorphosis, PL's must be reared separately in some sort of grow-out unit. Hedgecock et al. (1976) and Carlberg and Van Olst (1976) used special units with individual compartments to grow out juvenile and larval lobster, respectively. We have constructed a similar system using two large grow-out tables that can house nearly 800 individual PL's. One table will be used as a replicate for the other. We plan to use these units to derive h^2 estimates for *M. rosenbergii*.

When the genetic markers become available, then genetic groups may be stocked in the same uncompartmentalized unit because they can be identified by their electrophoretic patterns. This will greatly aid in genetic variance estimation in *M. rosenbergii*.

Genetic Markers

In a recent paper, Moav et al. (1976a) states:

"Consideration of marking and identification of individual organisms shows that fish, as well as most other aquatic organisms of commercial value, differ from the common farm livestocks--for which all the advanced animal breeding plans had been developed--in that during the hatching and early juvenile stages, individuals cannot be marked and identified. Hence the necessity for genetic marking. We pointed out the advantages, as well as the severe limitations of most morphological genetic mutants, and reached the conclusion that electrophoretic markers can be used to create the necessary basic tools for implementing advanced breeding programs of fish."

In this regard, genetic selection and variance estimation is very difficult and expensive without the use of genetic markers for progeny group and stock identification.

We have identified markers in *M. rosenbergii* using starch gel electrophoresis applied to various tissues: 18 systems give us clear and unambiguous results, and two isozyme systems (GOT [Glutamate-Oxaloacetate-Transaminase] and PGM [Phosphoglucomutasel]) are polymorphic in our Anuenue broodstock. A GOT locus has three, and a PGM locus has two putative co-dominant alleles in high frequency. We have electrophoretically phenotyped our entire Anuenue broodstock for these as well as for nonpolymorphic systems using muscle from *single* walking legs of animals--which remain alive and well. Using the markers as molecular tags in a mass selection experiment, we should be able to stock progeny groups from different parents in the same grow-out unit eliminating much of the expense and environmental error in statistical replication.

Corollary Studies

Since quantitative genetic traits are controlled by both genetic and environmental factors, it is also important to conduct studies regarding the factors which influence phenotypic variance of important traits in *M. rosenbergii*. This is done to achieve results, independent of those generated in formal genetic mating schemes, that will further aid in identifying the sources of phenotypic variance. In one study we sought to establish the relationship between adult size and the time of larval metamorphosis to the PL stage. It is common knowledge to hatchery aquaculturists that there is a wide range for the time the first batch of scout larvae metamorphosize into benthic PL's to the time that the last larvae do so (we will call this range the "PL drop time"). Most *M. rosenbergii* hatcheries use only the PL's that have "come down" during the first half of the PL drop time in stocking their grow-out ponds. If there is a high correlation between adult size (i. e., growth rate) and PL drop time, then this practice could constitute *de facto* genetic selection.

As part of this experiment, we took the PL's from a single female and divided them into experiment and control groups based on whether they settled during the first three days (Group I), days 4-6 (Group II), or days 6-8 (Group III). Group IV contained all PL's settling in the 9th day or later. A portion of each group was put into three control groups. All groups were placed in tanks and allowed to grow for 13 months at which time postorbit lengths were taken on all animals. Many animals in each group were of "market lengths" (9-10 cm orbital length) at this time. We found no differences in mean size between experimental and control groups indicating that early settling PL's will not necessarily be the largest adults. These results are similar to those obtained by Sandifer (pers. comm.) in a study his group conducted in South Carolina.

If prawns, which develop relatively "slowly" as larvae, can still reach market size in about the same time as those who developed "faster" as larvae, then are the "runts" in a juvenile population capable of doing the same and under what conditions?

We addressed this question in a study that focused on the effect of stocking density and homogeneity of initial individual size on grow-out of runted and bull animals. Our experimental design is shown in Figure XIV-5. We first determined the mean and standard deviation of the orbital length frequency distribution of juveniles in nine AFRC hatchery tanks. These juveniles were reared for about three months at various densities from PL's that were hatched as larvae on or about August 27, 1973 (± 5 days) from an unknown number of berried females taken from a commercial pond. At the beginning of our study, the well-known *M. rosenbergii* variance amplification was of sufficient magnitude that bull and runt animals were present in these juvenile populations as shown in length frequency histograms in the box in Figure XIV-4. We then regrouped the runted, bull, and "average" sized juveniles into four size classes (I, II, III, IV) based on the standard deviation units and grew them out for approximately six months at different densities (I won't discuss the effect of latter variable in this review). Our results are summarized in Figures XIV-6 and -7. First of all, size class populations displayed an amplification of the variance of the orbital length frequency distribution such that many animals in the runt classes (I and II in Figures XIV-5 and -6) displayed faster growth rates than animals in the bull class (IV in Figures XIV-5 and -6) in effect achieving larger sizes than animals that were 100% larger six months before. To be sure, the mean size of class I was lower than class IV, but our results demonstrated growth rate is labile in *M. rosenbergii* and "runtiness" is not necessarily a permanent condition. In this regard, the runt class populations amplified their orbital-length frequency variances at much faster rates than the bull classes as shown in Figure XIV-7 where the coefficient of variation of orbital length in the size classes is plotted against time for two densities. At 1X density (approximately 1 juvenile/ft^2), the values are initially very close as one would expect as the result of the regrouping. However, at the end of the study (6/7/74), the values are disparate for the different size classes.

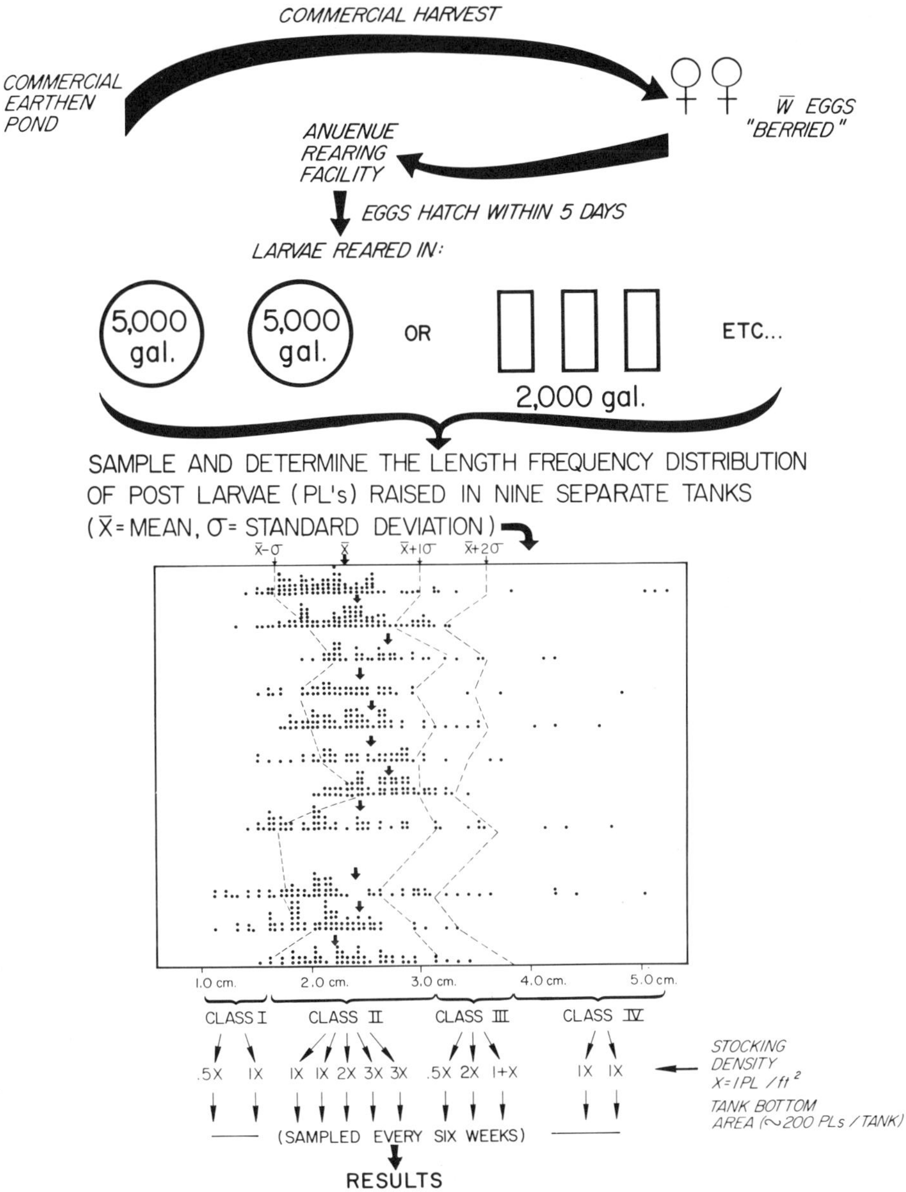

Figure XIV-5

Design of experiment to assess the effect of regrouping "runt", "bull", and "average" size juveniles on growth rate (see text).

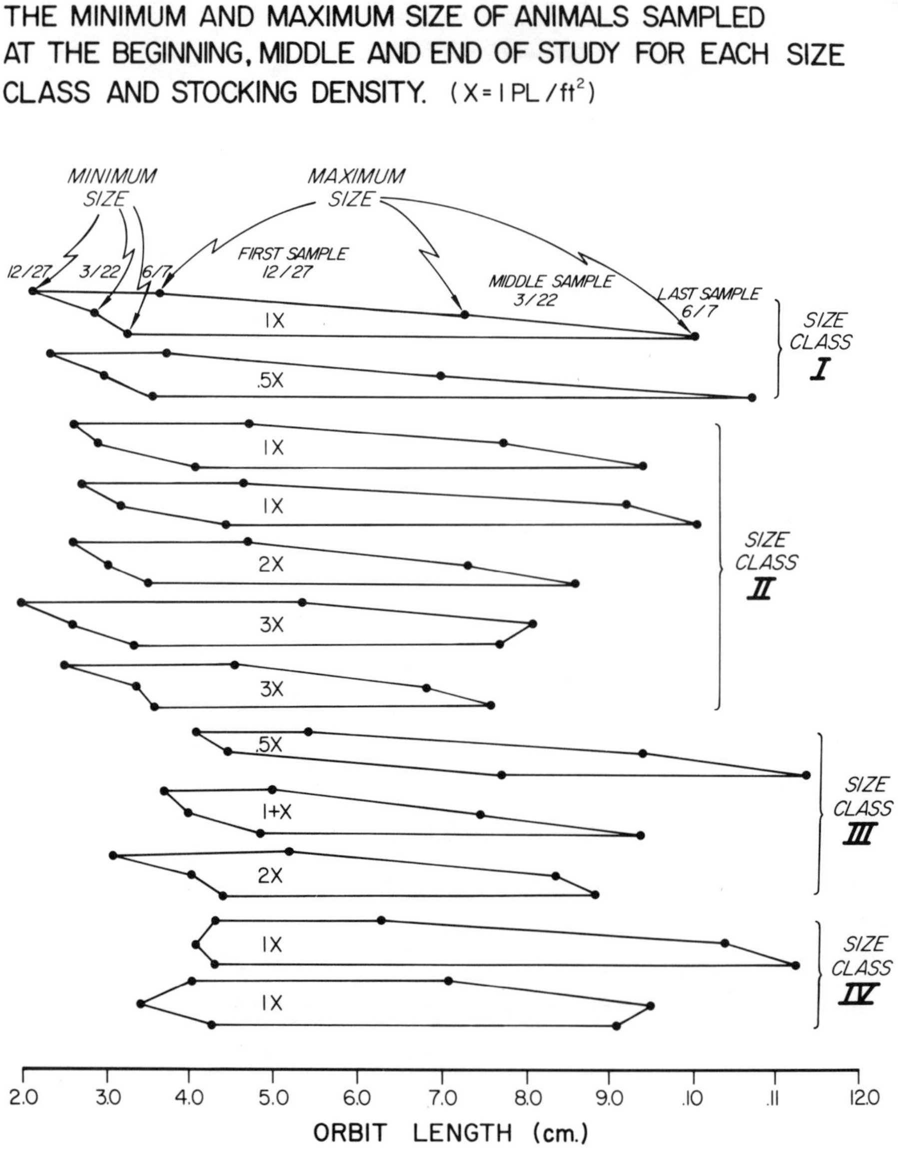

Figure XIV-6
The minimum and maximum size of animals sampled at the beginning, middle, and end of the study illustrated in Figure XIV-5. Size classes (I, II, III, and IV) are described in the text and in Figure XIV-4.

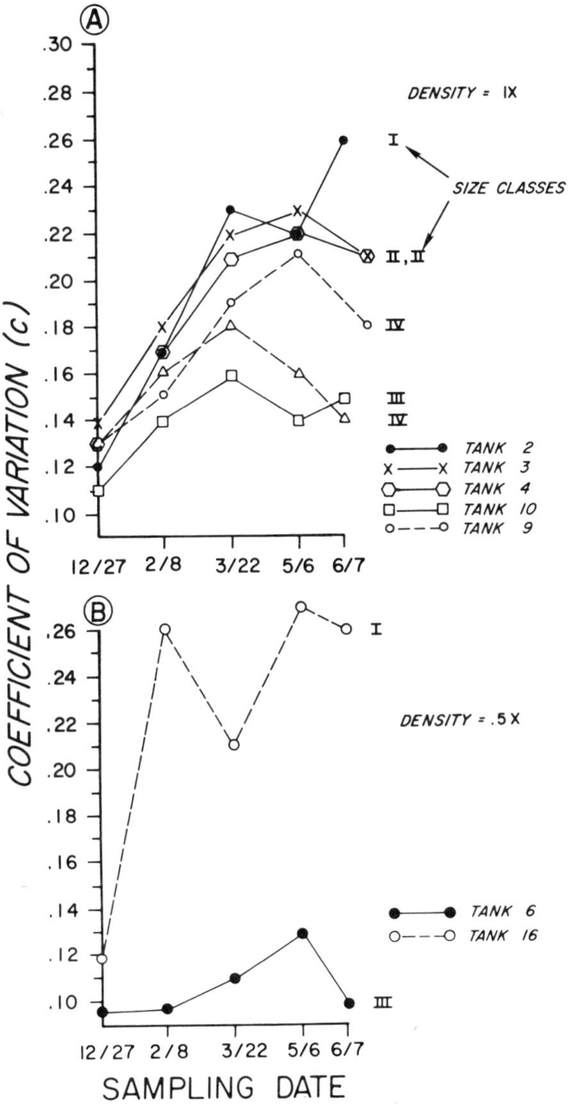

Figure XIV-7
Coefficient of variation (standard deviation/mean) for orbital length in size class populations discussed in text and Figure XIV-5, reared in tanks at two densities (X = 1 juvenile/ft^2). Figure demonstrates that runt classes (I, II) increase their orbital length frequency distribution variances at a faster rate than bull classes (III, IV).

Conclusion

In conclusion, I would like to offer a few highly personal comments.

Recently a small study group convened in Seattle (Anon., 1977) to evaluate the results of a survey (part of the NOAA National Aquaculture Plan) conducted by the University of Washington's College of Fisheries in cooperation with NOAA. They concluded that research priorities in the development of commercial prawn aquaculture should be given to genetics as well as other disciplines. But they also concluded:

> "Most believe that genetic improvement is needed and desirable in the long run. But there was agreement that the animals were good enough at the present time because of their inherent characteristics and could be used for a commercially viable aquaculture."

Indeed, it can be debated whether the lack of genetic knowledge of *M. rosenbergii* is a constraint to its commercial aquaculture at present. In the short term, I would agree that we don't have any major genetically mediated impediments to commercial viability. However, we must look down the road. As pointed out by Moav et al. (1976) and others, the genetic background (genotype) of an aquacultured group is closely tied with such things as nutritional requirements, disease resistance, behavior, and physiological responses. These factors contribute to the production performance of the group and the research to optimize them must take into account the group's genotype. In other words, genetic improvement has to be done in conjunction with improvement in other aspects of aquaculture management systems.

Generally, long-term, expensive, time-consuming selection and genetic engineering programs do not (and perhaps should not) materialize until a rather extensive commercial industry has been developed. In this regard, we must remember that we, as *M. rosenbergii* aquaculturists, are embarking on the first stages of a domestication process in a "wild" organism. Conversely, the application of agricultural genetic practices to other food crops and livestock, which began early in this century, built on existing stocks that had already become genetically distinct as the result of centuries of cultivation and domestication. Of course, we can look forward to compressing the time required for genetic refinement if Donaldson's (1970) work has any relevance to our efforts.

But what should our efforts be? First, I feel we must determine the extent of the genetic control of traits involved with the heterogenous individual growth rates that seem to occur in all our ponds and tanks. These traits will involve size and weight at certain grow-out times, growth rate *per se* and individual behavior. Next, we should determine the extent to which these traits vary throughout the species range so that intraspecific differences, if they prove valuable, can be utilized outright or within a few generations in our aquacultured groups without lengthy and expensive "genetic engineering" efforts involving long-term selection and gynogenesis.

Regarding gynogenesis, I would like to make one last point which may have relevance to other work of a "genetic engineering" nature in *M. rosenbergii*. In gynogenesis, maternal gametes (eggs) are induced to develop into offspring without a paternal set of chromosomes. Irradiation is used to kill sperm chromosomes, but not the capacity of the sperm to induce postfertilization cleavage. Usually, one tries to induce refusion of two of the maternal meiotic products to achieve diploidization so that offspring have a diploid set of only maternal chromosomes. Artificial gynogenesis speeds up the development of inbred lines because functionally the mother has "mated with herself" to produce the gynogenetic offspring. I personally think gynogenesis and other genetic engineering work should be pursued in *Macrobrachium*, but such work, I think, is completely in the realm of basic research and its payoff will be in the long term. Here we are talking about manipulating nature to the point of creating a variation that was not present in the first place while, with other genetic programs, we are using the natural species variation. We must remember that Mother Nature has a way of being resilient to our efforts to reconstruct her genetic work. Consequently, if we are to modify nature's work greatly, we should expect to pay dearly for it in both time and money.

Acknowledgements

Many thanks go to Mr. Takuji Fujimura and his staff at AFRC for their efforts in support of our work. Mr. Michael Tomihama compiled the data in Table XIV-2 and contributed to the section on genetic selection. Likewise, Mr. Alan Muraoka conducted the larval rearing of the ecotypes. Glen Pulianas and John Bishop did the chromosome work; John Wong, Debbie Stelmach, Gail Kitaji-Dennis, and Mark Muranaka were instrumental in the electrophoretic work; and Mary Ellen Lindenfelser helped with some aspects of the ecotype evaluation. Ms. Rumi Fujimura typed the early drafts of this manuscript. The help of all these people is gratefully acknowledged.

B. Discussion

Much to the credit of the audience, the question-and-answer period following Dr. Malecha's review involved "give and take" discussions which provided perspectives of the content that would not otherwise come from reading the formal narrative. Consequently, as much of the flair of original discussion as possible was preserved in this written summary.

Question: Why do all matings need to be done "simultaneously" in the variance-estimation experiment?

Answer: One wants to achieve simultaneous or near simultaneous matings to eliminate the environmental sources of variation which are time dependent since the culture conditions change even in a closed system. Simultaneity of mating and the subsequent brooding, hatching, and larvae, juvenile, and adult rearing is nearly impossible with *M. rosenbergii* if we use the natural reproductive cycle because, as I said in my discussion, at least one day/female is needed for sequentially mating a sire to a number of dams.

Question: Isn't the tank-to-tank (or system-to-system) variation so great as to mask the time-mediated variation you spoke of?

Answer: Yes, it could be. The variance experiments have to be done in *one* grow-out unit if possible (one tank, one table, or whatever) with the individuals somehow identified or housed in separate compartments. Genetic markers could greatly aid in this respect.

Question: How many replicate individual compartments or cells do you need?

Answer: No replications of individual cells.

Question: Why? What about the cell-to-cell variation?

Answer: That's the individual-to-individual (or animal-to-animal) variation you want to measure since there is one animal/cell or individual compartment in the grow-out unit.

Question: How many individuals/family do you need?

Answer: About 10-15 sibs/family is a good sample size. Also, a minimum of 20 fathers and 3-5 mothers/father seems like a good design.

Question: Why so many fathers and so few mothers?

Answer: Well, our ANOVA methods look at the paternal or sire component as the most important. The mothers are egg machines in this experimental design. What we wish to emphasize is the analysis of the paternal component of the variance. Look at the reverse--it's technically and biologically impossible to mate one mother against many fathers because of the constraints imposed by the frequency of the female premating molts. You could analyze the maternal component if you had *in vitro* fertilization and development the way they do in salmonoids and other species. A single female egg mass could be divided into groups, each fertilized by sperm from a different male. We very much need these *in vitro* methods in *M. rosenbergii*.

Question: If you're looking at the genetic component, though, can't you breed the same female to a number of males over a period of time and see the same genetic effect which is carried out in different broods? If your environmental effect is so strong that it masks this, then you're not going to see anything anyway.

Answer: If you had enough time, you could ignore the within year effects by randomizing them over a number of years, but you'll need a lot of years.

Question: Can you use preservation of the gametes to accomplish this?

Answer: Not really, because the cryogenic technology for preserving ova is not yet perfected as it is for spermatozoa. However, cryogenic preservation of spermatozoa is useful in another area. When a single male has been associated with high performance, you can keep his genetic material even after his death. This is done quite a lot in livestock breeding. *M. rosenbergii* males, by the way, live only about three years in our stocks.

Question: Are your genetic experiments similar to ones done in other aquaculture fields?

Answer: Yes, they are. We're really not pioneering anything new in the field of animal breeding. The various aspects of quantitative genetic theory that we are using have already been worked out and applied to a vast number of important food crops and livestock. Becker (1975) is a good source of review in this regard. This reference, by the way, discusses the applicability and limitations of various experimental designs to estimate heritability, especially in regard to estimating the standard error of the h^2 value and more especially with nested designs and unequal numbers in the half-sibs families. Two h^2 values have to be very different in many cases for this difference to be considered statistically significant.

Back to the original question, we are just using previously worked-out theories and methods and applying them to this crustacean. Of course, these methods will have to be modified to accommodate the biology of the animals, but in no way does the theory change. The theoretical backgrop to our ecotype work really comes under the aegis of zoogeography, evolutionary biology, or ecological genetics, or all three. However, useful genetic material brought in from the wild and incorporated into economically important stocks is common practice in botanical agriculture. For example, the high lysine gene in corn and various disease-resistant genes in cucumbers, peas, and tomatoes were handled in this manner.

Actually, Vavilov (1949-1950), the famous Russian plant breeder, can be credited with advancing, most formally, the concept that the "distribution centers" or centers of origin of cultivars contain useful genetic material that should be looked for, evaluated, and incorporated into domesticated stocks. In a similar way, we are attempting to do this as discussed in the "ecotype" collection and evaluation section in this review. Of course, M. rosenbergii is not yet a truly domesticated species so that our work doesn't strictly parallel that done in cultivars. But we certainly can identify a native distribution (see Figure XIV-1) of M. rosenbergii and make an attempt to characterize, evaluate, and possibly use the economically important intraspecific genetic variation we find in an effort to improve prawn aquaculture production.

Question: Are you attempting to estimate the heritability values from juvenile to market size? And, if so, when will you have that? I think that this is one of the very important things to do.

Answer: I agree it is very important, and our variance estimation will be done at different times during the grow-out of PL's, but I don't know if we can do an estimate on market-sized families. The size of the cells (4" x 4") in the grow-out table will limit the estimation to juvenile or subadult or younger animals. This is unfortunate, I know, because we really would like an h^2 estimate in market-size cohorts. When we finish the PL to subadult estimates, maybe we can attract the funds needed to design and construct a grow-out unit that will allow this. However, if we are going to go through all this trouble, we might as well put our financial resources into developing electrophoretic (genetic) markers as pointed out by Moav et al. (1976). This would allow the h^2 estimates to be done in the ponds, the actual production unit, because, with the markers, we could stock many electrophoretically marked families into one unit. Then we would have variance estimates more applicable to the "real world", i.e., in the production unit most frequently used in commercial prawn aquaculture (notwithstanding the admirable efforts of Steve Serfling and his coworkers).[a] In this regard, we must remember that the h^2 value is a ratio of variances and the denominator value (the total phenotypic variance) is very heavily influenced by the particular environment (and management system) in which the families are grown. Our laboratory estimates are only the beginning. Getting back to the h^2 measurements at different stages of growth, see Lannan (1972) and Reagan et al. (1976) where their h^2 values decreased through time. This means that environmental sources of variance are playing a greater role as time goes on. If we find this in M. rosenbergii, my guess is that it will be highly correlated to the bull-runt variance amplication I spoke of in the review. The take-home message could be that the market size of large bulls is not mediated by growth-rate polygenes. But we need to remember that genes controlling, in whole or in part, the behavior factors that

[a]High-density raceway rearing experiments and proprietary designs.

contribute to the "bull-runt" phenomenon may be important; and we should look at this. By the way, phenomena similar to the bull-runt thing in M. rosenbergii are seen also in carp where the large animals are called "jumpers" in Israel (Moav et al., 1973) and in Japanese koi where the "bulls" are called "Tobi koi" which roughly translates as "shoot carp" (Nakamura and Kasohara, 1955, 1957, 1961). One last point: you're not losing genetic variance or any genetic material if the h^2 values go down; it's just that the genetic variation will be harder to identify, making genetic selection based on size more "difficult" and also, selection responses will be lower.

Question: For our purposes, we have been raising larvae singly in vessels and a few hundred in batch culture in big aquaria. We have noticed that, when our larvae are grown in the aquaria, the variability of the individual sizes is tremendous, something like 300%. But, when we grew the larvae in individual vessels, the individual variation is more like 25% up to stage 11. This suggests that the larvae size variation is more environmental than genetic. What do you think?

Answer: That was just for the larvae, wasn't it?

Answer: Yes, larvae up to the 11th stage.

Answer: I would suspect that rearing juvenile M. rosenbergii individually would give you a similar result. Remember if animals are individually housed, then you can just let the animals develop normally and analyze the data by sex at the end of the experiment. It is well known, from work in other organisms, that the heritability values can differ in the two sexes.

CHAPTER XV
ABOUT STANDARDS FOR PRAWN RESEARCH

A problem for culturists which surfaced at several sessions was that of relating one datum to another. The problem was particularly acute in comparing animal densities. A. Farmanfarmaian of the Department of Physiology at Rutgers University, who is working with the Public Service Gas and Electric Company's prawn and trout project, suggested an approach to development of standard units and was requested by the chairman to put his ideas in writing. The following is presented for consideration by researchers, who are requested to send their comments to Dr. Farmanfarmaian for inclusion in the work of a standards committee of the World Mariculture Society.

A. Standard Units for the Expression of Organismal Density in Aquaculture Ponds, Raceways, and Tanks: A. Farmanfarmaian, Marine Sciences Center, Rutgers University

In recent literature and proceedings of several conferences on aquaculture [Bardach et al. (1972), Goodwin and Hanson (1974) and Anon. (1976d)] stocking and production densities are reported in a variety of units by different individuals. In some cases, densities are described in absolute totals at different stages of research or production in a particular facility (e.g., see pages 407-409 in Bardach et al. and pages 23-38 in Anon. (1976d). Such descriptive reports serve a useful purpose in the initial development of aquaculture technology. In more advanced stages where comparative quantitative data are needed for fixed and operating cost calculations, or market and economic projections, it is desirable to use standardized units for expressing density or other aquaculture indices (e.g., see hatchery constants, page 415 in Bardach et al.).

Most authors have used numbers of animals per unit volume or unit surface when expressing stocking densities of embryos or juveniles. Production densities are often reported in units of weight of animals per unit volume or unit surface and rarely per unit flow. In addition to such expressions of density, separate descriptions of shapes and sizes of ponds, raceways, and tanks; water flows; and sizes and total numbers of animals are provided. Thus, most aquaculturists do recognize the importance of the biomass (weight), static capacity (volume), dynamic capacity (volume flow or turnover time), and territoriality or individuality (numbers of animals). However, it is desirable to combine these factors into a single metric expression of density for a given aquaculture system (pond, raceway, tank) as follows:

$$D = \frac{W}{V \cdot Q}$$

Where:

D is density of the system
W is the total weight of the animals in grams in the system
V is the capacity or volume of the system in liters[a]
Q is the volume flow in liter-hours for once-through or recirculating system

This equation, however, does not take into account the fact that most of the physiological functions of an animal (e.g., O_2 consumption and CO_2 and NH_3 production) are not directly related to weight but to log of weight. Furthermore, the equation does not take territoriality into consideration; e.g., a 3 g animal requires less living space than 3 one-gram animals do. The problem of territoriality is directly related to the fixed costs of land and tanks and also to cannibalism in crustacea, and therefore, it must be included in the density index. Thus, the equation may be modified as follows:

$$D = \frac{\log W \cdot N}{V \cdot Q}$$

Where N is the total number of animals in the system.
All other terms are as defined above.
Log W · N is a combined physiological expression of biomass and territoriality.

In this equation, it is assumed that all the terms are positive. This is always true in production or experimental aquaculture. It is further assumed

---

[a]The use of V or capacity in liters avoids calculations based on dimensions of varied geometric shapes.

that the system for which the density is being calculated is essentially homogeneous, i.e., there are no significant unstirred layers or phase separations in the ponds, raceways, or tanks under consideration. This second assumption is true for any expression of density which is an average property of a system. A simple inspection of the second equation shows that for a given capacity (V) and flow (Q), density is highly sensitive to N at stocking time when the embryonic or juvenile weight is small. During grow-out and harvest, density is highly influenced by weight (W) since N remains constant or actually decreases. As V increases, the "spatial" or "structural" density falls and as Q increases, the "functional" or "physiological" density falls. Mathematically, V and Q can completely compensate for each other but in reality, an infinite increase in V can completely compensate for Q; i.e., a few animals in a large pond require little or no flow. However, Q cannot in reality completely compensate for V because each animal has a minimal spatial requirement imposed by territoriality. In the case of crustacea, this space is probably not less than the volume of a sphere whose radius equals half the total length of the animal, inclusive of appendages.

In conclusion, one can say that even though some aquaculturists strongly prefer to express density in terms of weight per unit area of land or tank surface, such expression has very limited usefulness because it ignores the fact that, fundamentally, all organisms express some form of territoriality as they live in three-dimensional space and the fourth dimension of time. Furthermore, most of their functions are related to the log of weight, not weight itself. The expression of density proposed in the second equation above is by no means perfect, but it does take into account all the mentioned factors and their practical limits in aquaculture technology.

Finally, in the case of animals such as crustaceans where, due to adherent water, the wet weight can vary considerably in the hands of different individuals, it is recommended that the weighing procedure be defined and validated by the constant dry weight method at least once in each facility over the full working range of wet weights.

(This paper was supported in part by Research Grant NSF/RANN GI-43925 and Rutgers Marine Sciences Center.)

APPENDIX A

ON THE AQUACULTURE OF FRESHWATER PRAWNS (*MACROBRACHIUM* SPECIES)

June 14-15, 1976

Charleston, South Carolina

Sponsored by:	The Oceanic Institute The National Sea Grant Program The Marine Resources Research Institute (MRRI), South Carolina Wildlife and Marine Resources Department
Host:	Paul A. Sandifer and the MRRI Directorate and Staff
Chairman:	Harold L. Goodwin
Analyst:	Joe A. Hanson
Registration:	Libby Goodwin and Doris Akers

AGENDA

Monday, June 14th

THE TECHNOLOGY OF PRAWN CULTURE

8:00 a.m.	Buses depart Riverview Holiday Inn for MRRI
8:30 a.m.	Registration at MRRI; coffee available
9:00 a.m.	Introduction to the workshop: Sandifer and Goodwin
9:15 a.m.	*Prawn Species for Culture*: Lester Miyajima, Aquaprawns, Inc. Discussant: George Kelley, Jacksonville University and Harrell International.

9:45 a.m.	Selective Breeding and Genetics: Spencer Malecha, University of Hawaii.
10:15 a.m.	Coffee Break
10:30 a.m.	Problems of Disease: Carl A. Sindermann, Mid-Atlantic Coastal Marine Fisheries Center, NMFS.
11:15 a.m.	Juvenile Production: Paul A. Sandifer, MRRI. Discussant: Michael Fujimoto, Anuenue Fisheries Research Center, Hawaii.
12:00 N	Lunch at MRRI
1:30 p.m.	Advances in Prawn Systems Engineering: Paul Zielinski, Clemson University.
2:15 p.m.	Pond Growout Systems: Kenneth Kato, Pacific Aquaculture, Inc. and Michael Fujimoto, Anuenue Fisheries Research Center. Discussant: Theodore Smith, MRRI.
3:00 p.m.	Intensive Culture Systems: Edward McSweeney, Syntex (USA). Discussants: Steve Serfling, Solar Aquafarms; Albert Eble, Trenton State University; J.B. Peebles, University of Hawaii.
3:45 p.m.	Feeds and Nutrition: Kirk Biddle, NMFS Southeastern Technology Utilization Center. Discussants: Jeanne Joseph, MRRI; Margo Stahl, University of Hawaii, Douglas Conklin, Bodega Bay Lab., University of California.
5:00 p.m.	Adjourn. Buses to Riverview Holiday Inn.
6:10 p.m.	Buses from Riverview to Downtown Holiday Inn.
6:30 p.m.	Attitude Adjustment and Social Lubrication.
8:00 p.m.	Dinner. Speaker: Harold H. Webber, President, Groton Bio Industries and President, World Mariculture Society
10:15 p.m.	Buses to Riverview Holiday Inn.

Tuesday, June 15th

PROCESSING, MARKETING AND ECONOMICS

8:00 a.m.	Buses to MRRI
8:30 a.m.	Registration at MRRI
9:00 a.m.	Problems of Prawn Processing: Virginia Sidwell, NMFS Southeastern Technology Utilization Center. Discussants: George Kelley, Jacksonville University and Kenneth Kato, Pacific Aquaculture.
9:45 a.m.	The Role of Industry in Macrobrachium Culture: John Bennett, Syntex (USA).
10:30 a.m.	Coffee Break
10:45 a.m.	The Problems of New Product Marketing When There Is No Real Definition of the Market: Ian Chaston, Booth Fisheries.
11:40 a.m.	Regulations Affecting the Prawn Processor and Marketer: Thomas J. Billy, NMFS Fishery Products Inspection and Safety Division.
12:30 p.m.	Lunch at MRRI
1:30 p.m.	The Economics of Prawn Aquaculture: Kenneth Roberts, Agricultural Economics Dept., Clemson University, on assignment to Sea Grant, NOAA.
2:15 p.m.	Priorities for Prawn Aquaculture Research and Development: Summary and discussion.
4:30 p.m.	Workshop ends.

APPENDIX B

ORGANIZATIONS IN PRAWN CULTURE

For this appendix, persons and organizations were invited to prepare a statement outlining the particulars of their prawn culture interests. The appendix includes only those who wished to be included. The submissions are presented as received, without editing.

In such a rapidly developing field, there are continuous changes in personnel, locations, and interests. This appendix was accurate at the time of submission by participants, but should be verified by persons interested in additional information about listed organizations and their personnel.

AQUACULTURE RESEARCH/ENVIRONMENTAL ASSOCIATES (AREA)

Address: P.O. Box 1303
Homestead, Florida USA 33030
Telephone: (305) 248-4205

Paul Mulvihill
Principal

Cecilia A. Mulvihill
Principal

In the early 1060's the organizers of AREA started studying potential organisms suitable for aquaculture. This study was directed toward existing research and development at that time, future potentials and economic returns. Catfish, trout oysters, clams, salmon, crabs, lobsters and many other organisms were rejected because of various reasons (i.e., research was nearing completion, larval rearing was too difficult, the period of growth was too long, etc.). Finally *Macrobrachium* became the primary target for development, hence Aquaculture Products and Research Inc. (APARI) was founded in 1968. APARI obtained assistance from many sources and finally a *Macrobrachium* hatchery was established in Homestead, Florida, in 1971. At that point APARI had been successful in the larval rearing of *Macrobrachium carcinus*, *M. acanthurus*, *M. americanium* and *M. rosenbergii*. It was found that *M. rosenbergii* offered the greatest potential for future development. A thorough analysis of the capital requirements showed that funding could only be forthcoming from a public stock issue or from some major corporation that had the assets to take this still unproven risk. At that time aquaculture ventures were glamour issues on the stock exchanges and a great number of firms who had obtained public financing were in serious trouble, many of which have since failed. APARI elected to sell post-larvae to firms that did not desire to go to the expense and technical problems involved in establishing a hatchery. As an example, Foods Multinational Limited, a subsidary of General Mills in Honduras; MARIBEC, a subsidiary of International Basic Economy Corporation in Jamaica; Grosstrol Inc. in Kansas; farmers in Texas and even Israeli government agencies, etc. Because of the extremely large capital requirements to further develop this industry discussions were held with many major firms. In June 1975 the Board of Directors of APARI voted to sell its assets to the Weyerhaeuser Company. The new firm AREA was established with the concept of continuing aquaculture research but not solely limited to *Macrobrachium*.

BOOTH FISHERIES (Division of Consolidated Foods)

Address: 2 North Riverside Plaza
Chicago, Illinois 60606
Telephone: (312) 346-3117

Dr. Ian Chaston
Aquaculture Manager

Booth Fisheries is probably the largest processor and marketer of shrimp products in the world. Active R&D in Aquaculture has terminated in 1974 and the company now merely is monitoring current progress, especially in the area of crustacean culture. Long-term, we expect to become re-involved in Aquaculture either through purchase of raw material from farm operations on a contractual basis or through active investment in joint venture/aquisition propositions.

BODEGA MARINE LABORATORY (University of California)

Address: P.O. Box 247
Bodega Bay, California 94923
Telephone: (707) 875-3662

Dr. Cadet Hand
Director

Associates: Dr. Douglas E. Conklin
Dr. Michael C. Hartman
Dr. Edgar H. Nilson
Dr. Dennis Hedgecock
Mr. William S. Fisher

Project goals are as follows: development of crustacean aquaculture using primarily lobsters as a test organism; understand of crustacean nutritional requirements and nutritional physiology; development of pelletized diets for lobsters; development of crustacean reference diets; development of larval rearing systems and larval feeds; understanding disease-related physiology and microbiology; crustacean disease identification and control; understanding of crustacean genetics; development of lobster broodstock management.

CARIBE KING SHRIMP, INC.

Address: Bo. Guaniquilla, Buzon Rural A603 Telephone: (809) 891-1510
Aguada, Puerto Rico 00602 891-1010

John W. Parker
President

Associates: Rafael J. Martinez
Hector Reichard
James G. Williams
Randolph Hagood

Caribe King Shrimp, Inc. began operations in Puerto Rico in 1971. Its facilities include a hatchery and laboratory in Aguada, Puerto Rico, for post-larvae (Juvenile) production on five acres of oceanfront property, together with 50 acres of freshwater ponds, laboratory and research ponds in Cabo Rojo, Puerto Rico.

Shrimp production averages 8 shrimp per pound (whole) and is sold to Puerto Rican and U.S. markets.

CENTRE NATIONAL POUR L'EXPLOITATION DES OCEANS (CNEXO)
CENTRE OCEANOLOGIQUE DU PACIFIQUE (COP)

Address: B.P. 7004 Taravao Telex: PAPEETE OCEANEX
Vairao, Tahiti 094 FP

M. de Chazeaux
Director

AQUACOP (Aquaculture Team of the COP)
Alain Michel
Jacques Calvas
Gerard Cuzon
Jean-Michel Griessinger
Philippe-Jacques Hatt
Jean-Francois le Bitoux
Jean-Louis Martin
Jean-Francois Virmaux

The development of commercial prawn culture in French Polynesia is made in cooperation with the Territory Fishery Service and began in October 1973 from 80 adults from Fujimura laboratory.

CENTRE OCEANOLOGIQUE DU PACIFIQUE (COP) (Continued)

The following fields are under investigation:

Larval Rearing: The technique developed is intensive. In a fiberglass tank with a conical bottom a high density of larvae (100/liter), in clear water, without phytoplankton, the food is composed of artemia nauplii, frozen artemia, bonito and octopus flesh. The best results were 62 post-larvae/liter in 0.8-cubic-meter tanks.

Grow-out Culture System: Two hectares of experimental earth ponds 500 m2 to 3,500 m2 are in operation in Tahiti and Moorea. Different experiments are in progress to determine the economics of this culture in Tahiti.

Nutrition: Research is conducted on local by-products to determine an artificial diet which could be produced locally. A formula with 40% of copra meal is now produced by the COP experimental unit in pellets. Until December 1975, two tons were produced with results of 300 grs/m2/year.

CENTRO DE INVESTIGACIONES DE BIOLOGIA MARINA (CIBIMA)

Address: Universidad Autonoma de Santo Domingo Telephone: 685-6682
Ave. George Washington, Guibia
Santo Domingo, Republica Dominicana

Idelisa Bonnelly de Calventi
Director

Associate: Sophie Jakowska Address: 27W 96th Street Apt. 80
New York, N.Y. 10025

Research: Biological and Ecological Studies of Fresh Water Shrimps; Pathology of Vertebrates and Invertebrates.

UNIVERSITY OF CALIFORNIA, DAVIS

Address: University of California, Hydrobiological Laboratory
Department of Land, Air and Water Resources
Water Science and Engineering Section
Davis, California 95616

Dr. Allen W. Knight
Project Director

Associates: Mr. Steve Nelson
Mr. Dave Armstrong
Ms. Marian Stephenson
Mr. Steve Serfling
Ms. Henrietta Stern
Ms. Debbie Chippendale
Ms. Mary Ann Simmons
Ms. Claude Nagamine

Through funding by the Davis Aquaculture Group, the hydrobiological laboratory is interested in the culture and potential use of the prawn (*Macrobrachium rosenbergii*) and other species for valley aquaculture in California. The primary objective is to obtain ancillary physiological ecological information that would assist those actually involved in mass culture. Specifically, we are presently developing suitable methodology and data base for simultaneously evaluating the relationship between un-ionized ammonia, pH, and water temperature. Standard bioassay as well as modified methods for tolerance and toxicity studies, developed in our laboratory, are employed. Respiration and growth studies will also be conducted evaluating the combined effect of the above factors and initial evaluations are currently underway. We propose to utilize empirical models (multivariate techniques), some presently under development for the opossum shrimp, to examine the problem of prawn tolerance where several factors vary simultaneously. Ultimately, the model would be utilized to describe the prawn's simultaneous tolerance and growth in response to numerous environmental factors. In addition, the energetics of the prawn is under evaluation. Ultimately, we will prepare a model describing the utilization of energy by prawns of selected size and exposed to numerous environmental factors and fed selected synthetic and natural foods.

CLEMSON UNIVERSITY

Address: Clemson, South Carolina 29631 Telephone: (803) 656-3000

Paul B. Zielinski, Associate Professor of Civil Engineering 116 Lowry Hall
Walter E. Castro, Professor of Mechanical Engineering 301D Riggs Hall

Activity: Prawn systems engineering including tank fluid flow studies, air lift pump capacities and aeration, pond oxidation and general engineering assistance to Marine Resources Research Institute.

Averett S. Tombes, Professor of Zoology 336 Long Hall

Activity: An observation of the external structure of the animal during growth to determine how early the animal can be sexed, which is preparatory to the study of varying growth rates in the male.

Arnold G. Eversole, Assistant Professor of Entomology and Economic Zoology 306 Long Hall

Activity: The determination of the effect of Mirex, a fire ant pesticide, on the fresh water stages of *Macrobrachium*.

UNIVERSITY OF COLORADO, Boulder

Address: Department of Environmental, Population, and Organismic Biology
Department of Psychology
University of Colorado
Boulder, Colorado 80302

Dr. David Chiszar
Projects Director, Animal Behavior Laboratories

Kent Scudder, Coordinator
Aquaculture Division

Team: Scott Metzger, M.A., behavioral dynamics
Dave Duvall, M.S., endocrinologist
Jay Krafsur, biologist
Ken Ross, biologist

Interest: Investigation of the ontogeny of prawn behavior and the various internal and external factors contributing to the agnostic behavior of *Macrobrachium* sp.. Development of appropriate controls of behavior to assist in the optimization of commercial prawn production in an intensive culture system.

ELSHAR
Aquaculture Development Services

Address: P.O. Box 341216, Coral Gables, Florida 33134 (Winter)
138 Belfort Street, Portland, Maine 04103 (Summer)
RR5, Joe Riche Road, Hiway 33, Kelowna, British Columbia VIX 4K4, Canada (Summer)

Elliott C.L. Maynard
President, Technical Director

Sharon M. Maynard
Vice-President, Coordinator

ELSHAR's interests lie in the areas of Administrative Coordination, Functional Culture Systems, Hatchery and Ponding Design based on individual company requirements, production cycles, financial gain, and sales promotion, in relation to the development of the giant fresh water prawn aquaculture industry.

As Coordinator, Sharon Maynard has worked with various companies to improve internal communications and to increase efficiency between Board of Directors and Investors, Administrative and Technical Divisions, and Hatchery and Ponding staffs. She has developed special Biological Routines and Training Programs, as well as Scheduling and Data Processing Systems. As Technical Director, Elliott Maynard has set the basic scientific coordinates for Two Fresh Water Shrimp Aquaculture Companies, one in South Florida, and the other in Puerto Rico. His early work in 1969-70 resulted in the first juveniles of *Macrobrachium carinus* and *M. acanthurus* ever produced under conditions of mass culture. The work with *Macrobrachium* species represents over seven years of applied experience with all aspects of open and closed hatchery culture systems, ponding designs, commercial production, and sales promotion.

ELSHAR is presently able to provide guidance, supervision, and exact administrative and technical coordinates to individuals and corporations to further the development of the giant fresh water prawn industry on an international basis, including areas of the continental United States, Central and South America, the Caribbean, Europe, and Africa.

FLORIDA ATLANTIC UNIVERSITY (Aquaculture Laboratory)

Address: Boca Raton, Florida 33431 Telephone: (305) 395-5100 Ext. 2701

Dr. Sheldon Dobkin
Interim Dean, College of Science

Associates: Norman Chwang
Denise Holland
Peggy Van Arman

Interests: Selective breeding and hybridization of *Macrobrachium acanthurus*, *M. rosenbergii* and other species of the genus.

The conslusion that we have come to based upon three and one-half years of study (1973-1976) are as follows:

1. *Macrobrachium acanthurus* can be easily reared in the laboratory and will grow readily in pond culture. In south Florida, this species grows faster than *M. rosenbergii* for approximately six months (total age), but thereafter, *M. rosenbergii* surpasses it.

2. Selective breeding could allow the production of two commercial crops of male *M. acanthurus* during the eight-month growing season in south Florida. Females grow considerably more slowly than males after they reach maturity.

3. Development of a cold-resistant hybrid *M. rosenbergii* would allow more efficiency of production of this species in marginal areas (Florida, Texas, etc.).

4. *M. rosenbergii* remains the best species for culture in terms of larval survival and survivial during the grow-out period, and hence, efficiency of culture. Hybridization experiments ought to be accelerated, however, to produce a market-sized animal in a shorter period of time.

V. L. GIANNINI & COMPANY

Address: 10717 Wilshire Blvd. Telephone: (213) 475-5881
Los Angeles, California 90024

Valerio L. Giannini
Principal

Techno-Economic business/market research of state of the art, investment analysis, financial porjections, market and feasibility studies of Macrobrachium for three major U.S. industrial firms and federal government agency, all since 1969.

HAWAII INSTITUTE OF MARINE BIOLOGY

Address: P.O. Box 1346, Coconut Island Telephone: (808) 247-6631
Kaneohe, Hawaii 96744

Dr. John Bardach
Director

Dr. Robert C. May, Aquaculture coordinator
Dr. Gregory A. Ahearn, Physiology of nutrient absorption by crustacean gastrointestinal tract
Leigh A. Maginniss, Intestinal transport physiology of Macrobrachium
Margo S. Stahl, Macrobrachium nutrition and feed formulation
J. B. Peebles, Macrobrachium behavior

UNIVERSITY OF HAWAII (Aquacultural Engineering Group)

Address: Agricultural Engineering Department Telephone: (808) 948-8154
University of Hawaii
3131 Maile Way
Honolulu, Hawaii 96822

Jaw-Kai Wang
Project Leader

Associates: Michael Williamson
Richard Gibson

Areas of interest include Automated Hatchery System Development; Grow-out Pond Design; Optimal Stocking Strategy; Harvesting Equipment; Overall Production Optimization.

INSTITUTO DE DESARROLLO DE LOS RECURSOS NATURALES RENOVABLES (INDERENA)
CENTRO DE INVESTIGACIONES PESQUERAS

Address: Apartado Aereo No. 2459/2895
Cartagena, Columbia
South America

Telephone: 43990-45790

Dr. Luis E. Martinez Silva, Marine Biologist
Principal Investigator

Activities include Experiments on larval rearing of Macrobrachium acanthurus; Pond culture of same species in Lomarenas, Bolivar, Colombia (Joint FAO+UNDP/INDERENA Project); Biology of indigenous species of Macrobrachium of the Atlantic Coast of Colombia.

Mario Pedini
Fishery Biologist

Address: Food and Agriculture Organisation of the United Nations
Inland Water Fishery Development Programme COL/552
Apartado Aereo 2459/2895
Cartagena, Columbia, South America

Telephone: 43990-45790

Interest: Macrobrachium

KELVIN HUGHES AQUACULTURE SERVICES

Address: 4, Beaufort Gardens, Marlow
Bucks SL7 1EL
England

Telephone: 062-84-5631

Michael B. New
Manager

Activities include Aquaculture consultancy including site selection; Practical intensive and pond experience with Macrobrachium rosenbergii, M. americanum and M. acanthurus in the U.S.A., U.K., Mexico and Colombia.

Kelvin Hughes Aquaculture Services was formed in 1973 as a new department of Kelvin Hughes, an international marine equipment and hydrographic survey company.

MARINE RESEARCH LABORATORY
State of Florida, Department of Natural Resources

Address: 100 Eighth Avenue, S.E. Telephone: (813) 896-8626
St. Petersburg, Florida 33701

Scott A. Willis
Project leader *Macrobrachium* research, pond manager, juvenile/nursery development, larval rearing.

Mark E. Berrigan
Macrobrachium research, hatchery manager, pond growout studies.

Team: William C. Trimble, Mariculture/Pathology section supervisor
Richard D. Haag, Student assistant, *Macrobrachium* culture
George E. Henderson, Pathology subsection leader
Daniel E. Roberts, Research biologist
Brian V. Harpster, Research biologist
Cary D. Burns, Research biologist
Walter K. Havens, Marine science technician
Kenneth R. Halscott, Marine science technician
Jean Williams, Histologist
Woodrow P. Arvanitas, Student assistant, histology

Interests include current Marine Research Laboratory research projects including pond growout, larval food studies, disease diagnosis and control, and predator control studies. Post larval and juvenile *M. rosenbergii* have been stocked in twelve 0.025 hectare earthen ponds at densities of 5, 10, and 20 prawns/m^2 to analyze production and pond management techniques. Studies have also begun to find a larval food supplement or replacement for *Artemia*. An atlas of disease diagnosis and control will be compiled during the coming year, and specimens for analysis and diagnosis are solicited. Studies will continue in identification and control of predators, especially aquatic insects, to reduce post-larval loss after stocking ponds in Florida.

MARINE RESOURCES RESEARCH INSTITUTE (MRRI)
South Carolina Wildlife and Marine Resources Department

Address: 217 Ft. Johnson Road (P.O. Box 12559)
Charleston, South Carolina 29412

Dr. Paul A. Sandifer
Prawn Culture, Crustacean Biology
Principal

MARINE RESOURCES RESEARCH INSTITUTE (MRRI) (Continued)

Associates: Dr. Walter E. Castro, culture systems engineering (Clemson University)

Jeanne D. Joseph, lipids of marine organisms, lipid and fatty acid requirements of *Macrobrachium* (MRRI)

Dr. John J. Manzi, ecology, taxonomy and culture of phytoplankton, culture of larval *Macrobrachium*, molluscan biology (MRRI)

Dr. D.U. Silverthorn, crustacean physiology (Medical University of South Carolina)

Dr. Theodore I.J. Smith, general aquaculture, all aspects of prawn culture, intensive and extensive culture system design and management (co-Principal Investigator for MRRI Core Projects)

Dr. Paul B. Zielinski, systems engineering and design (Clemson University)

MRRI Team: J.S. Hopkins, Research Biologist 2
M.B. Maddox, Research Biologist 1
D.E. Marchette, Research Biologist 1
M.H. Smith, Research Biologist 1
F.S. Taylor, Research Biologist 1
C.W. Waltz, Research Biologist 1
J.E. Williams, Research Biologist 1
V.M. Hargis, Laboratory Technician 2
G. A. Ward, Laboratory Technician 2

Facilities: Laboratories and research hatchery at Institute; 4-6 ponds (0.25 ha) at Dennis Wildlife Center, Bonneau, S.C.; other ponds (0.07 to 0.32 ha) available at Wadmalaw Island and Beaufort, S.C.; additional facilities are being developed at the Institute.

Interests: Development of a commercial prawn aquzculture industry in South Carolina. The South Carolina *Macrobrachium* Aquaculture Program centered at the Marine Resources Research Institute in Charleston is composed of two complementary portions, one supported by the Coastal Plains Regional Commission and the other by NOAA Office of Sea Grant. Both segments have the development of profitable *Macrobrachium* aquaculture in South Carolina as their basic goal. However, the approaches to this goal are different.

The aquaculture studies supported by the Coastal Plains Regional Commission are primarily concerned with the intensive culture of *Macrobrachium* in recirculating water management systems. The objectives of this portion of our studies are: (1) to evaluate the potential for use of recirculating systems in aquaculture

MARINE RESOURCES RESEARCH INSTITUTE (MRRI) (Continued)

in South Carolina; (2) to conduct demonstrations and pilot-scale production of Macrobrachium using recirculating systems; and (3) to develop guidelines for the commercial applications of recirculating water management systems in all phases of Macrobrachium culture, from larval rearing through grow-out and breeding.

NOAA Office of Sea Grant supports primarily the more extensive aspects of the South Carolina program and has pond cultivation of Macrobrachium as a central goal. The Sea Grant supported projects are (1) a core project (basic objectives - to initiate pilot studies in pond grow-out and hatchery production of Macrobrachium in South Carolina and to provide leadership and coordination for the Macrobrachium program); (2) an engineering project (Clemson University - Zielinski and Castro) (basic objectives - to provide engineering support and design research for the development of Macrobrachium aquaculture); (3) larval diets project (basic objective - to examine the value of algae and Artemia substitutes in the rearing of larval and juvenile stages of Macrobrachium; (4) a lipid project (basic objectives - to monitor lipid and fatty acid compositions of cultured M. rosenbergii and to determine and dietary lipid/fatty acid requirements of the prawn); and (5) a temperature project (Medical University - Silverthorn) (basic objective - to examine temperature tolerance of juvenile and adult Macrobrachium).

MEDICAL UNIVERSITY OF SOUTH CAROLINA

Address: 80 Barre Street Telephone: (803) 792-2978
Charleston, South Carolina 29401

Dr. S. Unglaub Silverthorn Telephone: (803) 792-2978
Assistant Professor of Physiology

Interests: Cold tolerance of Macrobrachium rosenbergii and effects of cold shock on growth

MICRONESIAN MARICULTURE DEMONSTRATION CENTER

Address: Box 359 Koror, Palau 96940
West Caroline Islands
U.S. Trust Territory of the Pacific

Dr. James P. McVey
Director of Laboratory

Associates: Howard Deese, Project Leader (*Macrobrachium*) Directs and performs research on *Macrobrachium* hatchery and pond production
Evelyn Oiterong, Hatchery Technician

The *Macrobrachium* culture research program was initiated at M. M. D. C. in the fall of 1973. It was supported solely by the laboratory budget from the Congress of Micronesia until June 1975 when it received matching funds from NOAA Office of Sea Grant. The basic goal of the project was to establish whether the existing techniques of *Macrobrachium rosenbergii* culture could be could be used to establish an economically feasible industry for an exportable cash crop.

During the first two years of work we established a basic working facility consisting of a small hatchery system, two cement growout tanks, and three small ponds. We imported stock from Hawaii and captured wild stock from our own streams. We cultured both stocks in our hatchery, tanks and ponds and have succeeded at crossing the two stocks. Investigations to date have indicated that yields equivalent to other well operated commercial operations are possible with proper management.

We are now moving into a new phase of our project the goal of which is to develop a number of culture systems to the best efficiency possible. We are establishing a larger hatchery and a pilot research farm in order to investigate in detail the feeding, harvesting, processing and marketing techniques on a scale large enough for economic analysis. We are beginning some investigations into polyculture using prawns and milkfish or prawns and taro. The possibility of brackish water pond rearing in mangrove ponds is also being investigated. We hope to conduct most of this work with comparisons between the two stocks if the facilities permit.

MINISTRY OF AGRICULTURE, FISHERIES AND FOOD

Address: Attn: J.F. Wickins Telephone: Conwy 2419
Ministry of Agriculture, Fisheries and Food,
Fisheries Experiment Station
Benarth Road
CONWY, LL32 8UB
Great Britain

Prawn Personnel	Function
J.F. Wickins, Senior Scientific Officer Project Leader	Water Quality
T.W. Beard, Higher Scientific Officer	Culture Requirements
R.W. Sedgwick, Higher Scientific Officer	Nutrition
G. Williams, Laboratory Attendant	Husbandryman

Interests and activities include the controlled indoor culture of warm water prawns in intensively stocked recirculation systems.

RUTGERS UNIVERSITY MARINE SCIENCES CENTER

Address: Doolittle Bldg., Busch Campus Telephone: (201) 932-3118, 2829, 3081
Rutgers University
New Brunswick, New Jersey 08903

A. Farmanfarmaian
Professor of Physiology, Dept. Physiology, Nelson Labs

Facilities: PSEG Mercer Power Plant, Trenton, New Jersey

Interests and activities include aquaculture of _Macrobrachium_, trout, and other species which my be employed in diseasonal thermal aquaculture using once-through cooling water discharges from power plants and other technologies. Our activity includes the generation of physiological matrices (temperature, Dissolved Oxygen, density ratios, light, feeding and nutrition, molt conditions, and pollutant tolerances) which would permit the selection of a set of conditions for the optimization of conversion ratios and growth rates for the specific species, e.g., _Macrobrachium rosenbergii_, at any aquaculture site.

SAN DIEGO STATE UNIVERSITY

Address: Center for Marine Studies
Department of Biology
Aquaculture Program
San Diego, California 92182

Jack Van Olst
Project Leader

Jim Carlberg
Project Leader

Intensive Culture Systems for Macrobrachium

San Diego State biologists have been involved with research on the culture of Macrobrachium since 1970. General techniques and systems for larval culture have been developed. Some basic nutrition and food preference trials have been conducted. Aquaculture research at San Diego State University has mainly involved the development of commercial culture techniques for the American lobster, Homarus americanus, utilizing thermal effluent seawater from several coastal power stations. Some of the technology developed for lobster culture seems to apply to Macrobrachium and we have therefore investigated such areas as cannibalism, substrate choice, and the use of thermal effluent to accelerate growth rates. Cannibalism was observed to increase dramatically at high culture densities (about $100/m^2$) when early juveniles attained a total length of approximately 4-5 cm. The same general types of substrates which we have shown to be effective in limiting cannibalism in Homarus were also effective in significantly reducing cannibalism in Macrobrachium. Additional information was obtained on the use of simple plastic heat exchangers to transfer heat from thermal effluent seawater to freshwater culture systems for Macrobrachium.

More recently, under funding from the Research and Development Program of the Southern California Edison Company, research has centered on the development of extremely high density culture systems, in which the use of individual rearing compartments would be necessary to avoid cannibalism. Initial cost estimates of this capital-intensive method of culture, similar to that which we have proposed for lobsters, are not too discouraging, but more research is necessary. One experiment in progress at our aquaculture laboratory at the Scripps Institution of Oceanography is designed to measure the effects of rearing compartment size on growth and survival. Post-larvae have been stocked in eight sizes of rearing compartments have bottom areas from 6 to 750 cm^2. After approximately two months of culture there are obvious differences in size, and it appears that the animals in the largest compartments are growing almost twice as fast as those in the smallest compartments. Similar results were obtained in an experiment with Homarus which has been in progress for over

SAN DIEGO STATE UNIVERSITY (Continued)

one year. These data will be used to decide what sizes of rearing compartments would be optimal for both species. It appears that considerable savings in construction and operating costs might be possible if some reduction in growth rate were permissible.

Other current research involves the testing and evaluation of several types of inexpensive heat exchangers in the culture of *Macrobrachium*. Additionally, San Diego State University scientists have assisted several local firms interested in *Macrobrachium*, such as Homarus, Inc., Solar Aquaforms and Malaysian Prawn Farms, by providing biological and technical advice.

Future research on *Macrobrachium* in conjunction with Southern California Edison Company will emphasize the use of thermal effluent from two freshwater-cooled power stations in southern California. Additionally, some cooperative work is planned with the Department of Mechanical Engineering at San Diego State University to develop techniques for the use of solar energy water heating systems for *Macrobrachium* culture.

SOLAR AQUAFARMS

Address: P.O. Box 109
Encinitas, California 92024
Telephone: (714) 753-0649

Steven A. Serfling
President and Director

Team: Dominick Mendola, Vice President and Director
Carolyn Sheehan, Aquatic Technician
Stanley Serfling, Biologist

During the past three years, Solar Aquafarms has been engaged in developing an intensive, controlled environment system and process for the commercial production of *Macrobrachium*. This process utilizes greenhouse-enclosed raceways with closed-cycle, biologically filtered water, and solar heat exchangers for maintaining required conditions year-round. Solar Aquafarms is presently setting up a new pilot-production facility near San Diego.

SOUTHEAST UTILIZATION RESEARCH CENTER
National Marine Fisheries Service, National Oceanic and Atmospheric Administration, U.S. Department of Commerce

Address: University of Maryland
Regents Drive
College Park, Maryland 20740

Telephone: (301) 344-4155

Dr. Garfield N. Biddle, "Kirk"
Supervisory Research Animal Husbandman and Project Leader for aquaculture nutrition research

Team:
Mr. Alan R. Fortner, Chemist, water chemistry in experimental, closed-system aquaculture
Ms. Patricia Fair, Biological Technician, heavy metal ingestion studies with crustaceans
Mr. Mark Millikin, Biological Technician, general laboratory studies, system maintenance
Mr. Thomas Siewicki, Biological Aide, systems fabrication/maintenance, general laboratory studies

The overall goals of the SURC aquaculture/nutrition program are to (a) carefully and systematically study those nutrients which are required for the optimal growth of the target species under consideration; (b) evaluate means by which such knowledge can be translated into the economical development of suitable feeds; and (c) encourage (through research) the utilization of various fishery products and by-products in the nutrition of cultured organisms. Our program has been involved in several major activities, including: (a) studies on the nutritional requirements of juvenile blue crabs reared in closed culture systems (protein/amino acid requirements, the influence of dietary carotene/other pigment precursors on animal appearance, significance of dietary ingestion of heavy metals (i.e., cadmium, etc.), chemical characterization of adult *Artemia salina*, etc., have been evaluated); (b) a cooperative project with Dr. C.E. Epifanio, CMS, U. of Del., on the effects of diet processing on the nutritive value and acceptability of formulated feeds for shellfish (oysters/clams) together with studies on the nutrient needs required for optimal growth during the "grow-out period;" (c) detailed examination of changes in water chemistry with time in closed recirculating systems as well as studies on the efficacy of various charcoals and resins in maintaining suitable water quality in closed systems; and (d) as a result of integrated planning with the South Carolina MRRI, initiation of a new project on *Macrobrachium* Prawn nutrition which is to receive primary focus following relocation of SURC (estimated for 10/77) to Charleston, S. C.

SOUTH FLORIDA FISH FARMS, INC.

Address: P.O. Box 503
Miami, Florida 33157

Harriette Loizou
President

Team: Ron Mattson, General Manager

Facilities: 37 acres--Borrow pits (two locations)

Interests: Currently involved in catfish cage culture. Interested in intensive culture and equipment for rearing *Macrobrachium rosenbergii* and *Macrobrachium carcinus*.

SOUTHERN UNIVERSITY, Department of Biology

Address: Box 10809, Department of Biology
Southern University
Baton Rouge, Louisiana 70813

Telephone: (AC504) 771-5210

Dr. Jay V. Huner
Assistant Professor
Southern University

James W. Avault, Jr.
Professor
Louisiana State University

Facilities: Southern University, Baton Rouge, La.: Laboratory. Louisiana State University, School of Forestry and Wildlife Management, Baton Rouge, La.: Experimental Ponds. (AC 504) 388-6051

Interests: *Macrobrachium ohione* supports a modest fisheries in the lower Mississippi River and its distributaries. We initiated our study to obtain information on its life history in its natural habitat and its potential for polyculture with fish and crawfish. Our efforts are basically exploratory in nature. We are encouraged by the fact that 100 count (100/lb.) *M. ohione* are retailed for a price of $2.00/lb. and that current wild catches do not meet the demand. During the 1930's 2 million pounds per year were harvested versus today's estimated one hundred thousand pounds (catch reduced by habitat changes and pollution). We expect to report results of our preliminary studies sometime in the next year.

UNIVERSIDAD DEL VALLE

Address: Division de Ciencias
Departamento de Biolgia
Apartado Aereo 2188 - Melendez
Cali - Valle - Colombia, S.A.

Telephone: 59-11-71
Ext. 170

Kevin P. Mullen
Professor

Miguel E. Cantillo
Student

Field studies include small pilot culture tanks on Rio Calima, near Buenaventura, Colombia, and larval migration studies on Rio San Juan, (Pacific Coast).

Laboratory studies on behavior in captivity and larval development of native species of *Macrobrachium americanum*, *M. acanthurus* and *M. tenellum*.

UNIVERSIDAD NACIONAL AUTONOMA DE MEXICO (UNAM)

Address: Centro de Ciencias del Mar y Limnologia, UNAM
Apartado Postal 70-305, Mexico 20, D. F. Mexico

Manuel Guzman
Ecology and Culture

Associate: Dr. Craig Kensler, Unesco Fisheries Biologist

Team: Gilberto Diaz, water chemistry
Walter Ritter, population dynamics
Pilar Negrete, student assistant
Hector Hardines, technical assistant

Facilities: Laboratory of Limnology and Cultivation at UNAM. Also use of field laboratories in the states of Campeche, Quintana Roo, Sinaloa and Guerrero, including about two hectares of various sized ponds.

Interests: Ecology, biology and population dynamics of *M. tenellum* as related to the commercial fishery. Also, laboratory culture, reproduction, fecundity and biometrics.

UNIVERSIDAD NACIONAL AUTONOMA DE MEXICO (UNAM) (Continued)

Address: Instituto de Biologiz, UNAM
Apartado Postal 20-239
Mexico 20, D. F. Mexico

Jorge Cabrera
Principal

Associate: Dr. Alejandro Villalobos
(Universidad Autonoma Metropolitana, Mexico City)

Team: Jose Luis Garcia-Calderon
Carlos Martinez
Cristina Chavez

Facilities: Aquaculture Laboratory at UNAM and two permanent field stations: one in the state of Veracruz and one in the state of Jalisco.

Interests: Taxonomy, distribution and natural history of Macrobrachium species in Mexico. Emphasis on M. tenellum; larval development, rearing of juveniles and ecological-physiological work.

WEYERHAEUSER COMPANY

Address:

Main Office	Florida Office
3400 13th Avenue SW	P.O. Box 1584
Seattle, Washington 98134	Homestead, Florida 33030
(206) 623-3913	(305) 257-5101 or 4330

Personnel

Development: Harlan G. Freeman, Aquaculture Manager, Seattle
Norman H. Moe, Project Manager, Prawn Culture, Florida
Michael P. Mulvihill, Aquaculturist, Prawn Culture, Florida
Robert D. Lawrence, Grow-out Technician, Prawn Culture Florida
James P. Mulvihill, Hatchery Technician, Prawn Culture Florida

Research: Dr. Brian J. Allee, Director, Aquaculture Research, Seattle
Timothy J. Hall, Aquatic Biologist, Prawn Research Leader, Florida
Robert D. Thomsen, Aquaculture Biologist, Prawn Research, Florida

WEYERHAEUSER COMPANY (Continued)

Activities: Operate a *Macrobrachium rosenbergii* Hatchery, Grow-out, and Research site in Florida. Evaluating the feasibility of a large Fresh Water Prawn business. Also involved with Pacific Salmon Culture and Ocean Ranching in the Northwest.

APPENDIX C

BIBLIOGRAPHY

THE BIOLOGY AND CULTURE OF THE FRESHWATER PRAWN

(Genus *Macrobrachium*)

By
William C. Trimble and Paul A. Sandifer

About their bibliography, the authors have this to say:

"This bibliography developed concurrently with our research needs and inquiries and the expansion of literature and interest concerning *Macrobrachium*.

"The subject index of the bibliography hopefully will be useful to both researchers and businessmen. However, we caution that unpublished papers, reports, etc., were not indexed in detail because such references are frequently unobtainable. Also, some papers were indexed by title alone, because they had not been obtained for review.

"Literature references reflect our species bias (*M. rosenbergii*) and undoubtedly are incomplete; we solicit additions and corrections.

"We thank Jan Z. Hiester, William S. Ravenel, and Kathy T. Trimble for assistance in compiling this bibliography, and friends and colleagues worldwide for suggesting and assisting us in obtaining numerous references."

Now I must add that the original bibliography was expanded after it was received from the authors. Each of the authors of papers in the freshwater prawn book included his own bibliography and it did not seem tidy to me to have a bibliography at the end of each chapter and another large one at the end of the book. Consequently, I took the liberty of shuffling the chapter references into the work of the authors to achieve a single bibliography. I hope that this has been more of a service than a disservice and recognise that I must now take full responsibility for any errors which may appear herein. (ed.)

Aside from editorial meddling, the preparation of this work was supported by the Coastal Plains Regional Commission, the South Carolina Sea Grant Program (NOAA Office of Sea Grant, Department of Commerce), and the State of South Carolina.

SUBJECT INDEX

11. Parasites, Diseases, Predators, Competitors, and Abnormalities: 30, 49, 53, 57, 62, 63, 111, 112, 132, 140, 162, 164, 165, 166, 175, 176, 177, 185, 190, 195, 236, 237, 257, 274, 354, 377, 394, 396, 415, 419, 420, 421, 422, 427, 460, 489.

12. Effects of Environmental Factors:
    a. On Embryos and Larvae: 47, 54, 55, 89, 90, 91, 92, 97, 119, 129, 130, 161, 162, 165, 166, 195, 220, 227, 252, 265, 271, 273, 274, 280, 307, 342, 344, 351, 384, 415, 428, 432, 451, 455, 481.

    b. On Postlarvae and Juveniles: 57, 129, 130, 142, 146, 147, 148, 158, 159, 161, 165, 166, 168, 169, 176, 177, 180, 190, 193, 194, 195, 227, 252, 265, 270, 307, 332, 336, 344, 349, 389, 392, 393, 405, 424, 426, 452, 481, 483, 488, 489, 490.

    c. On Adults: 57, 67, 114, 146, 161, 165, 166, 176, 177, 195, 220, 227, 234, 265, 344, 426, 452.

13. Mass Production of Postlarvae:
    a. General: 53, 62, 161, 164, 165, 166, 167, 168, 169, 170, 171, 172, 175, 176, 177, 189, 193, 195, 251, 257, 265, 274, 278, 279, 280, 283, 341, 350, 422.

    b. Brood Stock: 52, 53, 130, 135, 136, 159, 161, 166, 175, 176, 177, 193, 194, 195, 257, 270, 274, 278, 279, 280, 336, 394, 484.

    c. Larval Culture: 1, 11, 13, 14, 18, 47, 48, 50, 52, 53, 121, 123, 130, 135, 151, 160, 161, 162, 163, 165, 166, 167, 168, 169, 170, 171, 172, 175, 176, 177, 178, 179, 180, 189, 193, 195, 200, 251, 255, 256, 257, 265, 270, 271, 274, 278, 279, 280, 285, 287, 288, 292, 296, 307, 321, 367, 384, 390, 391, 392, 394, 396, 397, 415, 422, 427, 432, 435, 455, 468, 469, 470, 481, 491, 495, 496.

14. Pond Culture: 14, 50, 52, 53, 57, 62, 66, 95, 121, 122, 123, 134, 135, 136, 137, 153, 164, 165, 166, 167, 168, 171, 174, 175, 176, 177, 189, 190, 193, 194, 195, 197, 219, 226, 262, 265, 270, 271, 274, 278, 279, 283, 292, 298, 309, 322, 341, 352, 354, 358, 359, 384, 388, 394, 419, 420, 426, 428, 467, 471, 488, 490.

15. Tank Culture: 53, 64, 105, 106, 130, 135, 136, 141, 143, 144, 158, 159, 164, 165, 166, 175, 193, 265, 270, 271, 294, 358, 384, 388, 390, 391, 393, 394, 395, 424, 336, 337, 471, 481, 483, 484, 490.

16. Use of Recirculating Systems: 55, 64, 105, 106, 112, 130, 158, 159, 190, 287, 288, 294, 390, 391, 392, 393, 394, 395, 396, 397, 401, 402, 416, 417, 425, 427, 435, 481, 483, 484, 495, 496.

17. Polyculture: 57, 62, 66, 107, 151, 164, 165, 174, 175, 197, 271, 277, 278, 280, 384, 420.

18. Economics: 69, 74, 167, 169, 170, 171, 172, 174, 175, 176, 177, 190, 193, 194, 195, 200, 251, 278, 320, 324, 331, 384, 404, 420, 473.

19. Processing and Storage: 190, 278.

20. Marketing: 66, 176, 177, 190, 278, 404.

21. Legal Regulations, Problems, and Constraints: 36a, 71, 153, 210, 227, 257, 275, 276, 277, 279, 289.

22. Site Selection: 107, 134, 137, 189, 190, 278, 473.

23. Bibliographies: 8, 276, 281.

24. Research Programs: 9, 17, 22, 24, 28, 31, 32, 33, 39, 40, 44, 50, 51, 52, 134, 137, 152, 161, 174, 175, 189, 190, 193, 197, 277, 278, 329, 330, 341, 346, 384, 388, 407, 422, 444, 490.

25. Commercial Culture Efforts: 35, 50, 72, 95, 105, 106, 125, 135, 136, 188, 190, 232, 276, 277, 292, 324, 354, 404, 445a, 449, 473, 474.

26. Engineering and System Design: 96, 105, 106, 134, 137, 142, 188, 218, 219, 232, 309, 323, 397, 402, 403, 335, 468, 469, 470, 491, 495, 496.

27. General and Miscellaneous: 6, 10, 11, 12, 14, 15, 16, 17, 18, 19, 20, 21, 22, 23, 24, 25, 28, 29, 31, 34, 35, 36, 37, 38, 41, 42, 50, 51, 70, 82, 85, 86, 96, 99, 100, 102, 104, 108, 109, 110, 125, 126, 127, 128, 134, 137, 138, 155, 159, 173, 184, 190, 198, 205, 206, 210, 212, 222, 226, 232, 233, 254, 261, 262, 264, 275, 276, 277, 283, 289, 292, 300, 301, 317, 318, 324, 329, 330, 350, 352, 353, 360, 367, 376, 378, 382, 384, 386, 391, 410, 411, 423, 442, 443, 444, 446, 450, 457, 458, 464, 471, 473, 474, 476, 478, 479, 480, 490.

REFERENCES

1. Adisukresno, S., A. Poernomo, M. Budiono, and S. Busman. [1975?]. Preliminary experiments on mass rearing of Macrobrachium larvae. Shrimp Culture Research Center, Japara, Indonesia.

2. Allen, W.V. and J. Kilgore. 1975. The essential amino acid requirements of the Red Abalone, Haliotis refescens. Comp. Biochem. Physiol. 50A: 771-775.

3. Andrews, J.W., L.V. Sick and G.P. Baptist. 1972. The influence of dietary protein and energy levels on growth and survival of Penaeid shrimp. Aquacult. 1: 341-347.

4. Anonymous. 1963. An experiment on the culture of giant freshwater prawn [in Thai]. Rep. Bangkhen Fish. Stn.

5. Anonymous. 1964. A preliminary study on the history and ecology of giant freshwater prawn [in Thai]. Rep. Bangkhen Fish. Stn.

6. Anonymous. 1968. [Further progress in production of Macrobrachium rosenbergii in Thailand]. FAO Fish Culture Bull. 1(1): 15.

7. Anonymous. 1970. [Cyclops and copepods substituted for Artemia as feed for larval Macrobrachium rosenbergii reared in Thailand]. FAO Fish Culture Bull. 2(4): 17.

8. Anonymous. 1971a. Bibliography of aquaculture. Coastal Plains Center for Marine Development Services, Wilmington, N.C. 245 pp.

9. Anonymous. 1971b. [Preparation of a program to culture Macrobrachium rosenbergii in Mauritius]. FAO Aquaculture Bull. 4(1): 18.

10. Anonymous. 1972a. [Propagation of freshwater prawn]. FAO Aquaculture Bull. 4(2): 4-5.

11. Anonymous. 1972b. [Initial larval culture of Macrobrachium rosenbergii in Mauritius]. FAO Aquaculture Bull. 4(3): 16.

12. Anonymous. 1972c. [Culture of Macrobrachium rosenbergii planned for Ecuador]. FAO Aquaculture Bull. 4(3): 16.

13. Anonymous. 1972d. [Pilot hatchery and future plans for culture of Macrobrachium rosenbergii announced for Macritius]. FAO Aquaculture Bull. 4(4): 19.

14. Anonymous. 1972e. [Pilot hatchery and pond production of Macrobrachium rosenbergii at Mauritius]. FAO Aquaculture Bull. 5(1): 22-23.

15. Anonymous. 1972f. Super shrimp. The News & Courier (Charleston, S.C.) March 23: 5C.

16. Anonymous. 1973a. [Effluent from heated power plant utilized for production of Macrobrachium rosenbergii, Ictalurus punctatus and clams in Fremont, Nebraska]. FAO Awuaculture Bull. 5(2): 16.

17. Anonymous. 1973b. [Expansion of facilities for culturing Macrobrachium rosenbergii in Mauritius]. FAO Aquaculture Bull. 6(1): 25.

18. Anonymous. 1973c. [Hatchery in Mauritius planned for production of Macrobrachium rosenbergii]. FAO Aquaculture Bull. 5(3-4): 25.

19. Anonymous. 1973d. [New Macrobrachium rosenbergii production facilities in Malaysia]. FAO Aquaculture Bull. 6(1): 13-14.

20. Anonymous. 1973e. [New Macrobrachium rosenbergii production facilities in Singapore]. FAO Aquaculture Bull. 6(1): 14.

21. Anonymous. 1973f. [Rearing and culture experiments with Macrobrachium americanum in Mexico]. FAO Aquaculture Bull. 5(2): 12-13.

22. Anonymous. 1974a. Annual Report '74. Aquaculture Dept., Southeast Asian Fish. Development Center.

23. Anonymous. 1974b. [Aquarium Farms, Inc. at Omaha, Nebraska]. FAO Aquaculture Bull. 6(2-3): 7.

24. Anonymous. 1974c. First NSF Grant to PSE&G for culture of Macrobrachium rosenbergii. R&D Quarterly, Public Service Electric and Gas Co., Newark, N.J. Summer: 1.

25. Anonymous. 1974d. First shrimp (Macrobrachium rosengergii) harvested at Mercer. R&D Quarterly, Public Service Electric and Gas Co., Newark, N.J. Fall: 2-3.

26. Anonymous. 1974e. [Larval rearing of Macrobrachium rosenbergii in Tahiti]. FAO Aquaculture Bull. 6(2-3): 14.

27. Anonymous. 1974f. [Macrobrachium rosenbergii introduced into Malawi]. FAO Aquaculture Bull. 6(2-3): 22.

28. Anonymous. 1974g. Malaysian prawn is center of interest at Anuenue Fisheries. Univ. Hawaii, Sea Grant Newsl. 4(12): 2-3.

29. Anonymous. 1974h. Patent issued on system of tanks for culturing shrimp [Macrobrachium rosenbergii]. FAO Aquaculture Bull. 6 (4): 13.

30. Anonymous. 1974i. Treatment of infectious prawn [Macrobrachium rosenbergii] diseases. FAO Aquaculture Bull. 6(4): 15.

31. Anonymous. 1975a. Aquaculture Department 1975 Annual Report. SEAFDEC. Tigbauan, Iloilo, Phillippines.

32. Anonymous. 1975b. Commission Agenda Item, Briefing Session, Native freshwater shrimp studies. Texas Parks & Wildlife Dept. 3 pp. (Unpublished).

33. Anonymous. 1975c. Commission Agenda Item, Briefing Session, Status of South American freshwater shrimp [Macrobrachium amazonicum] at Heart of the Hills Fisheries Research Station. Texas Parks & Wildlife Dept. 3pp. (Unpublished).

34. Anonymous. 1975d. Fishfarms Hawaii produces and markets Macrobrachium and catfish successfully. Univ. Hawaii, Sea Grant Newsl. 5(1): 1,6.

35. Anonymous. 1975e. Fresh water prawn farming. Aquacult. Newsl. (14): 1-3.

36. Anonymous. 1975f. Genetics: Macrobrachium rosengergii. Univ. Hawaii, Sea Grant Newsl. 5(10): 5.

36a. Anonymous. 1975g. Federal food, drug and cosmetic act, as amended Supt. Doc. USGO.

37. Anonymous. 1976a. Giant Asian shrimp to be grown in desert. The Fish Boat 21(2): 19.

38. Anonymous. 1976b. Giant-prawn venture reaches a vital stage in Austrailia. Fish Farming Int. 3(1): 9.

39. Anonymous. 1976c. Priorities of research needs as suggested by Florida's Macrobrachium Industry. Florida Practicum Conf. on Macrobrachium. Fla. Dept. Nat. Resources, Mar. Res. Lab. 3/25-26/76, held in St. Petersburg, Fla.

40. Anonymous. 1976d. Power plant waste heat utilization in aquaculture. Public service Electric & Gas Co., 80 Park Place, Neward, N.J.

41. Anonymous. 1977a. Freshwater prawns: an emerging aquaculture enterprise for the Americas. Commer. Fish Farmer & Aquacul. News 3(2): 7-14.

42. Anonymous. [1977?]b. Aquaculture Department: descriptive pamphlets. SEAFDEC, Iloilo. Philippines.

43. Anonymous. 1977c. Genetic research aimed at improving Macrobrachium rosenbergii. Univ. Hawaii, Sea Grant Newsl. 7(1): 5.

44. Anonymous. 1977d. Session in Seattle. Research Priorities Evaluated for Commercial Prawn Aquaculture. The Commercial Fish Farmer and Aquaculture News. 3(2): 10-11. Jan. 1977.

45. Al-Murrani, W.K. 1974. The Limits to Artificial Selection. An. Breed Abs. Vol. 42, No. 12. pp 587-592.

46. Aquacop. 1976. Incorporation de proteines vegetales dans un aliment compose pour crevettes Macrobrachium rosenbergii [in French, English summary]. Aquaculture 8: 71-80.

47. Aquacop. 1977a. Macrobrachium rosenbergii (de Man) culture in Polynesia: observations on the water chemodynamism in an intensive larval rearing. Presented 8th Annual Meeting World Mariculture Society, San Jose, Costa Rica, 9-13 January 1977. (Manuscript).

48. Aquacop. 1977b. Macrobrachium rosenbergii (de Man) culture in Polynesia: progress in developing a mass intensive larval rearing in clear water. Presented 8th Annual Meeting World Macriculture Society, San Jose, Costa Rica, 9-13 January 1977. (Manuscript).

49. Aquacop. 1977c. Observations on deseases of crustacean cultures in Polynesia. Presented 8th Annual Meeting World Mariculture Society, San Jose, Costa Rica, 9-13 January 1977. (Manuscript).

50. Aquaculture Products and Research, Inc. (APARI). [1972]. Demonstration project to develop techniques for rearing fresh-water shrimp in controlled aquatic environment. Tech. Assist. Study, Economic Development Administration, Washington, D.C. 22 pp. (Unpublished).

51. Arana, M.F. 1974. Experiences sobre el cultivo del langostino Macrobrachium americanum (Bate) en el noroeste de Mexico. Simposio FAO/Carpas sobre acuiultura en America Latina: 1-9.

52. Ardill, D., R. Jenson, B. Jugnarain, and R.K. Thompson. 1973. The introduction of the fresh water prawn, Macrobrachium rosenbergii (de Man), into Mauritus. Rev. Agric. Sucr. Ile Maurice 52: 6-11.

53. Ardill, J.D. and R.K. Thompson. 1975. The freshwater prawn, Macrobrachium rosenbergii, in Mauritius. FAO/CIFA Symp. Aquaculture in Africa, Accra, Ghana. 14 pp.

54. Armstrong, D. 1976. Section III. Acute toxicity of nitrite to larval stages of Macrobrachium rosenbergii. Pages 22-29 in A.W. Knight, principal investigator, Laboratory studies on selected nutritional, physical, and chemical factors affecting the growth, survival, respiration, and bioenergetics of the giant prawn Macrobrachium rosenbergii. Water Sci. &Eng. Paper 4501, Univ. Calif., Davis.

55. Armstrong, D.A., M.J. Stephenson, and A.W. Knight. 1976. Acute toxicity of nitrite to larvae of the giant Malaysian prawn, Macrobrachium rosenbergii. Aquaculture 9: 39-46.

56. Baerends, G.P., R. Brojwer and H.T.J. Waterbolk. 1955. Ethological studies on Lebistes reticulatus (Peters). Behaviour. 8: 229-235.

57. Baissac, P. de B. 1976. Problems caused by the occurence of the filamentous algae, Spyrogyra sp., in culture ponds of Macrobrachium rosenbergii (de Man) in Mauritius. Rev. Agric. Sucr. Ile Maurice 53 (3&4): 206-211.

58. Balazs, G.H., S.E. Olbrich, and M.E. Tumbleson. 1974. Serum constituents of the Malaysian prawn (Macrobrachium rosenbergii) and pink shrimp (Penaeus marginatus). Aquaculture 3: 147-157.

59. Balazs, G.H. and E. Ross. 1976. Effect of protein source and level on growth and performance of the captive freshwater prawn, Macrobrachium rosenbergii. Aquaculture 7: 299-313.

60. Balazs, G.H., E. Ross, and C.C. Brooks. 1973. Preliminary studies on the preparation and feeding of crustacean diets. Aquaculture 2: 369-377.

61. Banerji, S.K. 1969. X. Crustacean production in India. Pages 258-272 in Prawn Fisheries of India, Bull. Cent. Mar. Fish. Res. Inst., No. 14.

62. Bardach, J.E., J.H. Ryther, and W.O. McLarney. 1972. Freshwater shrimp culture. Pages 619-632 in Bardach et al., Aquaculture: the farming and husbandry of freshwater and marine organisms. J. Wiley & Sons, Inc., N.Y.

63. Barkate, J.A. 1972. Preliminary studies of some shrimp diseases. Proc. 3rd Ann. Workshop, World Marie. Soc: pp 337-346.

64. Beard, T.W. 1972. A preliminary report on the growth and survival of Macrobrachium rosenbergii de Man, Penaeus aztecus Ives, P. indicus H. Milne Edwards, and P. monodon Fabricius in laboratory recirculation systems. Int. Counc. Explor. Sea (C.M. 1972/K:26) Shellfish and Benthos Committee. 13 pp.

65. Becker, W.A., 1975. Manual of Quantitative Genetics (3rd ed.). Published by Program in Genetics, Washington State University Press, Pullman Washington, 70 pages.

66. Bhimachar, B.S., ed. 1963. Information on prawns from Indian waters - synopsis of biological data. Indo-Pac. Fish. Counc. Proc. 10 (2): 124-133.

67. Bhimachar, B.S. 1965. Life history and behaviour of Indian prawns. Fish. Technol. (India) 2 (1): 1-12.

68. Biddle, G.N. 1976. The nutrition of Macrobrachium species. U.S. Dept. Commerce, NOAA, NMFS. Presented 1976 Workshop on the Aquaculture of Freshwater Prawns (Macrobrachium spp.), Charleston, S.C., 14-15 June 1976. 36 pp. (Manuscript).

69. Blommestein, E., H. Deese, and J.P. McVey. 1977. The social-economic impact of Macrobrachium rosenbergii farming in Palua (a preliminary report). Presented 8th Annual Meeting World Mariculture Society, San Jose, Costa Rica, 9-13 January 1977. (Manuscript).

70. Bonnelly de Calventi, I. and S. Jakowska. 1974. Contribution al conocimiento de los decapodos terrestres y de agua dulce de la Republica Dominicana [in Spanish]. Centro de Investigaciones de Biologia Marina Univ. Antonoma de Santo Domingo. 12 pp.

71. Booker, J.R. 1976. How to export fishery products to the United States. Food, Drug, Cosmetic Law Jout., May: pp 289-306.

72. Bowen, H.J. 1976. Fresh water power plant aquaculture. Pages 221-231 in Power Plant Waste Heat Utilization in Aquaculture - Workshop I Papers (Trenton State College, Trenton, N.J., 6-7 Nov. 1975). Public Service Electric and Gas co., Newark, N.J.

73. Broad, C. 1950. The shrimps in North Carolina. Pages 191-204 in Marine Fisheries of North Carolina: biology and natural history.

74. Brooks, M.J. 1976. Budget for Macrobrachium rosenbergii hatchery and farm, Colombia, South America. Student report, Department of Sisheries and Allied Aquacultures, Auburn University, Auburn. 31 pp. (Unpublished).

76. Campbell, K.C. 1969. Prawn farming in the Far East. World Fish. 18 (1): 41-43.

77. Carlberg, J.M. and J.C. Van Olst, 1976. Brine Shrimp (Artemia salina) Consumption by the Larval Stages of the American Lobster (Homarus americanus) in Relation to Food Density and Water Temperature. Proc. VII Ann. Meeting World Mariculture Soc. San Diego, Calif., Jan. 1976: 379-390.

78. Castell, J.D. and S.D. Budson. 1974. Lobster nutrition: The effect on Homarus americanus of dietary protein levels. J. Fish. Res. Bd. Can. 31: 1363-1370.

79. Castell, J.D. and J.F. Covey. 1975. The role of lipids in the nutrition of lobsters. Proc. World Maricult. Soc., 6 (In Press).

80. Castell, J.E., E.G. Mason, and J.F. Covey. 1975. Cholesterol requirements of juvenile American lobster (*Homarus americanus*). J. Fish. Res. Bd. Can. 32: 1431-1435.

81. Castro, W.E., P.B. Zielinski and P.A. Sandifer. 1975. Performance characteristics of air lift pumps of short length of small diameter. Proc. World Marie. Soc. 6th Ann. Meet.

82. Ceccaldi, H.J. 1968. General aspects of shrimp culture. Stud. Rev. GFCM (52): 97-104.

83. Chace, F.A., Jr. and H.H. Hobbs, Jr. 1969. The freshwater and terrestrial decapod crustaceans of the West Indies with special reference to Dominica. U.S. Nat'l. Museum Bull. 292, Smithsonian Institution Press, Washington, D.C. 258 pp.

84. Chaiyarat, C. and P. Intarachote. 1972. Study on stomach contents of giant fresh-water prawn (*Macrobrachium rosenbergii* de Man) from Songkhla Lake [Thailand]. Dept. Fish., Songkhla Mar. Fish. Sta., Annual Rep., 1971-72: 185-202. (Unpublished).

85. Charniaux-Cotton, H., 1960. *Sex Determination*. *in* the Physiology of Crustacea. Ch. 13. T.H. Waterman (ed.) Academic Press, N.Y.

86. Chen, T.P. 1972. [Attempts to breed *Macrobrachium rosenbergii* in Taiwan]. FAO Aquaculture Bull. 4 (2): 5.

87. Cherfas, B.I. (ed.), 1969. *Genetics, Selection, and Hybridization of Fish* translated from Russian by Israel Program for Scientific Translations (IPST cat. No. 600424), printed by Keter Press, Jerusalem.

88. Chopra, B.N. 1943. Prawn fisheries of India. Indian Sci. Cong. Assoc. Proc., 30th Session, pt. II: 153-173.

89. Choudhury, P.C. 1970. Complete larval development of the palaemonid shrimp *Macrobrachium acanthurus* (Wiegmann, 1836), reared in the laboratory. Crustaceana 18(2): 113-132.

90. Choudhury, P.C. 1971a. Complete larval development of the palaemonid shrimp Macrobrachium carcinus (L.) reared in the Labroatory (Decapoda, Palaemonidae). Crustaceana 20 (1): 51-69.

91. Choudhury, P.C. 1971b. Laboratory rearing of larvae of the palaemonid shrimp Macrobrachium acanthurus (Wiegmann, 1836). Crustaceana 21 (2): 113-126.

92. Choudhury, P.C. 1971c. Responses of larval Macrobrachium carcinus (L.) to variations in salinity and diet (Decapoda, Palaemonidae). Crustaceana 20 (2): 113-120.

93. Claybrook, D.L. 1976. Biosynthesis of amino acids from 3-[^{14}C]-pyruvate in the Fiddler Crab, Uca pugilator. Comp. Biochem. Physiol. 54B: 63-68.

94. Clausen, J. and W.M. Hiesey, 1958. Experimental Studies on the Nature of Species IV. Genetic Structure of Ecological Races. Carnegie Inst. Wash. Pub. No. 615: 1-312.

95. Clendenon, J. 1970. 1-pound shrimp make big splash [Aquaculture Products and Research, Inc.]. Ft. Lauderdale News and Sun-Sentinel (Ft. Lauderdale, Fla.) July 12, 8 (48): 1A, 12A.

96. Cohen, D. 1976. The introduction of M. rosenbergii into Israel's fish ponds: the use of geothermal springs in heating the larviculture and nursery. Hebrew Univ. Jerusalem, Israel. 6 pp. (Unpublished).

97. Cohen, D., A. Finkel, and M. Sussman. 1976. On the role of algae in larviculture of Macrobrachium rosenbergii. Aquaculture 8: 199-207.

98. Conklin, D. 1975. Nutritional studies of lobsters. Proc. First Intl. Conf. Aquacult. Nutr., 287-296.

99. Corbin, J.S. 1976. Aquaculture in Hawaii 1976: progress, resources and organization. Dept. Planning & Econ. Development, Hawaii. 107 pp. + xxx pp. Bibliography & Appendices.

100. Costello, T.J. 1971. Freshwater prawn culture techniques developed. Am. Fish Farmer & World Aquacul. News 2 (2): 8-10, 27.

101. Costello, T.J. 1972. Aquaculture of mollusks in the Golf Coast Region. Bull. Am. Malacological Union, Inc. (Symp. on Molluscan Aquaculture, 37th Annual Meeting, 1971): 22.

102. Costello, T.J. 1973. [Culture of shrimps and prawns (Macrobrachium rosenbergii) in Honduras]. FAO Aquaculture Bull. 5 (3-4): 10.

103. Cowey, C.B. and J.R.M. Forster. 1971. The essential amino acid requirements of the prawn Palaemon serratus. The growth of prawns on diets containing proteins of different amino acid compositions. Mar. Biol. 10: 77-81.

104. Cox, K.W. 1973. Mauritus. Aquaculture - oyster and prawn (Macrobrachium rosenbergii) culture. FAO Fish. Advisory Ser. Proj. Rep., Mauritius, FAO-FI-DP-MAR-72-004-1: 1-13.

105. Day, J.J. and P.S. Hirschman. 1970. Apparatus for growing crustaceans [including Macrobrachium rosenbergii], U.S. Pat. 3,724,423 April 3, 1973. U.S. Patent Office, Washington, D.C.

106. Day, J.J. and P.S. Hirschman. 1974. Apparatus for growing crustaceans. U.S. Pat. 3,797,458 March 19. Offic. Gaz. U.S. Patent Office 920 (3): 700. [Abstr.].

107. Dean, J.M. 1972. The use of old rice fields for aquaculture. (Report to the S.C. State Development Board). Tech. Rep. 2, Belle W. Baruch Coastal Res. Inst., Univ. S.C., Columbia. 9 pp. (Unpublished).

108. Dean, J.M., ed. 1975. The potential use of South Carolina rice fields for aquaculture. S.C. State Development Board, Columbia. 96 pp.

109. Deese, H. 1976. Freshwater prawn project. Micronesian Mariculture Demonstration Center, Newsl. (5): 3.

110. de la Cruz, M.C.R. 1968. Contribucion al conocimiento de los palemonidos de Mexico: III. Palemonidos del Golfo de California, con notas sobre la biologia de Macrobrachium americanum Bate [in Spanish, English abstr.]. FAO Fish. Rep. 57 (2): 373-380.

111. Delves-Broughton, J. 1974. Preliminary investigations into the suitability of a new chemotherapeutic, furanace, for the treatment of infectious prawn diseases. Aquaculture 3: 175-185.

113. de Man, J.R. 1910. Decapoda of the Siboga Expedition. Family Penaeidae Spence Bate. Siboga Expedition Monograph 39a: 1-131.

114. Denne, L.B. 1968. Some aspects of osmotic and ionic regulation in the prawns Macrobrachium australiense (Holthuis) and M. equidens (Dana). Comp. Biochem. Physiol. 26: 17-30.

115. Deshimaru, O. and K. Kuroki. 1974. Studies on a purified diet for prawn. I. Basal composition of diet. Bull, Jap. Sci. Fish., 40: 413-419.

116. Dobkin, S. 1969. Abbreviated larval development in caridean shrimps and its significance in the artificial culture of these animals. FAO Fish. Rep. 57 (3): 935-946).

117. Deshimaur, O. and K. Kuroki. 1975. Studies on a purified diet for prawn. IV. Evaluation of protein, free amino acids and their mixture as nitrogen source. Bull. Jap. Soc. Sci. Fish. 41: 101-103.

118. Deshimaru, O. and K. Shigeno. 1972. Introduction to the artificial diet for prawn Penaeus japonicus. Aquacult. 1: 115-133.

119. Dobkin, S. 1971. A contribution to knowledge of the larval development of Macrobrachium acanthurus (Wiegmann, 1836) (Decapoda, Palaemonidae). Crustaceana 21 (3): 294-297.

120. Dobkin, S. 1973. Selective breeding and hybridization of the shrimps Macrobrachium acanthurus and M. carcinus. Ann. Rept. Florida-Atlantic Univ., Boca Raton.

121. Dobkin, S., W.P. Azzinaro, and J. van Montfrans. 1974. Culture of Macrobrachium acanthurus and M. carcinus with notes on the selective breeding and hybridization of these shrimps. Proc. 5th Annual Workshop World Mariculture Society: 51-62.

122. Dobkin, S., N.L. Chwang, P.G. Van Arman, and D.H. Holland. 1977. Growth rates of Macrobrachium rosenbergii grown in ponds in south Florida. Presented 8th Annual Meeting World Mariculture Society, San Jose, Costa Rica, 9-13 January 1977. (Manuscript).

123. Dobkin, S., J. van Montfrans, and D.H. Holland. 1975. Selective breeding of the fresh and brackish-water shrimp Macrobrachium acanthurus. Proc. 6th Annual Meeting World Mariculture Society: 33-36.

124. Donaldson, L.R., 1970. Selective Breeding of Salmonid Fishes. in Marine Aquaculture, W.J. McNeil (ed.), Oregon State Univ. Press.

125. Dotter, B.M. 1975. Sun ventures into growing prawns. Our Sun 40(2): 10-12.

126. Doumange, F. 1972. [CNEXO constructs laboratory for culture of Macrobrachium, penaeid shrimp, and molluscs]. FAO Aquaculture Bull. 4 (2): 9.

127. du Barry, J. 1972. Strange but true: family sized shrimp. True (Sept.) 53 (424): 35.

128. Dugan, C.C. 1971. Florida research: freshwater shrimp. Am. Fish Farmer & World Aquacul. News 2 (10): 9.

129. Dugan, C.C. and T.A. Frakes. 1973. Culture of brackish-freshwater shrimp, Macrobrachium acanthurus, M. carcinus and M. ohione. Proc. 3rd (1972) Annual Workshop World Mariculture Society: 185-191.

130. Dugan, C.C., R.W. Hagood, and T.A. Frakes. 1975. Development of spawning and mass larval rearing techniques for backish-freshwater shrimps of the genus Macrobrachium (Decapoda, Palaemonidae). Fla. Mar. Res. Publ. No. 12, Fla. Dept. Nat. Res., Mar. Res. Lab., St. Petersburgh. 28 pp.

131. Dupree, H.K. 1975. Nutritional studies on warmwater fishes. Proc. First Int. Conf. Aquacult. Nutr., 65-84.

132. Dutt, S. and K. Ravindranath. 1974. Notes on abnormal shrimps and prawns. Curr. Sci. (Bangalore) 43 (21): 684-685.

133. Dutt, S. and K. Ravindranath. 1976. Changes in the relative growth rates of the inner pair of telson spines and the spinous end of the telson in three species of Macrobrachium Bate (Decapoda, Palaemonidae). Crustaceana 31 (3): 296-300.

134. Eble, A.F. 1975. Freshwater aquaculture of the tropical prawn, Macrobrachium rosenbergii and the rainbow trout, Salmo gairdnerii, using thermal effluent discharges from an electric generating station. Presented 67th Annual Cov., Nat'l. Shellfish Assoc., Charleston, S.C., 22-26 June 1975. 1 p. [Abstr.].

135. Eble, A.F. 1975. Integration of thermal and food processing residuals into a system for commercial culture of freshwater shrimp. Appendix VII, 30 pp. in Guerra, C.R., principal investigator, Integration of thermal and food processing residuals into a system for commercial culture of freshwater shrimp. Second Semiannual Rep. (July 1, 1975 - Dec. 31, 1975) NSF/RANN from PSE&G Co., R&D Dept., Newark, N.J. (Unpublished).

136. Eble, A.F. 1976. Integration of thermal and food processing residuals into a system for commercial culture of freshwater shrimp. Appendix VII, 40 pp. in Guerra, C.R., principal investigator, Integration of thermal and food processing residuals into a system for commercial culture of freshwater shrimp. Second Annual Rep. (July 1, 1975 - June 30, 1976) NSF/RANN from PSE&G Co., R&D Dept., Newark, N.J. (Unpublished).

137. Eble, A.F., N.E. Stolpe, M.C. Evans, and N.F. DeBlois. 1976. The use of thermal effluents of an electric generating station in New Jersey in the culture of the tropical prawn, Macrobrachim rosenbergii and the rainbow trout, Salmo gairdneri. Pages 47-65 in Power Plant Waste Heat Unilization in Aquaculture - Workshop I Papers, (Trenton State College, Trenton, N.J., 6-7 Nov. 1975). Public Service Electric and Gas Co., Newark, N.J.

138. Escritor, G.L. 1975. Status of the giant prawn resources of the Philippines. Contr. Internat'l. Conf. Prawn Farming, Vung-Tau, Vietnam. 1975. 11 pp.

139. Evangeline, G., V. Venkatesan, and K.P. Sambandam. 1975. Observations on the prawn fishery of the Killai backwaters. Pages 316-321 in R. Natarajan, ed., Recent Researches in Estuarine Biology, Hindustan Publishing Corp., Dehli, India.

140. Evans, I.H. 1976. Progress in indentification of potential pathogens of Macrobrachium rosenbergii (de Man). Appendix XV, 6 pp. in Guerra, C.R.. principal investigator, Integration of thermal and food processing residuals into a system for commercial culture of freshwater shrimp. Second Annual Rep. (July 1, 1975 - June 20, 1976) NSF/RANN from PSE&G Co., R&D Dept., Newark, N.J. (Unpublished).

141. Evans, M.C. 1975a. Fabricated substrates in the intensive culture of Macrobrachium rosenbergii (de Man). Appendix IX, 7 pp. in Guerra, C.R., principal investigator, Integration of thermal and food processing residuals into a system for commercial culture of freshwater shrimp. Second Semiannual Rep. July 1, 1975 - Dec. 31, 1975) NSF/RANN from PSE&G Co., R&D Dept., Newark, N.J. (Unpublished).

142. Evans, M.C. 1975b. Fabricated substrates - an approach to the intensive culture of Macrobrachium rosenbergii (de Man). Presented 67th Annual Conv., Nat'l. Shellfish Assoc., Charleston, S.C. 22-26 June 1975. 1 p. [Abstr.].

143. Evans, M.C. 1976a. Fabricated substrates in the intensive culture of Macrobrachium rosenbergii (de Man). Appendix X, 7 pp. in Guerra, C.R., principal investigator, Integration of thermal and food processing residuals into a system for commercial culture of freshwater shrimp. Second Annual Rep. (July 1, 1975 - June 20, 1976) NSF/RANN from PSE&G Co., R&D Dept., Newark, N.J. (Unpublished).

144. Evans, M.C. 1976b. A prediction curve for density vs size in Macrobrachium rosenbergii (de Man). Appendix XI, 10 pp. in Guerra, C.R., principal a system for commercial culture of freshwater shrimp. Second Annual Rep. (July 1, 1975 - June 30, 1976) NSF/RANN from PSE&G Co., R&D Dept. Newark, N.J. (Unpublished).

145. Falsoner, D.S., 1960. Quantitative Genetics, Ronald Press, New York

146. Farmanfarmaian, A. 1975. Integration of thermal and food processing residuals into a system for commercial culture of freshwater shrimp [July 1 - December 31, 1975]. Appendix VIII, iv + 46 pp. in Guerra, C.R., principal investigator, Integration of thermal and food processing residuals into a system for commercial culture of freshwater shrimp. Second Semiannual Rep. (July 1, 1975 - Dec. 31, 1975) NSF/RANN from PSE&G Co., R&D Dept., Newark, N.J. (Unpublished).

147. Farmanfarmaian, A. 1976. Integration of thermal and food processing residuals into a system for commercial culture of freshwater shrimp [January 1 - June 30, 1976]. Appendix IX, 35 pp. in Guerra, principal investigator, Integration of thermal and food processing residuals into a system for commercial culture of freshwater shrimp. Second Annual Rep. (July 1, 1975 - June 30, 1976) NSF/RANN from PSE&G Co., R&D Dept., Newark, N.J. (Unpublished).

148. Farmanfarmaian, A., R. Moore, and K. Volosin. 1976. Diseasonal thermal aquaculture - physiological and environmental factors influencing the survival, growth and pollution of giant shrimp and rainbow trout. Pages 199-211 in Power Plant Waste Heat Utilization in Aquaculture - Workshop I Papers, (Trenton State College, Trenton, N.J., 6-7 Nov. 1975). Public Service Electric and Gas Co., Newark, N.J.

149. Fielder, D.R. 1970. The larval development of Macrobrachium australiense Holthuis, 1950 (Decapoda, Palaemonidae), reared in the laboratory. Crustaceana 18(1): 60-74.

150. Figueroa, A.V. 1969. Problemas de especiacion en America de un grupo de Palaemonidae del genero Macrobrachium [in Spanish, English abstr.]. FAO Fish. Rep. 57 (3): 1055-1066.

151. Fisheries Division, Ministry of Agriculture and Co-operatives, Kuala Lumpur, Malaysia. 1972. Review of the status of coastal aquaculture in Malaysia. Pages 52-59 in T.V.R. Pillay, ed. Coastal aquaculture in the Indo-Pacific Region. Fishing News (Books) Ltd., Surrey, England.

152. Florida Department of Natural Resources. 1976. Priorities of research needs as suggested by Florida's Macrobrachium industry (Florida Practicum Conference on Macrobrachium, 25-26 March 1976). FDNR, Marine Research Laboratory, St. Petersburg, Fla. 1 p. (Unpublisehed).

153. Florida Game and Fresh Water Fish Commission. 1976. Guidelines for outdoor pool culture of Macrobrachium rosenbergii in Florida. Game and Fresh Water Fish Commission, Tallahassee, Fla. 2 pp.

154. Forster, J.R.M. and P.A. Gabbott. 1971. The assimilation of nutrients from compounded diets by the Prawns Palaemon serratus and Pandalus platyceros. J. Mar. Biol. Assn., U.K. 51: 943-961.

155. Forster, J.R.M. 1972. Research on intensive prawn rearing. Proc. Shellfish Assoc. Great Britain, 3rd Shellfish Conf.: 12-22.

156. Forster, J.R.M. 1973. Studies on compound diets for prawns. Proc. 3rd (1972) Annual Workshop World Mariculture Society: 389-402

157. Forster, J.R.M. and T.W. Beard. 1973. Growth experiments with the prawn Palaemon serratus (Pennant) fed with fresh and compounded diets. Fish. Invest., Ser. II. 27 (7): 16 pp.

158. Forster, J.R.M. and T.W. Beard. 1974. Experiments to assess the suitability of nine species of prawns for intensive cultivation. Aquaculture 3: 355-368.

159. Forster, J.R.M. and J.F. Wickins. 1972. Prawn culture in the United Kingdom: its status and potential. Lab. Leaflet (new series) No. 27, Fish. Expt. Stat., Conway, Caernarvonshire. 32 pp.

160. Fugimura, T. 1966. Notes on the development of a practical mass culturing technique of the giant prawn Macrobrachium rosenbergii. Indo-Pac. Fish. Counc. Proc., 12th Session, Honolulu, Hawaii (IPFC/C66/WP47). 3 pp.

161. Fugimura, T. 1967a. Development of a prawn industry: development of a rearing technique for the giant long-legged prawn, Macrobrachium rosenbergii. Quarterly Progress Report: Jan. 1 to June 30, 1966. U.S. Dept. Interior, Fish & Wildlife Ser., Bur. Comm. Fish. 3 pp.

162. Fugimura, T. 1967b. Quarterly Progress Report: July 1 to Sept. 30, 1966. 4 pp.

163. Fugimura, T. 1967c. Quarterly Progress Report: Oct. 1 to Dec. 31, 1966. 2 pp.

164. Fugimura, T. 1967d. Quarterly Progress Report: Jan. 1 to Mar. 31, 1967. 3 pp.

165. Fugimura, T. 1967e. Final Report: July 1, 1966 to June 30, 1967. 3 pp.

166. Fugimura, T. 1968a. Development of a prawn industry: development of commercially applicable mass rearing technique for the giant prawn Macrobrachium rosenbergii. Quarterly Report: July 1 to Sept. 30, 1967. U.S. Dept. Interior, Fish & Wildlife Ser., Bur. Comm. Fish. 6 pp.

167. Fugimura, T. 1968b. Quarterly Report: Oct. 1 to Dec. 31, 1967. 3 pp.

168. Fugimura, T. 1968c. Quarterly Report: Jan. 1 to Mar. 31, 1968. 3 pp.

169. Fugimura, T. 1968d. Final Report: July 1, 1967 to June 30, 1968. 6 pp. + iv pp. appendix.

170. Fugimura, T. 1969a. Quarterly Progress Report: July 1 to Sept. 30, 1968. 4 pp. + ii pp. appendix.

171. Fugimura, T. 1969b. Quarterly Progress Report: Oct. 1 to Dec. 31, 1968. 4 pp. + iv pp. tables.

172. Fugimura, T. 1969c. Quarterly Progress Report: Jan. 1 to Mar. 31, 1969. 2 pp.

173. Fugimura, T. 1970. The giant prawn, aloha aina sp. Dept. Land Nat. Res. Hawaii 1 (1): 9-12.

174. Fugimura, T. 1972. Development of a prawn culture industry. Annual Report: July 1, 1970 to June 30, 1971. U.S. Dept. Interior, Fish & Woldlife Ser., Bur. Comm. Fish. 6 pp. + ii pp. tables.

175. Fugimura, T. 1974. Development of a prawn culture industry in Hawaii. Job Completion Report: July 1, 1969 to June 30, 1972. U.S. Dept. Commerce, NOAA, NMFS. 21 pp. + vii pp. figures.

176. Fujimura, T. and H. Okamoto. 1970. Notes on progress made in developing a mass culturing technique for Macrobrachium rosenbergii in Hawaii. Indo-Pac. Fish. Counc. Proc., 14th Session, Bangkok, Thailand. Sym. 53. 17 pp.

177. Fugimura, T. 1972. Notes on progress made in developing a mass culturing technique for Macrobrachium rosenbergii in Hawaii. Pages 313-327 in T.V.R. Pillay, ed. Coastal Aquaculture in the Indo-Pacific Region. Fishing News (Books) Ltd., Surrey, England.

178. George, A.I. 1970. [Artificial propagation of Macrobrachium rosenbergii by the Department of Fisheries, Edathua, Kerala State, India]. FAO Fish Culture Bull. 2 (2): 5.

179. George, A.I. and M.J. Sebastian. 1970a. Review of the backwater fisheries and brackish water fish culture in Kerala State. Indo-Pacific Fish. Counc. Proc., 14th Session, Bangkok, Thailand, Sym. 19. 11 pp.

180. George, A.I. and M.J. Sebastian. 1970b. The prawn fisheries of Kerala backwaters: problems and prospects. Indo-Pac. Fish. Counc. Proc., 14th Session, Bangkok, Thailand, Sym. 20. 6 pp.

181. George, A.I. and M.J. Sebastian. 1972. The prawn fisheries of Kerala backwaters - problems and prospects with special reference to aquaculture (abst.). Page 118 in T.V.R. Pillay, ed. Coastal Aquaculture in the Indo-Pacific Region. Fishing News (Books) Ltd., Surrey, England.

182. George, M.J. 1969a. II. Systematics-taxonomic consideration and general distribution. Pages 5-48 in Prawn Fisheries of India. Bull. Cent. Mar. Fish. Res. Inst. 14.

183. George, M.J. 1969b. VII. Genus Macrobrachium Bate 1868. Pages 178-216 in Prawn Fisheries of India. Bull. Cent. Mar. Fish. Res. Inst. 14.

184. George, M.J. 1972. Brief review of the progress of prawn mariculture. Seminar on Mariculture & Mechanised Fishing, Madras, 28-29 Nov. 1972: 77-83.

185. Ghosh, A. 1972. An abnormal rostrum of the commercial prawn Macrobrachium rosenbergii (de Man). Sci. Cult. (Bangalore) 38 (2): 98-99.

186. Gjedrem, T., 1975. Possibilities for Genetic Gain in Salomonids. Aquaculture: 6 (1975): 23-29.

187. Glude, J.B. 1975. Nutritional consideration in the culture of tropical species. Proc. 1st Int. Conf. Aquaculture Nutrition (Univ. Delaware): 107-117.

188. Godfriaux, B.L. 1976. Aquaculture - future research [Mercer Generating Station, Public Service Electric and Gas Co.]. Pages 213-220 in Power Plant Waste Heat Utilization in Aquaculture - Workshop I Papers (Trenton State College, Trenton, N.J., 6-7 Nov. 1975). Public Service Electric and Gas Co., Newark, N.J.

189. Godfriaux, B.L., H.J. Valkenberg, A. Van Riper, C.R. Guerra, A.F. Eble, and P. Campbell. n.d. Power plant heated water use in aquaculture. 18 pp. (Unpublished).

190. Goodwin, H.L. and J.A. Hanson. [1975]. The aquaculture of freshwater prawns (Macrobrachium species). (Augmented summary of Proc. Workshop on Culture of Freshwater Prawns, St. Petersburg, Fla.). The Oceanic Inst., Waimanalo, Hawaii. 95 pp.

191. Greenwood, J.G., D.R. Fielder, and M.J. Thorne. 1976. The larval life history of Macrobrachium novaehollandiae (de Man, 1908) (Decapoda, Palaemonidae), reared in the laboratory. Crustaceana 30 (3): 252-286.

192. Gregor, J.W. & P. J. Watson, 1961. Ecotypic Differentiation: Observations and Reflections. Evolution 15: 166-173.

193. Guerra, C.R., principal investigator. 1975a. Integration of thermal and food processing residuals into a system for commercial culture of freshwater shrimp. Second Semiannual Rep. (July 1, 1975 - Dec. 31, 1975) NSF/RANN from PSE&G Co., R&D Dept., Newark, N.J.

194. Guerra, C.R. 1975b. Integration of thermal and food processing residuals into a system for commercial culture of freshwater shrimp. First Annual Rep. NSF/RANN from PSE&G Co., R&D Dept., Newark, N.J. (Unpublished).

195. Guerra, C.R. 1976a. Integration of thermal and food processing residuals into a system for commercial culture of freshwater shrimp. Second Annual Rep. (July 1, 1975 - June 30, 1976) NSF/RANN from PSE&G Co., R&D Dept., Newark, N.J. (Unpublished).

196. Guerra, C.R. 1976b. Thermal aquaculture using power plant effluents - a synopis. Pages 191-197 _in_ Power Plant Waste Heat Utilization in Aquaculture - Workshop I Papers (Trenton State College, Trenton, N.J., 6-7 Nov. 1975). Public Service Electric and Gas Co., Newark, N.J.

197. Guerrero, L.A. and R.D. Guerrero, III. 1976. Culture of freshwater shrimps in fertilized ponds. FAO Tech. Conf. Aquaculture, Kyoto, Japan 1976. Exp. pap. 16. 5 pp.

198. Guerrero, R.D., III L.A. Guerro, and J.H. Grover. 1975. Notes on the culture of freshwater shrimps in Central Luzon. Presented Int. Conf. Prawn Farming, Vung-Tau, Vietnam, 1975. 6 pp. (Unpublished).

199. Gunter, G. 1937. Observations on the river shrimp, _Macrobrachium ohionis_ (Smith). Amer. Midl. Nat. 18 (6): 1038-1042.

200. Hagood, R.W. and S.A. Willis. 1976. Cost comparisons of rearing larvae of freshwater shrimp, _Macrobrachium acanthurus_ and _M. rosenbergii_, to juveniles. Aquaculture 7: 59-74.

201. Halver, J.E. 1975. Nutritional studies of salmonid fishes. Proc. First Intl. Conf. Aquacult. Nutr., 25-37.

202. Halver, J.E. 1976. Formulating practical diets for fish. J. Fish. Res. Bd. Can. 33: 1032-1039

203. Harrison, C. 1975. The essential amino acids of Mytilus californianus. Veliger 18: 189-193.

204. Hart, C.W., Jr. 1961. The freshwater shrimps (Atyidae and Palaemonidae) of Jamaica, W.I. Proc. Acad. Nat. Sci., Phila. 113 (4): 61-80.

205. Hechtman, E.S. 1974a. [Use of Macrobrachium rosenbergii, Tilapia catfish, and clams as part of a waste disposal system for raceways in Nebraska]. FAO Aquaculture Bull. 6 (2-3): 17.

206. Hechtman, E.S . 1974b. A new system of shrimp (Macrobrachium rosenbergii) culture in the United States. FAO Aquaculture Bull. 6 (4): 13.

207. Hedgecock, D., K. Nelson, R. Shleser, 1976. Growth Differences Among Families of the Lobster, Homarus americanus. Proc. VII Ann. Meeting World Mariculture Soc., San Diego, Calif. Jan. 1976.

208. Hedgepeth, J.W. 1947. River shrimps: interesting crustaceans about which little has been written. Prog. Fish-Cult. 9 (4): 181-184.

209. Hedgepeth, J.W. 1949. The North American species of Macrobrachium (river shrimp). Texas J. Sci. 1 (3): 28-38.

210. Hendricks, P. 1975. Aquaculture: an overview. University Hawaii, Sea Grant Newsl. 5 (12): 4.

211. Hillman, R.E., 1964. Chromatographic Evidence of Intraspecific Genetic Differences in the Eastern Oyster, Crossostrea virginca. Sys. Zool. 13: 12-18.

212. Hinds, V.T. 1971. [Culture of Macrobrachium lar in Tahiti]. FAO Aquaculture Bull. 3(4): 11.

213. Hobbs, H.H., Jr. and W.H. Massman. 1952. The river shrimp, Macrobrachium ohione (Smith), in Virginia. Va. J. Sci. 3 (3): 206-207.

214. Holthuis, L.B. 1950. The palaemonidae collected by the Siboga and Snellius Expeditions with remarks on other species. I. Subfamily Palaemonidae. The Decapoda of the Siboga Expedition,

Part X. Siboga Expedition Monograph 39a9: 1-268.

215. Holthuis, L.B. 1952. A general revision of the palaemonidae (Crustacea Decapoda Natantia) of the Americas. II. The subfamily palaemonidae. Allan Maneock Found. Publ., Occ. Paper 12: 11-132.

216. Holthuis, L.B. and A.J. Provenzano, Jr. 1970. New distribution records for species of Macrobrachium with notes on the distribution of the genus in Florida (Decapoda, Palaemonidae). Crustaceana 19 (2): 211-213.

217. Holthuis, L.B. and H. Rosa, Jr. 1965. List of species of shrimps and prawns of economic value. FAO Fish. Tech. Pap. No. 52. 21 pp.

218. Hoyt, J.W. 1966. The effects of algal contaminants on frictional resistance. Proc. 11th intern. Towing Tank Conf. Tokyo.

219. Huang, W.Y., J.K. Wang, and T. Fujimura. 1976. A model for estimating prawn populations in ponds. Aquaculture 8: 57-70.

220. Hughes, D.A. and J.D. Richard. 1973. Some current-directed movements of Macrobrachium acanthurus (Wiegmann, 1836) Decapoda, Palaemonidae) under laboratory conditions. Ecology 54 (4): 927-929.

221. Ibrahim, K.H. 1962. Observations on the fishery and biology of the freshwater prawn, Macrobrachium malcolmsonii Milne-Edwards of River Godavary. Indian J. Fish. 9 (2): 433-487.

222. Ingle, R.M. and B. Eldred. 1960. Notes on the artificial cultivation of freshwater shrimp. The West Indies Fish. Bull (4): 1-5.

223. Iwai, T.Y. 1976. Energy transformation and nutrient assimilation by the freshwater prawn Macrobrachium rosenbergii under controlled laboratory conditions. M.Sc. Theisis, Dept. Anim. Sci., U of Hawaii. 128 pp.

224. Jalihal, D.R. and K.N. Sankolli. 1975a. On the abbreviated metamorphosis of the freshwater prawn Macrobrachium hendersodayanum (Tiwari) in the laboratory. Karnatak Univ. J. Sci. 20: 283-291.

225. Jalihal, D.R. and K.N. Sankolli. 1975b. On the palaemonid prawn Macrobrachium hendersodayanum (Tiwari) from the Malaprabha River. Karnatak Univ. J. Sci. 20: 297-304.

226. Jhingran, V.G. and V. Gopalakrishnan. 1974. Catalogue of cultivated aquatic organisms. FAO Fish. Tech. Pap. No. 130. 83 pp.

227. John, M.C. 1957. Bionomics and life history of Macrobrachium rosenbergii (de Man). Bull. Cent. Res. Inst. Univ. Kerala, India, Ser. C, Nat. Sci. 15: 93-102.

228. Johnson, D.S. 1960a. Some aspects of the distribution of freshwater organisms in the Indo-Pacific area, and their relevence to the validity of the concept of an oriental region in zoogeography. Proc. Cent. Bicent. Congr. Biol., Singapore, 1958: 170-181.

229. Johnson, D.S. 1960b. Sub-specific and infra-specific variation in some freshwater prawns of the Indo-Pacific region. Proc. Cent. Bicent. Congr. Biol., Singapore, 1958: 259-267.

230. Johnson, D.S. 1962. A synopsis of the Decapoda, Caridea, and Stenopodidae of Singapore, with notes on their distribution and a key to the genera of Caridea occurring in Malayan wters. Bull. Natl. Mus. Singapore 30: 44-79.

231. Johnson, D.S. 1964. Distribution and other notes on some freshwater prawns (Atyidae and Palaemonidae) mainly from the Indo-West Pacific region. Bull. Natl. Mus. Singapore 32: 5-30.

232. Johnson, D.S. 1966. Edible crustaceans. Malay. Nat. J. 19: 275-282.

233. Johnson, D.S. 1967a. Non-penaeid prawns of inland waters, including brackish waters, in western Malaysia and Signapore. Sym. on Scientific & Technological Res. in Malaysia & Singapore, Kuala Lumpur.

234. Johnson, D.S. 1967b. Some factors influencing the distribution of freshwater prawns in Malaya. Proc. Sym. Crustacea, Mar. Biol. Assoc. India 1: 418-433.

235. Johnson, D.S. 1968. Biology of potentially valuable freshwater prawns with special reference to the riceland prawn Cryphiops Macrobrachium lanchesteri (de Man). FAO Fish. Rep. 57 (2): 233-241.

236. Johnson, S.K. 1975. Handbook of shrimp diseases. TAMU-SG-75-603, Center for Marine Resources, Texas A & M University, College Station, Texas. 19 pp.

237. Johnson, S.K. 1977. Diseases of shrimps similar to those of aquatic insects. The Commer. Fish Farmer & Aquacul. News 3 (2): 15-17.

238. Jones, D.A., J.G. Munford, and P.A. Gobbott. 1974. Microcapsules as artificial food particles for aquatic filter feeders. Nature 247: 233-235.

239. Jones, S. 1969a. Introduction. Pages 1-4 in Prawn Fisheries of India. Bull. Cent. Mar. Fish. Res. Inst. 14.

240. Jones, S. 1969b. XI. General observations. Pages 273-279 in Prawn Fisheries of India. Bull. Cent. Mar. Fish. Res. Inst. 14.

241. Jones, S. 1969c. The prawn fishery resources of India. FAO Fish. Rep. 57 (3): 735-747.

242. Joseph, J.D. 1977. Assessment of the nutritional role of algae in the culture of larval prawns (*Macrobrachium rosenbergii*). Presented 8th Annual Meeting World Mariculture Society, San Jose, Costa Rica, 9-13 January 1977. (Manuscript).

243. Joseph, J.D. and S.P. Meyers. 1975. Lipid fatty acid composition of shrimp meals and crustacean diets. Feedstuffs 47 (35): 28-29.

244. Joseph, J.D. and J.E. Williams. 1975. Shrimp head oil: a potential feed additive for mariculture. Proc. 6th Annual Meeting World Mariculture Society: 147-155.

245. Kanazawa, A., M. Shimaya, M. Kawasaki and K. Kashiwada. 1970. Nutritional requirements of prawn. I. Feeding an artificial diet. Bull. Jap. Soc. Sci. Fish 36 (9): 949-954.

246. Kanazawa, A., N. Tanaka, S. Teshima, and K. Kashiwada. 1971. Nutritional requirements of prawn. II. Requirement for sterols. Bull. Jap. Soc. Sci. Fish. 37: 211-215.

247. Katre, S. and T.J. Pandian. 1972. On the hatching mechanism of a freshwater prawn *Macrobrachium idae*. Hydrobiologica 40 (1): 1-17.

248. Kemp, S. 1915. Fauna of the Chilka Lake: Crustacea Decapoda. Mem. Indian Mus. 5 (3): 199-325.

249. Ketola, H. G. 1976. Quantitative nutritional requirements of fishes for vitamins and minerals. Feedstuffs, 48 (8): 42.

250. Kitayama, T., T. Kamata, M. Shimaya, O. Deshimauru, and C.O. Chichester 1972. The biosynthesis of astaxanthin. VIII. The conversion of B carotene - 15, 15'-$^3H_2$ into astaxanthin in the prawn, Penaeus japonicus. Bull. Jap. Soc. Sci. Fish 38: 1171-1175.

251. Kloke, C.W. and M. Potaros. 1975. The technology and economics of small-scale commercial prawn (Macrobrachium rosenbergii) hatcheries: a case study in Thailand. FAO Indo-Pac. Fish. Counc. Occas. Pap. 1975/1. 18 pp.

252. Knight, A.W., principal investigator. 1976. Laboratory studies on selected nutritional, physical, and chemical factors affecting the growth, survival, respiration, and bioenergetics of the giant prawn, Macrobrachium rosenbergii. Water Sci. & Eng. Paper 4501, Univ. Calif., Davis. 87 pp.

253. Kow, T. A. 1968. Unit stocks of shrimps and prawns in the IPFC Region and unit fisheries exploiting them. FAO Fish. Rep. 57 (2): 205-217.

254. Kuronuma, K. 1971. The giant freshwater prawn [Macrobrachium rosenbergii] in Japan. FAO Aquaculture Bull. 4 (1): 15.

255. Kuwabara, J.S. 1975. The application of a statistical optimization technique for the culture of Macrobrachium rosenbergii larvae (giant Malaysian prawn). Dept. of Agric. Engineering, Univ. Hawaii, Honolulu, Working Paper No. 9. 37 pp. (Unpublished).

256. Kuwabara, J., J.K. Wang, and T. Fujimura. 1974. First report of Malaysian giant prawn larvae culture. State Anuenue Fish. Res. Center, Hawaii Dept. Fish and Game, Honolulu. (Unpublished).

257. Lagler, K.F. [1975?]. Natural populations of the giant freshwater prawn, Macrobrachium rosenbergii (de Man). Int. Conf. Prawn Farming, Vung-Tau, Vietnam, 1975. (Unpublished).

258. Lasser, G.W. and W.V. Allen. 1976. The essential amino acid requirements of the Dungene ss Crab, Cancer magister. Aquacult. 7: 235-244.

259. Lannan, J., 1972. Estimating Heritability and Predicting Response to Selection for the Pacific Oyster Crassostrea gigas. Proc. Nat. Shellfish Ass. 62: 62-66.

260. Lasley, J.F., 1972. Genetics of Livestock Improvement. Prentice-Hall, Inc. New Jersey.

261. Leandri, R. 1974. [Macrobrachium rosenbergii rearing in Tahiti]. FAO Aquaculture Bull. 6(2-3): 14.

262. Lee, C.L. 1974. A Guide for the development of an udang galah Macrobrachium rosenbergii industry in Malaysia. Malaysian Agriculture Research Development Institute, Report No. 19. 22 pp.

265. Lewis, J.B. 1961. Preliminary experiments on the rearing of the freshwater shrimp Macrobrachium carcinus (L.). Gulf Caribb. Fish. Inst. Proc. 14: 199-201.

266. Lewis, J.B. and J. Ward. 1965. Developmental stages of the palaemonid shrimp Macrobrachium carcinus (Linnaeus, 1958). Crustaceana 9(2): 137-148.

267. Lewis, J.B., J. Ward, and A. McIver. 1966. The breeding cycle, growth and food of the freshwater shrimp Macrobrachium carcinus (Linnaeus). Crustaceana 10(1): 48-52.

268. Lewontin, R.C., 1965. Selection for Colonizing Ability in the Genetics of Colonizing Species. H.G. Baker & G.L. Stebbins (eds.), Academic Press, N.Y.

269. Liao, I.C., N.H. Chao, and L.S. Hsieh. 1973. Preliminary report of the experiments on the propagation of the giant freshwater prawn, Macrobrachium rosenbergii, in Taiwan [in Chinese, English abstr.]. J. Fish. Soc. Taiwan 2(2): 48-58.

270. Ling, S.W. [1962?]a. A general account on the biology of the giant freshwater prawn, Macrobrachium rosenbergii, and methods for its rearing and culturing. FAO (Fisheries Research Institute, Glugor, Penang, Malaysia). 21 pp. (Mimeo).

271. Ling, S.W. 1962b. Studies on the rearing of larvae and juveniles and culturing of adults of Macrobrachium rosenbergii (de Mann). FAO Indo-Pac. Fish. Counc., Curr. Aff. Bull. 35: 1-11.

272. Ling, S.W. 1968. A brief working bibliography on shrimp culture with particular reference to Macrobrachium spp. FAO Indo-Pac. Fish. Counc., Occas. Paper 68/1. 4 pp.

273. Ling, S.W. 1969a. The general biology and development of Macrobrachium rosenbergii (de Man). FAO Fish. Rep. 57(3): 589-606.

274. Ling, S.W. 1969b. Methods of rearing and culturing Macrobrachium rosenbergii (de Man). FAO Fish. Rep. 57(3): 607-619.

275. Ling, S.W. 1971. Some brief notes on the status and problems of shrimp and prawn farming development in Asia. Indo-Pac. Fish. Counc., 48th Exec. Comm. Meeting. 14 pp. (Unpublished).

276. Ling, S.W. 1972. A review of the status and problems of coastal aquaculture in the Indo-Pacific region. Pages 2-25 in T.V.R. Pillay, ed. Coastal Aquaculture in the Indo-Pacific Region. Fishing News (Books) Ltd., Surrey, England.

277. Ling, S.W. 1973. Status, potential and development of coastal aquaculture in the countries bordering the South China Sea. FAO Indo-Pac. Fish. Counc., SCS/DEV/73/5/Rome.

278. Ling, S.W. and T.J. Costello. 1976a. Review of culture of freshwater prawns. FAO Tech. Conf. on Aquaculture, Kyoto, Japan, 1976. Ref. pap. 29. 12 pp.

279. Ling, S.W. and T.J. Costello. 1976b. Status and problems of Macrobrachium farming in Asia. Proc. Food-Drugs from the Sea Conf., Mayaguez, Puerto Rico, 1974 (Mar. Technol. Soc.): 66-71.

280. Ling, S.W. and A.B.O. Merican. 1961. Notes on the life and habits of the adults and larval stages of Macrobrachium rosenbergii (de Man). Indo-Pac. Fish. Counc. 9(2): 55-61.

281. Longwell, A.C. 1975. Review of Genetic and Related Studies on Commercial Oysters and Other Pelecypod Mollusks. Mimeo of paper presented at the Thirteenth Pacific Science Congress, Vancouver, Canada, August 18-30, 1975.

282. Loosanoff, V. L. and C. A. Nomenjko. 1951. Existence of physiologically Different Races of Oysters, Crassostrea virginica. Biol. Bull. (Woods Hole): 2:151-156.

283. Lui, L. C. 1970. [Hatchery and pond production of Macrobrachium rosenbergii in Malacca]. FAO Aquaculture Bull. 3(1): 4.

284. Macagno, E. and D. R. Friedlander. 1976. Demonstration of a synctium by the migration of horse radish peroxidase between neurons of the prawn, Macrobrachium rosenbergii. Neurosci. Abstr. 2(1&2). 1 p.

285. Maddox, M. B. and J. J. Manzi. 1976. The effects of algal supplements on static system culture of Macrobrachium rosenbergii (de Man) larvae. Proc. 7th Annual Meeting World Mariculture Society: 677-698.

286. Maginniss, L. A. 1974. Glucose transport by the isolated midgut ot the Malaysian prawn, Macrobrachium rosenbergii. Am. (Soc.) Zool. 14(4): 1292. (abstr.).

287. Manzi, J. J. and M. B. Maddox. 1976. Algal supplement enhancement of static and recirculating system culture of Macrobrachium rosenbergii larvae. Helgol. Wiss. Meeresunters. 28: 447-455.

288. Manzi, J. J., M. B. Maddox, and P. A. Sandifer. 1977. Algal supplement enhancement of Macrobrachium rosenbergii (de Man) larviculture. Presented 8th Annual Meeting World Mariculture Society, San Jose, Costa Rica, 9-13 January 1977. (Manuscript).

289. Mardela Corporation. 1973. NOAA Aquaculture Survey 1972. Mardela Corp., Burlingame, Calif. 46 pp.

290. Maugle, P. 1974. Preliminary survey of the Macrobrachium and shrimp fisheries resources on the west coast of Columbia. Rept. cent. de Cooperativas de la Reforma Agraria.

291. Mayer, E. 1964. Systematics and the origin of the species. Dover Publ. Inc., N. Y., 334 pp.

292. Maynard, E. C. L., A. Riollano, and P. G. Rios. 1974. La crianza de camarones en agua dulce del genero Macrobrachium [in Spanish]. Agric. Al Dia (Puerto Rico) 19(1&2): 9-23.

293. McCormick, R.N. 1934. Macrobrachium ohionis, the large freshwater shrimp. Proc. Indiana Acad. Sci. 43: 218-224.

294. McSweeny, E.S. In press. Intensive culture systems. Proc. Workshop on the Aquaculture of Freshwater Prawns (Macrobrachium supp.), Charleston, S.C., 1976.

295. McVey, J.P. 1975. New record of Macrobrachium rosenbergii (de Man) in the Palau Islands (Decapoda, Palaemonidae). Crustaceana 29(1): 31-32.

296. Meeran, N.K. and M.J. Sebastian. 1972. On rearing the larvae of the giant freshwater prawn, Macrobrachium rosenbergii (de Man). Pages 94-95 in Proc. Seminar on Mariculture & Mechanised Fishing, Madras, India.

297. Menon, M.K. 1938. Early stages of two species of Palaemon. Proc. Indian Acad. Sci. 8(4): 288-294.

298. Menon, V.R., A. Rajagopal, and R. Srinivasan. 1972. A preliminary report on pond culture and breeding of the freshwater prawn, Macrobrachium malcolmsoni in Tamil Nadu. Pages 97-100 in Proc. Sem. on Mariculture & Mechanised Fishing, Madras, India.

299. Merican, A.B.O. 1970. [Food for shrimps and prawns (including Macrobrachium rosenbergii)]. FAO Aquaculture Bull. 3(1): 5.

300. Mermilliod, W. and F.M. Truesdale. 1976. River shrimp [Macrobrachium ohione]: something special. (Center for Wetland Resources, Louisiana State University) Aquanotes 5(3): 3-4.

301. Meyers, S.P. 1974. Cultivation of freshwater prawn Macrobrachium. Feedstuffs 46(49): 25-26.

302. Meyers, S.P. and Z.P. Zein-Eldin. 1972. Binders and pellet stability in development of crustacean diets. Proc. World Maricult. Soc. 3: 351-364.

303. Meyers, S.P. and C.W. Brand. 1975. Experimental flake diets for fish and crustacea. Prog. Fish-Cult. 37(2): 67-72.

304. Miller, D. 1973. The composition of various fishery products used in the production of animal feeds. Feedstuffs, 45(50): 29.

305. Miller, G.C. 1971. Commercial fishery and biology of the freshwater shrimp, Macrobrachium, in the lower St. Paul River, Liberia, 1952-53. U.S. Dept. Comm., National Oceanic and Atmospheric Administration, National Marine Fisheries Service, Special Sci. Rep., Fish. No. 626. 13 pp.

306. Milligan, H. 1976. A Method for Obtaining Metaphase Chromosome Spreads from Marine Shrimp with Notes on the Karyotype of Penaeus aztecus, Penaeus setiferus, and Penaeus duorarum. Proc. VII World Mariculture Society, San Diego, Calif. 1976: 327-332.

307. Minamizawa, A. and T. Morizane. 1970. Report on study about cultivation techniques for freshwater shrimp. Ehime Prefecture Fish. Lab., Japan (Transl. from Japanese, Bur. Comm. Fish., U.S. Dept. Interior). 55 pp.

308. Misharev, Y.Y. 1970. Molting and mating in freshwater shrimp Macrobrachium asperulum von Martens). All-Union Research Institute for Fish. and Ocean. 65: 424-428 (From Problems of commercial hydrobiology, Fish. Res. Board Canada, Transl. Series No. 1516. 11 pp.).

309. Mitchell, R.E. and A.M. Kirby, Jr. 1976. Performance characteristics of pond aeration devices. Proc. 7th Annual Meeting World Mariculture Society: 561-581.

310. Miyajima, L.S. 1975. Amino acid assay: an analytical approach to basic nutritional studies in freshwater shrimp. Proc. 1975 Fish Farming Conf. & Annu. Conv. Catfish Farm. Texas: 46-50.

311. Miyajima, L.S., G.A. Brokerick, and R.D. Reimer. 1975. Indentification of the essential amino acids of the freshwater shrimp, Macrobrachium ohione. Proc. 7th Annual Meeting World Mariculture Society: 699-704.

312. Moav, R. and Wohlfarth G.W., 1973. Carp breeding in Israel. in Aquacultural Genetics, Selected Topics, R. Moav (ed.), John Wiley & Sons, New York.

313. Moav, R., G. Hulata, G. Wohlfarth, 1975. Genetic Differences Between the Chinese and Duropean Races of the Common Carp. 1. Analysis of Gnotype-Environment Interactions for Growth Rate. Heredity 34(3): 323-340.

314. Moav, R., 1976. Genetic Improvement in Aquaculture Industry. Research report No. 9, F.A.O. Technical Conference on Aquaculture, Kyoto, Japan, May 26-June 2, 1976. FIR: AG/Conf./ 76/R9.

315. Moav, R.T. Brody, G. Wohlfarth, & G. Hulata, 1976a. Applications of Electrophoretic Genetic Markers to Fish Breeding. I. Advantages & Methods. Aquaculture 9: 217-228.

316. Moav, R.M. Soller, & G. Hulata, 1976. Genetic Aspects of the Transition from Traditional to Modern Fish Farming. Theor. App. Genetics 47: 285-290.

317. Mock, C.R. 1973. Shrimp culture in Japan. Mar. Fish. Rev. 35(3-4): 71-74.

318. Mock, C.R. 1974. Crustacean culture. NOAA Tech. Rep., NMFS Circ-388: 111-113.

319. Monaco, G. 1975. Laboratory rearing of larvae of the palaemonid shrimp Macrobrachium americanum (Bate). Aquaculture 6: 369-375.

320. Morita, S.K. 1977. An Econometric model of prawn pond production. Presented 8th Annual Meeting World Mariculture Society, San Jose, Costa Rica, 9-13 January, 1977. (Manuscript).

321. Morizane, T., M. Nakamura, and T. Yamada. 1973. Studies on the massrearing method of larval freshwater prawn Macrobrachium rosenbergii [in Japanese]. Aquiculture 21(3): 110-116.

322. Morizane, T. and T. Yamada. 1973. Trial paddy-field culture of freshwater prawn Macrobrachium rosenbergii [in Japanese]. Aquaculture 21(3): 117-118.

323. Muir, J.F. 1975. Water re-cycling systems in fish farming. Fish Farming Int. 2(2): 14-15, 48.

324. Nagdeman, J.J. 1976. No fish story: aquaculture may still be the wave of the future. Barron's National Business and Financial Weekly 56(3, Jan. 19): 9,10,12,14,16.

325. Nakamura, N. and S. Kasahara, 1955. A study on the Phenomenen of the Tobi koi or Shoot Carp. I. On the earliest stage at which the shoot carp appears. Bull. Jap. Soc. Sci. Fish. 21(2): 73-76 (in Japanese with English summary).

326. Nakamura, N. and S. Kasahara, 1957. A Study on the Phenomenon of the Tobi kói or Shoot Carp. III. On the Result of Culturing the Modal Group and the Growth of Carp Fry Reared Individually. Bull, Jap. Soc. Sci. Fish. 22(11): 674-678 (in Japanese with English summary).

327. Nakamura, N. and S. Kasahara, 1961. A Study on the Phenomenon of the Tobi koi or Shoot Carp. IV. Effects of Adding a Small Number of Larger Individuals to the Experimental Batches of Carp Fry and of Culture Density Upon the Occurence of Shoot Carp. Bull. Jap. Soc. Sci. Fish. 27(11): 958-962. (in Japanese with English summary).

328. Nakamura, R. 1975. A preliminary report on the circadian rhythmicity in the spontaneous locomotor activity of *Macrobrachium rosenbergii* and its possible application to prawn culture. Proc. 6th Annual Meeting World Mariculture Society: 37-41.

329. National Oceanic & Atmospheric Administration. 1973. Aquaculture in the National Oceanic and Atmospheric Administration. NOAA, Special Emphasis Document. (Unpublished).

330. National Oceanic & Atmospheric Administration. 1975. NOAA Aquaculture Plan. Draft for Review, NOAA, Washington, D.C. 157 pp. (Unpublished).

331. Nelson, R.D. 1972. Commercial feasibility of freshwater shrimp aquaculture in South Carolina: a preliminary study. Development Tech., Inc., St. Simons Island, GA. 34 pp. (Unpublished).

332. Nelson, S.G., D.A. Armstrong, A.W. Knight, and H.W. Li. 1976. The effects of temperature and salinity on the metabolic rate of juvenile *Macrobrachium rosenbergii* (Crustacea: Palaemonidae).

Comp. Biochem. Physiol. 54A: 1-5.

333. Neal, R.A. 1975. Alternatives in aquaculture development: consideration of extensive versus intensive methods. Jour. Fish. Res. Bd. Can. 30: 2118-2222.

334. Nelson, S.G. and A.W. Knight. 1977. Ecological energetics and its application to the evaluation of diets for aquatic species, with special regard to the giant Malaysian prawn, Macrobrachium rosenbergii. Presented 8th Annual Meeting World Mariculture Society, San Jose, Costa Rica, 9-13 January 1977. (Manuscript).

335. Nelson, S.G., A.W. Knight, and H.W. Li. Submitted for publication. The metabolic cost of food utilization and ammonia production by juvenile Macrobrachium rosenbergii (Crustacea: Palaemonidae).

336. Nelson S.G. and H.W. Li. 1976. Section VII. Bioenergetics of juvenile Macrobrachium rosenbergii, with brief notes on behavior and reproduction. Pages 60-78 in A.W. Knight, principal, and chemical factors affecting the growth, survival, respiration, and bioenergetics of the giant prawn Macrobrachium rosenbergii. Water Sci. & Eng. Paper 4501, Univ. Calif., Davis.

337. New, M.B. 1976a. A review of dietary studies with shrimp and prawns. Aquaculture 9: 101-144.

338. New, M.B. 1976b. A review of shrimp and prawn nutrition. Proc. 7th Annual Meeting World Mariculture Society: 277-287.

339. Newell, S.Y. and J.W. Fell. 1975. Preliminary experimentation in the development of natural food analogues for culture of detritivorous shrimp. Tech. Bull. 30, Univ. of Miami, Coral Gables, Fla. 115 pp.

341. Ong, K.S. [1975?]. A brief review of the Macrobrachium culture project in Penang, Malaysia. Contr. Int. Conf. Prawn Farming, Vung-Tau, Vietnam, 1975. (Unpublished).

342. Pandian, T.J. and S. Katre. 1972. Effect of hatching time on larval mortality and survival of the prawn Macrobrachium idae. Mar. Biol. 13: 330-337.

343. Panikkar, N.K. 1937. The prawn industry of the Malabar coast. J.

Bombay Nat. Hist. Soc. 39(2): 343-353.

344. Panikkar, N.K. 1968. Osmotic Behavious of shrimps and prawns in relation to their biology and culture. FAO Fish. Rep. 57(2): 527-538.

345. Pantulu, V.R. 1965. Inland prawn fisheries of India and their development. Fish. Technol. (India) 2: 54-58.

346. Patwardhan, S.S. 1937. Palaemon. Indian Zool. Mem. 6. 120 pp.

347. Peebles, J.B. 1974. Shelter usage and movement in Macrobrachium rosenbergii. Student Report, Univ. Hawaii. 25 pp. (Unpublished).

348. Peebles, J.B. 1976. The roles of prior residence and relative size in competition for shelter by Macrobrachium rosenbergii (Decapoda)., unpubl. m.s.

349. Perdue, J.A. and R. Nakamura. 1976. The effect of salinity on the growth of Macrobrachium rosenbergii. Proc. 7th Annual Meeting World Mariculture Society: 647-654.

350. Perrot, J. 1976. Progress in shrimp culture techniques and the production of juveniles [in French, English abstr.]. FAO Tech. Conf. Aquaculture, Kyoto, Japan. Res. pap. 12.

351. Pillai, N.N. and K.H. Mohamed. 1973. Larval history of Macrobrachium idella (Hilgendorf) reared in the laboratory. J. Mar. Biol. Assoc. India 15(1): 359-385.

352. Price, V. 1975. Studies on the commercial rearing of the giant prawn, Macrobrachium rosenbergii (de Man) in Puerto Rico: stocking densities in earthen ponds and nutritional studies. College of Fisheries, Univ. of Washington, Seattle. 27 pp. (Unpublished).

353. Pridgen, M. 1975. South Carolina's "Big Mac." Sea Grant 70's 5(5): 3-4.

354. Provenzano, A.J., Jr. 1973. Some results of a pilot project on freshwater prawn culture in Jamaica. Proc. 4th Annual Workshop Mariculture Society: 57-61.

355. Provosoli, L. 1975. Nutritional aspects of curstacean aquaculture. Proc. First Intl. Conf. Aquacult. Nutr., 13-21.

356. Rajyalakshmi, T. 1961. Studies on maturation and breeding in some estuarine palaemonid prawns. Proc. Natl. Inst. Sci. India, Part B, Biol. Sci. 27(4): 179-188.

357. Rajyalakshmi, T. 1966. On the age and growth of some estuarine prawns. Indo-Pac. Fish. Counc. Proc., Session 11(II): 52-83.

358. Rajyalakshmi, T. 1968. Try prawn [Macrobrachium malcolmsonii] in tanks and inland waters. Indian Farming (Nov.): 37-43.

359. Rajyalakshmi, T. 1974. The freshwater prawn, Macrobrachium malcolmsonii, for use in pond culture. Seafood Export J. 6(3): 1-5.

360. Rajyalakshmi, T. 1975, Freshwater caridean prawns (Macrobrachium malcolmsonii) are generally sedentary. Seafood Export J. 7(2). 4 pp.

361. Rajyalakshmi, T., T.S. Ramaraju, and D.R. Rao. 1968. On the marking of freshwater prawn, Macrobrachium malcolmsonii, in River Godvari. Indian J. Fish. 15: 61-67.

362. Rajyalakshmi, T. and M. Ranadhir. 1969. The commercial prawn Macrobrachium malcolmsonii (H. Milne Edwards) of the River Godavary, a discussion on the trend and characteristics of the population during 1963-1966. FAO Fish. Rep. 57(3): 903-921.

363. Ramamurthy, S. and M.S. Muthu. 1969. IX. Prawn fishing methods. Pages 235-257 in Prawn Fisheries of India. Bull. Cent. Mar. Fish. Res. Inst. No. 14.

364. Raman, K. 1964. On the location of a nursery ground of the giant prawn, Macrobrachium rosenbergii (de Man). Curr. Sci. (Bangalore) 33(1): 27-28.

365. Raman, K. 1967. Observations on the fishery and biology of the giant freshwater prawn, Macrobrachium rosenbergii (de Man). Proc. Sym. Crustacea, Mar. Biol. Assoc. India, Part II: 649-669.

366. Ramirios, C.I.G. 1975. Estudio de los camarones de agua dulce de El Salvador, con notas taxonomicas y ecologicas de cada una de las especies econtradas [in Spanish]. Thesis. Ciudad Univ.,

San Salvador, C.A. 69 pp. (Unpublished).

367. Ranade, M.R. 1972. [Larval production and rearing of jubenile Macrobrachium rosenbergii in India]. FAO Aquaculture Bull. 4(2): 4-5.

368. Rao, K.V. and K. Dorairaj. 1969. VIII. Prawn grounds on the continental shelf fished by trawlers. Pages 217-234 in Prawn Fisheries of India. Bull. Cent. Mar. Fish. Res. Inst., 14.

369. Rao, R.M. 1965. Breeding behaviour in Macrobrachium rosenbergii (de Man). Fish. Technol. (India) 2(1): 19-25.

370. Rao, R.M. 1967. Studies on the biology of Macrobrachium rosenbergii (de Man) of the Hooghly Estuary with notes on its fishery. Proc. Natl. Inst. Sci. India, Part B, Biol. Sci. 33: 252-279.

371. Rao, R.M. 1969. Studies on the prawn fisheries of the Hooghly Estuarine system. Proc. Natl. Inst. Sci. India 33(5&6): 252-279.

372. Rasalan, S.B., N.N. Delmendo, and T.G. Reyes. 1969. Some observations on the biology of the freshwater prawn Macrobrachium lanceifrons (Dana), with notes on the fishery. FAO Fish. Rep. 57(3): 923-933.

373. Reagan, R, G.B. Pardue, E.J. Eisen. 1976. Predicting Selection Responses for Growth of Channel Catfish. J. Hered. 67: 49-53.

374. Record, P. 1969. [Run of giant freshwater shrimp, mayacaste, in southeastern Mexico]. Pages 183-189 in P. Record, Tropical Frontier. Alfred A. Knopf, New York.

375. Reimer, R.D., K. Strawn, and A. Dixon. 1974. Notes on the river shrimp, Macrobrachium ohione (Smith) 1874, in the Galveston Bay System of Texas. Trans. Am. Fish. Soc. 103(1): 120-126.

376. Rivera, E.D. [1975?]. Brief note on the preliminary culture of Macrobrachium rosenbergii in the Phillippines. Contr. Int. Conf. Prawn Farming, Vung-Tau, Vietnam, 1975. 3 pp. (Unpublished).

377. Rigdon, R.H. and K.N. Baxter. 1970. Spontaneous necrosis in the brown shrimp Penaeus astecus caught in trawles in the Gulf of

Mexico. Trans. Am. Fish. Soc. 99: pp 533-536.

378. Roegge, M.A., W.P. Rutledge, and W.C. Guest. 1975. Freshwater shrimp studies. Texas Parks & Wildlife Dept., Inland Fish. Res., Heart of the Hills Fish. Res. Stat., Ingram. 21 pp.

379. Roman, A. 1976. Effects of two diets on growth and survival of post-larval Macrobrachium rosenbergii (de Man). Appendix XII, 8 pp. in Guerra, C.R., principal investigator, Integration of thermal and food processing residuals into a system for commercial culture of freshwater shrimp. Second Annual Rep. (July 1, 1975 - June 30, 1976) NSF/RANN from PSE&G Co., R&D Dept., Newark, N.J. (Unpublished).

380. Roux, J. 1935. New freshwater decapod crustaceans from the Malay Penninsula. Bull. Raffles Mus. 9: 28-33.

381. Roux, J. 1936. Second notes upon freshwater decapod crustaceans from the Malay Penninsula. Bull. Raffles Mus. 12: 29-43.

382. Rouzard, R. 1973. [A new aquaculture demonstration station for culture of Macrobrachium rosenbergii, Penaeus merguiensis, and P. aztecus]. FAO Aquaculture Bull. 6(1): 11-12.

383. Ruello, N.V., P.F. Moffitt, and S.G. Phillips. 1973. Reproductive behaviour in captive freshwater shrimp Macrobrachium australiense Holthuis. Aust. J. Mar. Freshwater Res. 24: 197-202.

384. Ryther, J.H. 1968. Freshwater shrimp culture in Malaya. Pages 214-222 in J.H. Ryther, Vol. I. The status and potential of aquaculture, particularly invertebrate and algae culture. Part II. Invertebrate and algae culture. National Tech. Information Service, Springfield, Va.

385. Sanchez, C. 1976. Larval development of Macrobrachium tenellum [in Spanish, English abstr.]. FAO Tech. Conf. Aquaculture, Kyoto, Japan, 1976. Exp. pap. 57. 8 pp.

386. Sandifer, P.A. 1975. Macrobrachium aquaculture in the continental United States: a research and development plan. Marine Resources Research Institute, Charleston, S.C. 63 pp. (Unpublished).

387. Sandifer, P.A. and J.D. Joseph. 1976. Growth responses and fatty acid composition of juvenile prawns (Macrobrachium rosenbergii)

fed a prepared ration augmented with shrimp head oil. Aquaculture 8: 129-138.

388. Sandifer, P.A. and T.I. Smith 1975. Experimental Aquaculture of the Malaysian Prawn, Macrobrachium rosenbergii (de Man) in South Carolina. U.S.A. FAO Tech. Conf. on Aquacult., Doc. FIR: AQ/Conf. 76/E. 3, 7 pp.

389. Sandifer, P.A., J.S. Hopkins, and T.I.J. Smith. 1975. Observations on salinity tolerance and osmoregulation in laboratory-reared Macrobrachium rosenbergii postlarvae (Crustacea: Caridea). Aquaculture 6: 103-114.

390. Sandifer, P.A. and T.I.J. Smith. 1974a. Decapod mariculture in closed recirculating systems. Annual Report: April 1973-April 1974. Coastal Plains Regional Commission, Washington, D.C. 16 pp. (Unpublished).

391. Sandifer, P.A. and T.I.J. Smith. 1974b. Development of a crustacean mariculture program at South Carolina s Marine Resources Research Institute. Proc. 5th Annual Workshop World Mariculture Society: 431-439.

392. Sandiver, P.A. and T.I.J. Smith. 1975a. Decapod mariculture in closed recirculating systems. Annual Report: April 1974 - April 1975. Coastal Plains Regional Commission, Washington, D.C. (Unpublished).

393. Sandifer, P.A. and T.I.J. Smith. 1975b. Effects of population density on growth and survival of Macrobrachium rosenbergii reared in recirculating water management systems. Proc. 6th Annual Meeting World World Mariculture Society: 43-54.

394. Sandifer, P.A. and T.I.J. Smith. 1976. Experimental aquaculture of the Malaysian prawn, Macrobrachium rosenbergii (de Man) in South Carolina, U.S.A. FAO Tech. Conf. Aquaculture, Kyoto, Japan, 1976. Exp. pap. 3. 9 pp.

395. Sandifer, P.A. and T.I.J. Smith. 1977. Intensive rearing of post-larval Malaysian prawns, Macrobrachium rosenbergii, in a closed cycle nursery system. Presented 8th Annual Meeting World Mariculture Society, San Jose, Costa Rica, 9-13 January 1977. (Manuscript).

396. Sandifer, P. A., T. I. J. Smith, and D. R. Calder. 1974. Hydrozoans as pests in closed-system culture of larval decapod crustaceans. Aquaculture 4(1): 55-59.

397. Sandifer, P. A., P. B. Zielinksi, and W. E. Castro. 1974. A simple airlift-operated tank for closed system culture of decapod crustacean larvae and other small aquatic animals. Helgol. Wiss. Meeresunters. 26(1): 82-87.

398. Sandifer, P. A., P. B. Zielinski and W. E. Castro. 1975. Enhanced survival of larval grass shrimp in dilute solutions of synthetic polymer, polyethyline oxide. Fish. Bull. 73(3): 618-680.

399. Schmitt, W. L. 1933. Notes on shrimps of the genus Macrobrachium found in the United States. J. Wash. Acad. Sci. 23(6): 312-317.

400. Segal, E. and A. Roe. 1975. Growth and behavior of post-juvenile Macrobrachium rosenbergii (de Man) in close confinement. Proc. 6th Annual Meeting World Mariculture Society: 67-88.

401. Serfling, S. A. 1975. U.S. Patent No. 3,916,833.

402. Serfling, S. A. 1976. Section II. Recycling of nitrogenous wastes and role of detritus in a closed-cycle polyculture ecosystem designed to maximize productivity of the freshwater prawn, Macrobrachium rosenbergii. Pages 12-21 in A. W. Knight, principal investigator, Laboratory studies on selected nutritional, physical, and chemical factors affecting the growth, survival, respiration, and bioenergetics of the giant prawn Macrobrachium rosenbergii. Water Sci. & Eng. Paper 4501, Univ. Calif., Davis.

403. Serfling, S. A. and S. G. Serfling. 1976. Section IV. Effect of light intensity and photoperiodism on growth rate of juvenile prawns, Macrobrachium rosenbergii. Pages 30-37 in A. W. Knight, principal investigator, Laboratory studies on selected nutritional, physical, and chemical factors affecting the growth, survival, respiration, and bioenergetics of the giant prawn Macrobrachium rosenbergii. Water Sci. & Eng. Paper 4501, Univ. Calif., Davis.

404. Shang, Y. C. 1974. Economic Feasibility of freshwater prawn farming in Hawaii. Economic Res. Center, Univ. Hawaii, Honolulu, 49 pp.

405. Sharp, J. 1976. Section I. Effects of dissolved oxygen, temperature, and weight on respiration of Macrobrachium rosenbergii. Pages 4-11 in A.W. Knight, principal investigator, Laboratory studies on selected nutritional, physical, and chemical factors affecting the growth, survival, respiration, and bioenergetics of the giant prawn Macrobrachium rosenbergii. Water Sci. & Eng. Paper 4501, Univ. Calif., Davis.

406. Sheik, M.A. 1976. The fishpond project. Trust. Terr. Pacific Islands, Mar. Res. Div., Micronesian Mariculture Demonstration Center, Newsl.: 11-12.

407. Shell, E.W. and R.T. Lovell. 1973. Inland fisheries progress in Thailand: 1972. Project AID/csd - 2870, T.O.9; USOM/Thailand PIOT 493-180/7-3-20047. International Center for Aquaculture, Auburn Univ., Auburn, Ala. 69 pp. (Unpublished).

408. Shewbart, K.L. and W.L. Mies. 1973. Studies on nutritional requirements of Brown Shrimp. The effect of linolenic acid on growth of Penaeus aztecus. Proc. World Maricult. Soc. 4: 277-287.

409. Shewbart, K.L., W.L. Mies and P.D. Ludwig. 1973. Nutritional requirements of brown shrimp Penaeus aztecus. Texas A & M University Sea Grant Publ. 73-205: 53 pp.

410. Shokita, S. 1973. Abbreviated larval development of the freshwater prawn, Macrobrachium shokitai Fujino et Baba (Decapoda, Palaemonidae) from Iriomote Island of the Ryukyus. Annot. Zool. Jpn. 46(2): 111-126.

411. Shokita, S. 1975. The distribution and speciation of the inland water shrimps and prawns from the Ryukyu Islands - I [in Japanese, English summary]. Bull. Sci. & Eng. Div. No. 18, Univ. Ryukyus, Naha, Okinawa, Japan: 115-136.

412. Sick, L.V. 1975. Selected studies of protein and amino acid requirements for Macrobrachium rosenbergii larvae fed neutral density formula diets. Proc. First Internatl. Conf. Aquaculture Nutrition (Univ. Delaware) : 215-228.

413. Sick, L.V. 1976. Selected studies of protein and amino acid requirements for Macrobrachium rosenbergii larvae fed neutral

density formula diets. Proc. First Intl. Conf. Aquacult. Nutr., 215-228.

414. Sick, L.V. and J.W. Andrews. 1973. The effect of selected dietary lipids, carbohydrates and proteins on the growth, survival and body composition of Penaeus duorarum. Proc. World Maricult. Soc., 4: 263-276.

415. Sick, L.V. and H. Beaty. 1974. Culture techniques and nutrition studies for larval stages of the giant prawn, Macrobrachium rosenbergii. Tech. Rep. Ser. No. 74-5, Ga. Mar. Sci. Center, Univ. Ga. (Skidaway Island), Savannah. 30 pp.

416. Siddall, S.E. 1974. Studies of closed marine culture systems. Prog. Fish Cult. 36(1): 8-15.

417. Sick, L.V. and H. Beaty. 1975. Development of formula foods designed for Macrobrachium rosenbergii larval and juvenile shrimp. Proc. 6th Annual Meeting World Mariculture Society: 89-102.

418. Sick, L.V., D. White and G. Baptist. 1973. The effect of duration of feeding amount of food, light intensity and animal size on rate of ingestion of pelleted food by juvenile Penaeid shrimp. Prog. Fish. Cult. 35(1): 22-26.

419. Sidthimunka, A. and B. Choapaknam. 1968. A preliminary report on pond culture of the giant freshwater prawn, Macrobrachium rosenbergii de Man. FAO Fish. Rep. 44(5): 205-212.

420. Silva, L.E.M., M. Pedini, and M.B. New. 1977. Experimental culture of freshwater prawns (Macrobrachium acanthurus) in the Atlantic coast of Colombia. Presented 8th Annual Meeting World Mariculture Society, San Jose, Costa Rica, 9-13 January 1977. (Manuscript).

421. Sinderman, C.J., ed. 1974. Diagnosis and control of mariculture diseases in the United States. U.S. Dept. of Commerce, N.M. F.S., Middle Atlantic Coastal Fisheries Center, Highlands, N.J., Tech. Ser. No. 2. 306 pp.

422. Singholka, S. and K. Pawaputanon. [1975?]. Macrobrachium rosenbergii culture at Bangken Fisheries Station, Thailand. 7 pp.

(Unpublished).

423. Smith, S.I. 1874. Synopsis of the higher Crustacea of the northern United States. Pages 637-661 in Report of the Freshwater Fisheries of the United States, Report U.S. Fish Comm., 1872-1873.

424. Smith, T.I.J. and P.A. Sandifer. 1975. Increased production of tank-reared Macrobrachium rosenbergii through use of artificial substrates. Proc. 6th Annual Meeting World Mariculture Society: 55-66.

425. Smith, T.I.J. and P.A. Sandifer. 1976. Decapod mariculture in closed recirculating systems. Annual Report: April 1975 - April 1976. Coastal Plains Regional Commission, Washington, D.C. 52 pp.

426. Smith, T.I.J., P.A. Sandifer, and W.C. Trimble. 1976a. Pond culture of the Malaysian prawn, Macrobrachium rosenbergii (de Man), in South Carolina, 1974-1975. Proc. 7th Annual Meeting World Mariculture Society: 625-645.

427. Smith, T.I.J., P.A. Sandifer, and W.C. Trimble. 1976b. Progress in developing a recirculating synthetic seawater hatchery for rearing larvae of Macrobrachium rosenbergii. Proc. Food-Drugs from the Sea Conf., Mayaguez, Puerto Rico, 1974 (Mar. Technol. Soc.): 167-181.

428. Smitherman, R.O., D.D. Moss, and E.L. Diaz. 1974. Observations on the biology of Macrobrachium americanum from a pond environment in Panama. Proc. 5th Annual Workshop World Mariculture Society: 29-40.

429. Smitherman, R.O., O.L. Green, and G.B. Pardue. 1974. Genetic Experiments with Channel Catfish. Catfish Farmer & World Aquaculture News. Vol. 6(3), June, 1974.

430. Soo, K.C. and Y. Uno. 1969. The larval development of Macrobrachium nipponense (De Haan) reared in the labortory. La Mer (Bull. Soc. Fran.-Jap. d'Oceanogr.) 7(4): 278-294.

431. Sorgeloos, P. 1976. The brine shrimp. Artemia salina: a bottleneck

in mariculture? FAO Tech. Conf. on Aquac., 26 May - 2 June 1976. Kyoto, Japan. FIR: AQ/Conf./76/E. 77.

432. Stephenson, M. and M.A. Simmons. 1976. Section V. Respiration and growth of larval Macrobrachium rosenbergi. Pages 38-46 in A.W. Knight, principal investigator, Laboratory studies on selected nutritional, physical, and chemical factors affecting the growth, survival, respiration, and bioenergetics of the giant prawn Macrobrachium rosenbergii. Water Sci. & Eng. Paper 4501, Univ. Calif., Davis.

433. Stern, H. 1976. Section VI. Natural foods for prawns. Pages 47-59 in A.W. Knight, principal investigator, Laboratory studies on selected nutritional, physical, and chemical factors affecting the growth, survival, respiration, and bioenergetics of the giant prawn Macrobrachium rosenbergii. Water Sci. & Eng. Paper 4501, Univ. Calif., Davis.

434. Stern, H.L., D.A. Armstrong, A.W. Knight, and D.J. Chippendale. 1976. Survival and growth of juveniles of the giant Malaysian prawn, Macrobrachium rosenbergii, fed natural plant diets. Proc. 7th Annual Meeting World Mariculture Society: 677-675.

435. Stolpe, N.E. 1975a. Recirculating systems for embryo incubation and larval rearing of the freshwater prawn Macrobrachium rosenbergii (de Man). Presented 67th Annual Conv. Nat'l. Shellfish Assoc., Charleston, S.C., June 22-26, 1975. 1 p. [Abstr.].

436. Stolpe, N.E. 1975b. Prawn compartmentalization experiments. Appendix X, 8 pp. in Guerra, C.R., principal investigator, Integration of thermal and food processing residuals into a system for commercial culture of freshwater shrimp. Second Semiannual Rep. (July 1, 1975 - Dec. 31, 1975) NSF/RANN from PSE&G Co., R&D Dept., Newark, N.J. (Unpublished).

437. Stolpe, N.E. 1976. Note on prawn compartmentalization experiments. Appendix XIII, 4 pp. in Guerra, C.R., principal investigator, Integration of thermal and food processing residuals into a system for commerical culture of freshwater shrimp. Second Annual Rep. (July 1, 1975 - June 30, 1976) NSF/RANN from PSE&G Co., R&D Dept., Newark, N.J. (Unpublished).

438. Strenth, N.E. 1976. A review of the systematics and zoogeography

of the freshwater species of Palaemonetes Heller of North America (Crustacea: Decapoda). Smithsonian Contributions to Zoology, No. 228. Smithsonian Inst. Press, Washington, D.C. 27 pp.

439. Strong, D.R. 1972. Life History Variations Among Populations of an Amphipod (Hyllella azteca) Ecology 53(6): 1103-1111.

440. Strong, D.R. 1973. Amphipod Amplexus: The Significance of Ecotypic Variation Ecology 54(6): 1383-1388.

441. Subrahmanyan, C.B. and C.H. Oppenheimer. 1969. Food preferences and growth of grooved Penaeid shrimp. Proc. Food-Drugs from the Sea Conf., pp 65-75, H.W. Youngken, Jr. (Ed.) Mar. Tech. Soc., Wash., D.C.

442. Swingle, H.S. and R. Allison. 1971. The inland fisheries progress in Thailand. Project AID/csd-2270; USOM/Thailand PIOT 493-3-80413. International Center for Aquaculture, Auburn Univ., Auburn, Ala. 106 pp. (Unpublished).

443. Swingle, H.S. and E.W. Shell. 1972. Inland fisheries progress in Thailand: 1971. Project AID/csd-2870, T.0.7; USOM/Thailand PIOT 493-180/7-3-10095. International Center for Aquaculture, Auburn Univ., Auburn, Ala. 110 pp. (Unpublished).

444. Swingle, H.S. and R.O. Smitherman. 1969. The marine and coastal fisheries stations of Thailand. Project AID/csd-1581. International Center for Aquaculture, Auburn Univ., Auburn, Ala. 39 pp. (Unpublished).

445. Sze, C.P. 1973. The biochemical composition of juveniles of Macrobrachium rosenbergii (de Man) Crustacea, Palaemonidae). Malay. Agric. J. 49: 8-10.

445a. Takata, M. 1974. Development of the prawn culture industry in Hawaii. NTIS Ser. COM-75-10112, 26 pp.

446. Tarby, M.J. 1974. Characteristics of yellow perch cannibalism in Oneida Lake and the relation to first year survival. Trans. Amer. Fish. Soc. 103(3): 462-471.

447. Thompson, J.N. & J.M. Thoday, 1974. A Definition and Standard Nomenclature for "Polygenic loci", Heredity 33(3): 430-437.

448. Tiwari, K.K. (no date). Distribution of the Indo-Burmese Freshwater Prawns of the Genus Palaemon (Fabr.) and its Bearing on the Satpura Hypothesis. Bull. No. VII: Symposium on Organic Evolution. National Institute of Sciences of India, New Delhi.

449. Toner, M. 1970. Science may convert field to shrimp farm [Aquaculture Products and Research, Inc.]. The Miami Herald (Miami, Fla.) Dec. 14: 1A, 28A.

450. Uchida, R.N. 1972. Review of recent progress in coastal aquaculture in the United States. Pages 84-104 in T.V.R. Pillay, ed. Coastal Aquaculture in the Indo-Pacific Region. Fishing News (Books) Ltd., Surrey, England.

451. Uno, Y. 1971. Studies on the aquaculture of Macrobrachium nipponense (de Haan) with special reference to breeding cycle, larval development and feeding ecology. La Mer (Bull. Soc. Fran. - Jap. d'Oceanogr.) 9(2): 123-128.

452. Uno, Y., A.B. Bejie and Y. Igarashi. 1975. Effects of temperature on the activity of Macrobrachium rosenbergii. La Mer (Bull. Soc. Fran.-Jap. d'Oceanogr.) 13(3): 150-154.

453. Uno, Y. and M. Fujita. 1972. Studies on the experimental hybridization of freshwater shrimps, Macrobrachium nipponense and M. formosense. Int. Ocean Develop. Conf. & Exhib., Keidanren Kaikan, Tokyo. Vol. 2, 1 p. (Abstr.).

454. Uno, Y. and K.C. Soo. 1969. Larval development of Macrobrachium rosenbergii (de Man) reared in the laboratory. J. Tokyo Univ. Fish. 55(2): 179-190 + xii figures.

455. Uno, Y. and C.S. Kwon. 1969. Larval Development of Macrobrachium rosenbergii (de Man) reared in the laboratory. J. Tokyo Univ. Fish. 55(2): 179-190.

456. Van Den Oord, A. 1964. The absence of cholesterol synthesis in the crab, Cancer pagurus L. Comp. Biochem. Physiol. 13: 461-467.

457. Varikul, V. 1960. Some important brackish-water shrimps in Thailand. Thai Fish. Gazette 13(1): 49-55.

458. Varikul, V. 1961. The cultivation of brackish-water shrimp in Thailand. [in Thai]. Thai Fish. Gazette 14(2): 129-143.

459. Venkataramiah, A., G.J. Lakshmi and G. Gunter. 1975. Effect of protein level and vegetable matter on growth and food conversion efficiency of brown shrimp. Aquacult. 6: 115-125.

460. Venkataramaiah, A. 1971. Spontaneous necrosis in shrimp. FAO Aquacult. Bull. 3(3): 11.

461. Vavilov, N.J. 1949-1950. Thr Origin, Variation, Immunity, and Breeding of Cultivated Plants: Translated from the Russian by K. Starr Chester, Chronica Botanica. 13(1-6): 1949-1950.

462. Ville, J. 1970a. Recherches sur la reproduction des Macrobrachium des lagunes ivoiriennes. I.-La fecondite precoce chez les Macrobrachium de Cote-d'Ivoire. Ann. Univ. Abidjan, ser. E, ecologie, III, (1): 253-261.

463. Ville, J. 1970b. Recherches sur la reproduction des Macrobrachium des lagunes ivoiriennes. II.-Migrations liees a la reproduction chez les Macrobrachium de Cote-d'Ivoire. Ann. Univ. Abidjan, ser. E, ecologie, III, (1): 263-268.

464. Viosca, P., Jr. 1957. The Louisiana shrimp story. Louisiana Wild Life and Fisheries Commission, Baton Rouge, La., Wildlife Education Bull. 40. 15 pp.

465. Watanabe, W.O. 1975. Identification of the essential amino acids of the freshwater prawn, Macrobrachium rosenbergii. M. Sc. Thesis, Dept. Zool., U of Hawaii, 26 pp.

466. Watanabe, W.O., G.A. Ahearn and M.S. Stahl. 1976. Nutritional requirements of Macrobrachium rosenbergii. (Unpublished Report).

467. Wang, J.K., W.Y. Huang, and T. Fujimura. 1975. Some characteristics of earthen pond prawn production in Hawaii. Working Paper #10, Sea Grant Coll. Program, Univ. Hawaii, Honolulu. 20 pp. (Unpublished).

468. Wang, J.K. and J.S. Kuwabara. 1975. Engineering analysis of prawn larvae culture. Presented at 1975 Annual Meeting Amer. Soc. Agricult. Eng., Paper No. 75-5015. 8 pp. + v figures.

469. Wang, J.K. and M.R. Williamson. 1976a. Automated prawn larvae/juvenile separation. Oceans '76: 13F-1 - 13F-3.

470. Wang, J.K. and M.R. Williamson. 1976b. The design of a continous juvenile production system for Macrobrachium rosenbergii. Presented at 1975 Annual Meeting Amer. Soc. Agricult. Eng.

471. Walker, A. 1975. Crustacean aquaculture. Proc. Nutr. Soc. 34: 65-73.

472. Webber, H.H. 1975a. The Design of an aquaculture enterprise. Proc. Gulf Carib. Fish. Inst. 24th Ses.: 117-125.

473. Webber, H.H. 1975b. Crustacean aquaculture in middle America. Mar. Fish. Rev. 37(1): 24-30.

474. Webber, H.H. 1975c. Shrimp farming: president-elect of World Mariculture Society discusses his company's Costa Rica project. Commer. Fish Farmer & Aquaculture News 1(4): 8-12.

475. White, M.J.D. 1973. The Evolution of Sex Determination. I. Simple Systems (Ch. 16) and II. Multiple System (Ch. 17). Animal Cytology and Evolution, Cambridge Univ. Press.

476. Webber, H.H. 1976. Criteria for candidate species for aquaculture. Aquaculture 7: 107-123.

477. Whitney, J.O. 1970. Absence of sterol biosynthesis in the blue crab Callinectes sapidus (Rathbun) and in the barnacle Balanus nubilus (Darwin.) J. Exp. Mar. Biol. Ecol. 4: 229-237.

478. Wickins, J.F. 1969. Preliminary experiments in the culture of the prawn Pandalus platyceros (Brandt) and the giant prawn Macrobrachium rosenbergii (de Man). Int. Counc. Explor. Sea (C.M. 1969/E.11), Fish. Improvement Committee. 8 pp.

479. Wickins, J.F. 1970a. Bigger prawns for your table. Central Office of Information. Features Newsl. 54, No. 0704/7. 9 pp.

480. Wickins, J.F. 1970b. Rearing prawns. Proc. Shellfish Assoc. Great Britain, 6-7 April: 61-72.

481. Wickins, J.F. 1972. Experiments on the culture of the spot prawn Pandalus platyceros Brandt and the giant freshwater prawn Macrobrachium rosenbergii (de Man). Fish. Invest. Minist. Agric., Fish. Food (G.B.), Ser. II, 27(5). vii + 23 pp.

482. Wickins, J.F. 1974. The use of ecdysed carapace for measurements of the growth of prawns. Int. Counc. Explor. Sea (C.M. 1974/ K:13), Shellfish & Benthos Committee. 6 pp.

483. Wickins, J.F. 1976. The tolerance of warm-water prawns to recirc- lated water. Aquaculture 9(1): 19-37.

484. Wickins, J.F. and T.W. Beard. 1974. Observations on the breeding and growth of the giant freshwater prawn *Macrobrachium rosenbergii* (de Man) in the laboratory. Aquaculture 3(2): 159-174.

485. Williamson, D.I. 1968. The type of development of prawns as a factor determining suitability for farming. FAO Fish. Bull. 57(2): 77-84.

486. Williamson, D.I. 1972. Larval development in a marine and a freshwater species of *Macrobrachium* (Decapoda, Palaemonidae). Crustaceana 23(3): 282-298.

487. Williamson, D.I. 1974. Notes on Caridea from Lake Rudolf, East Africa (Decapoda, Natantia). Crustaceana 27(3): 308-310.

488. Willis, S.A. and M.E. Berrigan. 1977. Effects of stocking size and density on growth and survival of *Macrobrachium rosenbergii* (de Man) in ponds. Presented 8th Annual Meeting World Mariculture Society, San Jose, Costa Rica, 9-13 January 1977, (Manuscript).

489. Willis, S.A., R.W. Hagood, and G.T. Eliason. 1976. Effects of four stocking densities and three diets on growth and survival of postlarval *Macrobrachium rosenbergii* and *M. acanthurus*. Proc. 7th Annual Meeting World Mariculture Society: 655-665.

490. Wong, N. and H. Deese. [1974?]. *Macrobrachium* project. Trust Terr. Pacific Islands, Mar. Res. Div., Micronesian Mariculture Demonstration Center, Newsl.: 2-6.

491. Zielinski, P.B., W.E. Castro, and P.A. Sandifer. 1974. The evaluation and optimization of *Macrobrachium* shrimp larvae tank designs and support systems. Proc. 5th Annual Workshop World Mariculture Society: 41-50.

492. Zandee, D.I. 1962. Lipid metabolism in *Astacus astacus* L. Nature, Lond. 195: 814-815.

493. Zandee, D.I. 1964. Absence of sterol synthesis in some Arthropods. Nature, Lond. 202: 1335-1336.

494. Zandee, D.I. 1966. Metabolism in the crayfish, *Astacus astacus* L. I. Biosynthesis of amino acids. Arch. Int. Physiol. Biochem. 74: 34-44.

495. Zielinski, P.B., W.E. Castro, and P.A. Sandifer. 1976a. Engineering consideration in the aquaculture of *Macrobrachium rosenbergii* in South Carolina. Presented Winter Meeting, Amer. Soc. Agriculture Engineers, Chicago, Ill.

496. Zielinski, P.B., W.E. Castro, and P.A. Sandifer. 1976b. Model flow studies of a rectangular shrimp larval tank. Proc. 7th Annual Meeting World Mariculture Society: 583-605.

The Crawfish